AF411408

Natural Background Levels in Groundwater

Natural Background Levels in Groundwater

Editors

Elisabetta Preziosi
Marco Rotiroti
M. Teresa Condesso de Melo
Klaus Hinsby

MDPI • Basel • Beijing • Wuhan • Barcelona • Belgrade • Manchester • Tokyo • Cluj • Tianjin

Editors

Elisabetta Preziosi
Water Research Institute
Consiglio Nazionale delle
Ricerche
Monterotondo (Rome)
Italy

Marco Rotiroti
DISAT - Department of Earth
and Environmental Sciences
University of Milano-Bicocca
Milano
Italy

M. Teresa Condesso de Melo
CERIS - Instituto Superior
Técnico
Universidade de Lisboa
Lisboa
Portugal

Klaus Hinsby
Department of Hydrology
Geological Survey of Denmark
and Greenland
Copenhagen
Denmark

Editorial Office
MDPI
St. Alban-Anlage 66
4052 Basel, Switzerland

This is a reprint of articles from the Special Issue published online in the open access journal *Water* (ISSN 2073-4441) (available at: www.mdpi.com/journal/water/special_issues/Background_Levels_ Groundwater).

For citation purposes, cite each article independently as indicated on the article page online and as indicated below:

LastName, A.A.; LastName, B.B.; LastName, C.C. Article Title. *Journal Name* **Year**, *Volume Number*, Page Range.

ISBN 978-3-0365-3724-5 (Hbk)
ISBN 978-3-0365-3723-8 (PDF)

Contents

About the Editors

Elisabetta Preziosi

Elisabetta Preziosi is a researcher at the Water Research Institute of the National Research Council of Italy with long-term experience in groundwater resources management. With a Master in Geology (1989) and Ph.D. in hydrogeology (1997) at the University Sapienza of Rome (Italy), she spent two years at the Ecole des Mines of Paris (ENSMP) in Fontainebleau (France) as a post-doctoral fellow (1994–1995) and as a researcher (1997–1998). Currently, she holds a class on groundwater monitoring at the Sapienza University of Rome. She has authored or co-authored more than 30 papers published in international ISI journals, as well as book chapters, and she is an Associate Editor of the ISI journal *"Acque Sotterranee - Italian Journal of Groundwater"*. Her research interests include water–rock interactions and natural background levels; the effects of climate changes on the hydrologic cycle; groundwater flow modeling at the regional scale; sustainable water resources management; water scarcity and drought issues; and EU Groundwater Directive implementation.

Marco Rotiroti

Marco Rotiroti is an Assistant Professor at the University of Milano-Bicocca, Department of Earth and Environmental Sciences. His research interests include hydrogeology and hydrogeochemistry, aiming to support the sustainable management of the quantity and quality of groundwater resources. His research activities focus on the geochemical evolution of groundwater along flow paths, groundwater pollution by nitrate and arsenic, natural background levels, contaminated sites, groundwater/surface–water interactions, groundwater dating, the impacts of climate change on groundwater resources, groundwater flow and reactive transport modeling.

M. Teresa Condesso de Melo

M. Teresa Condesso de Melo is a Hydrogeologist with an MSc Degree in Groundwater Hydrology (UPC Barcelona, Spain) and a PhD in Hydrogeology (UA Aveiro, Portugal), with a 3-year doctoral research training with the Hydrogeology Group (Wallingford, UK) of the British Geological Survey. She is a senior lecturer and co-coordinator of the Erasmus Mundus Joint Master Degree Programme on Groundwater and Global Change - Impacts and Adaptation (GroundwatCh) at Instituto Superior Técnico (ULisboa). She has more than 25 years of international professional experience in hydrogeology, groundwater quality and tracers, hydrogeophysics, groundwater monitoring and modelling. She has participated in several EU funded projects—PALAEAUX, BASELINE and BRIDGE—which were important supports for the implementation of EU Water Framework Directive (WFD) and Groundwater Directive (GWD).

Klaus Hinsby

Klaus Hinsby has been a Sr. Research Hydrogeologist at GEUS, the Geological Survey of Denmark and Greenland, since 1988. He is the chair of the Water Resources Expert Group of the EuroGeoSurveys and the Danish chapter of the International Association of Hydrogeologists. His research activities and interests include: 1) the derivation of natural backgrounds and threshold values for groundwater; 2) climate change impacts on groundwater quantity and quality, including sea-level rise and seawater intrusion; 3) groundwater age distributions and contaminant trends; and 4) the science–policy interface, e.g., in relation to EU directives and the UN sustainable development goals. He has more than 100 publications including 40 peer-reviewed journal papers and 14 book chapters.

Preface to "Natural Background Levels in Groundwater"

The need for establishing a formal limit between the concentration of potentially toxic inorganic compounds in groundwater due to natural processes or to anthropogenic pollution has prompted researchers to develop methods to derive this boundary and define the "Natural Background Level" (NBL). NBLs can be used as screening levels to define the good chemical status of groundwater bodies, as well as to fix the remediation target in polluted sites.

The book "Natural Background Levels in Groundwater" brings together a set of case studies across Europe and worldwide where the assessments and identification of this boundary are performed with different methodologies. It provides an overview of the approaches and protocols applied and tested in different states for NBL assessment, ranging from well-known methods, such as component separation or cumulative probability plot methods, to new computer-aided protocols. The main objective of this book is to bring together and discuss different methodological approaches and tools to improve the assessment of groundwater NBLs. The overview, discussion and comparison of different approaches and case histories for NBL calculation can be useful for scientists, water managers and practitioners.

The Guest Editors wish to thank all authors who submitted very interesting papers, the numerous reviewers for very constructive opinions about the submitted papers and the MDPI Editors for their useful support during the publication process.

Elisabetta Preziosi, Marco Rotiroti, M. Teresa Condesso de Melo, and Klaus Hinsby
Editors

water

MDPI

Editorial

Natural Background Levels in Groundwater

Elisabetta Preziosi [1], Marco Rotiroti [2,*], M. Teresa Condesso de Melo [3] and Klaus Hinsby [4]

1   IRSA-CNR, Water Research Institute—National Research Council, Strada Provinciale 35d, 9, Montelibretti, 00010 Rome, Italy; elisabetta.preziosi@cnr.it
2   Department of Earth and Environmental Sciences, University of Milano-Bicocca, Piazza della Scienza 1, 20126 Milano, Italy
3   CERIS, Instituto Superior Técnico, University of Lisbon, Av. Rovisco Pais, 1049-001 Lisbon, Portugal; teresa.melo@tecnico.ulisboa.pt
4   Geological Survey of Denmark and Greenland (GEUS), Øster Voldgade 10, 1350 København K, Denmark; khi@geus.dk
*   Correspondence: marco.rotiroti@unimib.it

Citation: Preziosi, E.; Rotiroti, M.; Condesso de Melo, M.T.; Hinsby, K. Natural Background Levels in Groundwater. *Water* **2021**, *13*, 2770. https://doi.org/10.3390/w13192770

Received: 23 September 2021
Accepted: 5 October 2021
Published: 6 October 2021

Publisher's Note: MDPI stays neutral with regard to jurisdictional claims in published maps and institutional affiliations.

High levels of inorganic compounds in groundwater represent a significant problem in many parts of the world, with major economic, social and environmental drawbacks. The natural composition of groundwater derives mainly from the water–rock interactions, both in the vadose and saturated zone, but also depends on the biological and physico-chemical processes, the residence time and the initial composition of the recharge water. Contamination from industrial, agricultural and urban areas often overlaps with the natural features of groundwater, and the assessment of the impact of anthropogenic activities might be challenging.

The distinction between natural and anthropogenic components that determine groundwater chemistry is a fundamental issue in groundwater management, particularly when the concentration of inorganic compounds exceeds the threshold values set for the evaluation of the good chemical status of groundwater bodies, as requested by many environmental regulations, including the European Water Framework and Groundwater Directives. In order to reach the good status of groundwater, natural background levels (NBLs) assessment of chemical species in groundwater is needed to clearly define the environmental objectives for groundwater bodies as well as the remediation targets for contaminated sites.

A variety of methodologies have been adopted by different countries to assess the NBLs of chemical species in groundwater that are potential pollutants and may be hazardous for human health and groundwater-dependent ecosystems (GDE). In Europe, the research efforts over the last 20 years have been dedicated to the elaboration of common standardized methodologies for the derivation of groundwater NBLs through the development of dedicated joint projects: from BASELINE [1] (*"Natural Baseline Quality in European Aquifers: a basis for aquifer management"*) to BRIDGE [2] (*"Background criteria for the identification of groundwater thresholds"*), arriving to the ongoing project HOVER [3] (*"Hydrogeological processes and Geological settings over Europe controlling dissolved geogenic and anthropogenic elements in groundwater of relevance to human health and the status of dependent ecosystems"*).

The continuous improvement in methodologies, tools and approaches for estimating groundwater NBLs has inspired this Special Issue, which aims to provide new insights into how NBLs are defined in different regions of the world, and to provide an update on the methods and approaches used to derive NBL at different spatial scales: site-specific, catchment, regional, national or transboundary scales.

The main goal of this Special Issue of *Water* is to bring together and discuss different methodological approaches and tools to improve the assessment of groundwater NBLs. From its first announcement, and after being thoroughly peer reviewed, nine papers have been accepted for publication [3–11]. To provide an overview of the experiences collected by this Special Issue, a brief summary of each published paper is reported below. A variety of spatial scales is encountered, ranging between national [3,11], regional [3,7,8,10], meso-[6,9] and site-specific [4,5] scales, and different methodologies are applied.

For example, Lions et al. [3] describe the methodology developed within the EU project HOVER through its application to six study areas for the calculation of the NBLs of trace elements on a broad scale (regional or national). This method uses a statistical approach and is based on the lithological classification and on the land-use analysis, leading to the estimation of a NBL value for each class of lithology and geochemical conditions (e.g., pH, redox) identified. The authors point out that this method provides consistent results when large datasets are available, and the aquifers have fairly homogeneous hydrogeological and hydrogeochemical conditions. Where local-scale variability is dominant or specific anthropogenic pressures are relevant, local investigations are needed. The work by Voutchkova et al. [11] compares the results of NBL estimation obtained from different HOVER and BRIDGE (preselection) applications for trace metals nationwide (Denmark). Pros and cons of the different methods used are highlighted. This allows the authors to propose a roadmap for NBL estimation at the national scale in countries with variable data availability. The roadmap provides *"a systematic way of selecting an appropriate method or combination of methods to assure that NBLs are calculated based on groundwater data representing no or only very minor anthropogenic alterations to undisturbed conditions"*.

Another comparison of different methods is presented by Nakić et al. [8] for the estimation of the ambient background levels (ANLs) in four groundwater bodies in Croatia. The concept of ABL differs from that of NBL in that it addresses long-term human impacts (e.g., agriculture, urbanization) that have irreversibly altered the natural background hydrochemistry. The methods compared are (a) the BRIDGE preselection, (b) the probability plot and (c) the modified Lepeltier. Interestingly, the authors conclude by recommending the use of the probability plot and modified Lepeltier methods only when the censored data (i.e., concentrations below the limit of quantification) are less than 30%, and the use of preselection with the censored data greater than 30%.

Masciale et al. [7] and Parrone et al. [9] apply the Italian national guidelines for the assessment of the NBL in groundwater bodies, based on the preselection of data, enriched by temporal trend analyis and outliers treatment. The Italian Guidelines involve different procedures for the assessment of NBLs according to the spatial and temporal dimension of the sample and on the type of distribution of the pre-selected dataset, also considering the redox conditions of groundwater. The resulting NBLs are provided with their respective confidence levels, assessed on the basis of the total number of observations/monitoring sites, extent of the groundwater body and the type of aquifer. Both papers rely on a dedicated tool for the estimation of NBLs according to the national guidelines: eNaBLe [9]. The case studies are at the mesoscale (Vulsini volcanic district in central Italy) [9] and regional scale (Apulia region) [7].

Chidichimo et al. [5] also present a tool for calculating groundwater NBLs: GuEstNBL. This tool is able to assess groundwater NBLs by applying BRIDGE preselection and component separation methods. GuEstNBL is applied for the calculation of NBLs of trace elements on a site-specific scale (an industrial site located in Lamezia Terme, southern Italy). Here, the calculated NBLs support the definition of specific environmental cleanup goals.

Sacchi et al. [4] present another case of NBL assessment in groundwater on a site-scale case study: a former asbestos mine in Serpentinite in northern Italy. In mining environments, anthropogenic activities boost water–rock interactions, further increasing the concentration of potentially toxic elements, thus making the estimation of their groundwater NBL a challenging task. The authors use a statistical approach, based on the Italian Guidelines, to assess the NBLs in the former mining site. Interestingly, they conclude by giving a warning about the use of the median as a representative value for concentration time series of each monitoring station, as recommended by several guidelines (including BRIDGE); the median results in the elimination of half of the measurements and, in particular, higher concentration values, which could lead to overly conservative NBL estimates.

Filippini et al. [6] focus on the problem that national/regional groundwater monitoring networks may fail to capture hydrochemical heterogeneities on a local scale, possibly leading to misleading NBL values on a meso scale. In this work, the lack of information provided by the regional monitoring is filled with site-specific monitoring networks in contaminated/polluted sites under remediation. Since the common procedures for the assessment of NBLs cannot be applied to this type of dataset due to the limited homogeneity of the data, an *"unorthodox"* method is proposed based on the definition of a consistent working dataset, followed by a statistical identification and a critical analysis of the outliers.

Finally, Shahzad et al. [10] report a case study analysing the spatial (regional) and temporal variability of groundwater quality data using statistical and geostatistical techniques. Although NBLs are not calculated, this work highlights the importance of exploratory analysis in supporting the definition of hydrogeochemical conceptual models, further supporting the assessement of groundwater NBLs.

The articles presented in this Special Issue provide an overview of the different approaches and protocols in use for NBL assessment, ranging from well-known methods such as component separation or cumulative probability plot methods, to new computer-aided protocols or the recent "HOVER methodology" [3], based on the exploitation of large datasets in regions where groundwater monitoring networks have been operating for many years.

A variety of case studies describe different situations where NBL is needed to properly assign good or poor quality status to groundwater bodies. Key issues are highlighted, such as the correct definition of conceptual models, the necessity to clearly differentiate between pre-selection criteria and thresholds for NBL definition, the link between NBL assessment and groundwater monitoring networks and the need for targeted analytical methods to achieve lower LOQs.

Current trends in NBL assessment focus on methods for assessing the spatial distribution and temporal variation of NBLs that could impact the legal assessment of polluted sites or water bodies. Overall, this collection of papers will be a source of inspiration for future research and development on this topic. It is not yet possible to make a recommendation for a preferred method to derive NBLs, which would be robust enough to be applied in most hydrogeological and land use settings across Europe and worldwide, providing comparable NBLs. The selection of published articles provides examples covering a range of different settings that will be relevant to many, but not all settings commonly found in Europe and worldwide. In all cases, an individual assessment is required to decide whether one or more of the methods presented in this Special Issue will meet the needs of NBL derivation in other settings.

Author Contributions: Writing—original draft, E.P. and M.R.; writing—review and editing, E.P., M.R., M.T.C.d.M. and K.H. All authors have read and agreed to the published version of the manuscript.

Funding: This research received no external funding.

Acknowledgments: The authors appreciate the efforts of the *Water* editors and the publication team at MDPI, and the anonymous reviewers for their invaluable comments.

References

1. Nieto, P.; Custodio, E.; Manzano, M. Baseline Groundwater Quality: A European Approach. *Environ. Sci. Policy* **2005**, *8*, 399–409. [CrossRef]
2. Hinsby, K.; Condesso de Melo, M.T.; Dahl, M. European Case Studies Supporting the Derivation of Natural Background Levels and Groundwater Threshold Values for the Protection of Dependent Ecosystems and Human Health. *Sci. Total Environ.* **2008**, *401*, 1–20. [CrossRef] [PubMed]

3. Lions, J.; Devau, N.; Elster, D.; Voutchkova, D.D.; Hansen, B.; Schullehner, J.; Petrović, P.T.; Samolov, K.A.; Camps, V.; Arnó, G.; et al. A Broad-Scale Method for Estimating Natural Background Levels of Dissolved Components in Groundwater Based on Lithology and Anthropogenic Pressure. *Water* **2021**, *13*, 1531. [CrossRef]
4. Sacchi, E.; Bergamini, M.; Lazzari, E.; Musacchio, A.; Mor, J.-R.; Pugliaro, E. Natural Background Levels of Potentially Toxic Elements in Groundwater from a Former Asbestos Mine in Serpentinite (Balangero, North Italy). *Water* **2021**, *13*, 735. [CrossRef]
5. Chidichimo, F.; Biase, M.D.; Costabile, A.; Cuiuli, E.; Reillo, O.; Migliorino, C.; Treccosti, I.; Straface, S. GuEstNBL: The Software for the Guided Estimation of the Natural Background Levels of the Aquifers. *Water* **2020**, *12*, 2728. [CrossRef]
6. Filippini, M.; Zanotti, C.; Bonomi, T.; Sacchetti, V.G.; Amorosi, A.; Dinelli, E.; Rotiroti, M. Deriving Natural Background Levels of Arsenic at the Meso-Scale Using Site-Specific Datasets: An Unorthodox Method. *Water* **2021**, *13*, 452. [CrossRef]
7. Masciale, R.; Amalfitano, S.; Frollini, E.; Ghergo, S.; Melita, M.; Parrone, D.; Preziosi, E.; Vurro, M.; Zoppini, A.; Passarella, G. Assessing Natural Background Levels in the Groundwater Bodies of the Apulia Region (Southern Italy). *Water* **2021**, *13*, 958. [CrossRef]
8. Nakić, Z.; Kovač, Z.; Parlov, J.; Perković, D. Ambient Background Values of Selected Chemical Substances in Four Groundwater Bodies in the Pannonian Region of Croatia. *Water* **2020**, *12*, 2671. [CrossRef]
9. Parrone, D.; Frollini, E.; Preziosi, E.; Ghergo, S. eNaBLe, an On-Line Tool to Evaluate Natural Background Levels in Groundwater Bodies. *Water* **2021**, *13*, 74. [CrossRef]
10. Shahzad, H.; Farid, H.U.; Khan, Z.M.; Anjum, M.N.; Ahmad, I.; Chen, X.; Sakindar, P.; Mubeen, M.; Ahmad, M.; Gulakhmadov, A. An Integrated Use of GIS, Geostatistical and Map Overlay Techniques for Spatio-Temporal Variability Analysis of Groundwater Quality and Level in the Punjab Province of Pakistan, South Asia. *Water* **2020**, *12*, 3555. [CrossRef]
11. Voutchkova, D.D.; Ernstsen, V.; Schullehner, J.; Hinsby, K.; Thorling, L.; Hansen, B. Roadmap for Determining Natural Background Levels of Trace Metals in Groundwater. *Water* **2021**, *13*, 1267. [CrossRef]

Article

Roadmap for Determining Natural Background Levels of Trace Metals in Groundwater

Denitza D. Voutchkova [1,*], Vibeke Ernstsen [2], Jörg Schullehner [1,3], Klaus Hinsby [2], Lærke Thorling [1] and Birgitte Hansen [1]

1   Geological Survey of Denmark and Greenland (GEUS), C.F. Møllers Allé 8, 8000 Aarhus C, Denmark; jsc@geus.dk (J.S.); lts@geus.dk (L.T.); bgh@geus.dk (B.H.)
2   Geological Survey of Denmark and Greenland (GEUS), Øster Voldgade 10, 1350 København K, Denmark; ve@geus.dk (V.E.); khi@geus.dk (K.H.)
3   Department of Public Health, Research Unit for Environment, Work and Health, Aarhus University, Bartholins Allé 2, 8000 Aarhus C, Denmark
*   Correspondence: dv@geus.dk

Abstract: Determining natural background levels (NBLs) is a fundamental step in assessing the chemical status of groundwater bodies in the EU, as stipulated by the Water Framework and Groundwater Directives. The major challenges in deriving NBLs for trace metals are understanding the interaction of natural and anthropogenic processes and identifying the boundary between pristine and polluted groundwater. Thus, the purpose of this paper is to present a roadmap guiding the process of method selection for setting meaningful NBLs of trace metals in groundwater. To develop the roadmap, we compared and critically assessed how three methods for excluding polluted sampling points affect the NBLs for As, Cd, Cr, Cu, Ni, and Zn in Danish aquifers. These methods exclude sampling points based on (1) the primary use of the well (or sampling purpose), (2) the dominating anthropogenic pressure in the vicinity of the well, or (3) a combination of pollution indicators (NO_3, pesticides, organic micropollutants). Except for Ni, the NBLs derived from the three methods did not differ significantly, indicating that the data pre-selection based on the primary use of the wells is an important step in assuring the removal of anthropogenically influenced points. However, this pre-selection could limit the data representativity with respect to the different groundwater types. The roadmap (a step-by-step guideline) can be used at the national scale in countries with varying data availability.

Keywords: groundwater; natural background levels; arsenic; cadmium; chromium; copper; nickel; zinc; Denmark

Citation: Voutchkova, D.D.; Ernstsen, V.; Schullehner, J.; Hinsby, K.; Thorling, L.; Hansen, B. Roadmap for Determining Natural Background Levels of Trace Metals in Groundwater. *Water* **2021**, *13*, 1267. https://doi.org/10.3390/w13091267

Academic Editor: Alexander Yakirevich

Received: 3 March 2021
Accepted: 28 April 2021
Published: 30 April 2021

Publisher's Note: MDPI stays neutral with regard to jurisdictional claims in published maps and institutional affiliations.

1. Introduction

Trace metals are ubiquitous in groundwater in concentrations that are determined primarily by natural processes, but in addition, they may be affected by anthropogenic factors. Significant exceedances of the natural backgrounds are a question of anthropogenic pollution or other human activities changing the natural geochemistry that may eventually require the development of action plans and remediation measures. Therefore, the assessment of natural background levels (NBLs) for trace metals is a prerequisite to determine whether observed concentrations are affected by human activities. In that case, remediation measures need to be introduced to protect the legitimate groundwater uses and groundwater-dependent terrestrial and associated aquatic ecosystems. The remediation measures then aim at reducing the groundwater concentrations below a specific threshold value (TV) for the pollutant, defined by the member states based on the estimated NBL for the pollutant according to EU legislation and guidance [1–3].

1.1. Derivation of NBLs and TVs for Assessment of Groundwater Chemical Status in EU

The European Water Framework Directive (WFD) [1] and Groundwater Directive (GWD) [2] stipulate that groundwater status must be assessed by the member states and that good status must be achieved to protect human health and groundwater-dependent or associated ecosystems, e.g., wetlands, and transitional and coastal waters. According to the GWD, TVs are groundwater quality standards for pollutants or groups of pollutants established by the individual member states for ensuring compliance with the definition of good chemical status of groundwater bodies. The WFD defines a groundwater body as a distinct volume of groundwater within an aquifer or system of aquifers. "Aquifer" is then defined as *"a subsurface layer or layers of rock or other geological strata of sufficient porosity and permeability to allow either a significant flow of groundwater or the abstraction of significant quantities of groundwater"* [1]. The national authorities of the member states derive these TVs as the basis for the groundwater status assessments. Annex II.A of GWD (see Appendix A for direct quote) provides a general guideline on the TV derivation, while more details can be found in Guidance Document No. 18 [3]. Hinsby et al., 2008 [4] presented a selection of application examples.

The groundwater chemical status provisions of the GWD only apply for anthropogenically altered conditions. Thus, a fundamental first step in establishing TVs is assessing the NBLs, i.e., the naturally occurring concentrations of substances for different hydrogeological conditions. The definition provided in the GWD (Article 2.5) states that *" . . . 'background level' means the concentration of a substance or the value of an indicator in a body of groundwater corresponding to no, or only very minor, anthropogenic alterations to undisturbed conditions"*. Member States are free to apply their own approach for identifying these NBLs, depending on existing studies and conceptual models of the groundwater bodies [3]. According to the widely used BRIDGE methodology [4], NBLs are derived as the 90th (or 97th) percentiles of a pre-selected dataset, which should approximate a natural groundwater composition of a given aquifer type [4] (see details in Section 3.2). The 90th percentile was suggested for small datasets (<~60 sampling points) or datasets where human impact cannot be excluded, while the 97th percentile is for groundwater bodies where all data points represent groundwater with a natural composition [4]. Guidance Document No. 18 [3] also refers to the BRIDGE methodology and mentions the 90th percentile as a practical criterion for setting the NBLs.

The NBL derivation in this context is similar to determining groundwater baseline quality with statistical methods, which aim at distinguishing anomalies from typical values [5]. Even though sometimes NBL and baseline quality are used as synonyms, there is an important distinction to be made. Groundwater baseline quality is *"the range of concentrations derived entirely from natural, geological, biological or atmospheric sources, under conditions not perturbed by anthropogenic activity"* [5], while the NBLs could include minor anthropogenic influences (see definition above) and is represented by a single value from which TV can be derived. Identifying groundwater baseline properties could rely on various other approaches (next to the purely statistical ones), e.g., use of historical data (pre-industrial conditions), down-gradient profiles, extrapolation from adjacent areas with similar geology, groundwater dating, and geochemical modeling [5]. Determining the groundwater baseline quality is beyond the scope of this study, where we focus on NBLs.

1.2. Purpose of This Study

The major challenges in deriving NBLs are (1) understanding how the interaction of natural and anthropogenic processes affects groundwater quality and (2) identifying the boundary between pristine (or nearly pristine) and polluted groundwater. EU member states need to set NBLs to be able to define TVs for the chemical status assessment of groundwater bodies, stipulated in the EU legislation. It was previously demonstrated that a very wide range of TVs is derived and reported by the member states, partly because of the NBLs derivation, and that further harmonization is warranted [6,7]. Therefore, the general objective of this study is to present a roadmap—a guideline on how to derive meaningful NBLs for trace metals, which can be an especially challenging task in the context of intensive and widespread agriculture, and extensive groundwater pumping for drinking water supply in urban areas. Our specific objectives are:

(1) To apply and compare three different methods for excluding anthropogenically influenced points when calculating the NBLs for trace metals in Denmark. These methods rely on the exclusion of water sampling points from the datasets, based on:

- Primary use of the well (and/or the sampling purpose);
- Dominating land-use (thus, potential anthropogenic pressure);
- Combination of pollution indicators.

(2) To critically assess, i.e., discuss requirements, advantages, and disadvantages of the individual methods, and on that basis to develop a universally applicable roadmap for NBLs derivation at the national scale.

Through this process, we aim to demonstrate also how combining several methods, using several types of data may compensate for the individual limitations of the methods.

2. Study Setting

2.1. Denmark—A Case Study with Widespread and Intensive Agricultural Pressure

Denmark is an EU country with an area of ~43,000 km^2 and a population of ~5.8 million (2018, Eurostat). Agricultural land covers 61% of Denmark (Figure 1a), a major part of which is used for the annual cultivation of grains, grass in crop rotation, and rapeseed. Forests cover 13%, other nature (wet or dry, incl. meadows and heaths) 9%, and the buildings and built-up areas cover 7% (the land-cover statistics isfrom Statistic Denmark https://www.dst.dk (accessed on 15 October 2020)). This places Denmark in third place in Europe, after Ireland and U.K., on percent used agricultural area, according to Eurostat (see Farms and farmland statistics, https://ec.europa.eu/eurostat/statistics-explained/ (accessed on 1 January 2020)). In 2019, there were about 33,600 farms with a total number of 1.49 million cattle and 12.3 million pigs. Denmark also has a very high production number of intensively reared pigs (>30 million/year) [8] and relatively high cow milk production (5693 million kg for 2019). About 11% of the agricultural land was either cultivated organically or is being converted to organic farming in 2019.

The Danish landscape was shaped by a sequence of Pleistocene glaciations and post-glacial processes, and the resulting topography is flat to gently undulating with a maximum elevation of 170 m [9].

The drinking water supply in Denmark is highly decentralized (~2600 active public waterworks) and relies 100% on groundwater. The groundwater for drinking water purposes is extracted from aquifers primarily composed of: (1) Quaternary glacio-fluvial sand and gravel deposits, (2) Upper Cretaceous and Danian limestone and chalk, and (3) Miocene quartz sand and micaceous sand [10]. The island of Bornholm is an exception, as there are various older fractured aquifers (see Figure S1). A brief introduction to the Quaternary and pre-Quaternary geology [10,11] of Denmark is provided in Supplementary Materials.

Based on the National Water Resources Model for Denmark, DK model (https://vandmodel.dk/ (accessed on 29 April 2021)), 2050 groundwater bodies have been delineated in Denmark [12]. The WFD definition of groundwater body (see Section 1.1) was followed, and the specific delineation criteria included [12]: (1) a minimum thickness of 2 m and minimum extend of 25 ha, (2) aquifers of the same geological type were grouped together only if there was a hydraulic contact between them (existing low-permeability layers were <2 m thick). In some cases, large groundwater bodies (>100 ha) were subdivided based on their hydrological and geomorphological characteristics, or differences in aquifer thickness, to limit the heterogeneity within the groundwater bodies. The majority of the 2050 Danish groundwater bodies do not have any water analyses. In this study, we investigated analyses of trace metals from drinking water wells, tapping into 451 of them (Figure 1b). Because the 451 groundwater bodies have varying area and volume (and data availability), for the purposes of deriving NBLs here, we grouped them according to four main aquifer types: (1) Quaternary sand, (2) Carbonate fractured rocks, (3) pre-Quaternary sand, and (4) the diverse geological units on the island of Bornholm (Figure 1b).

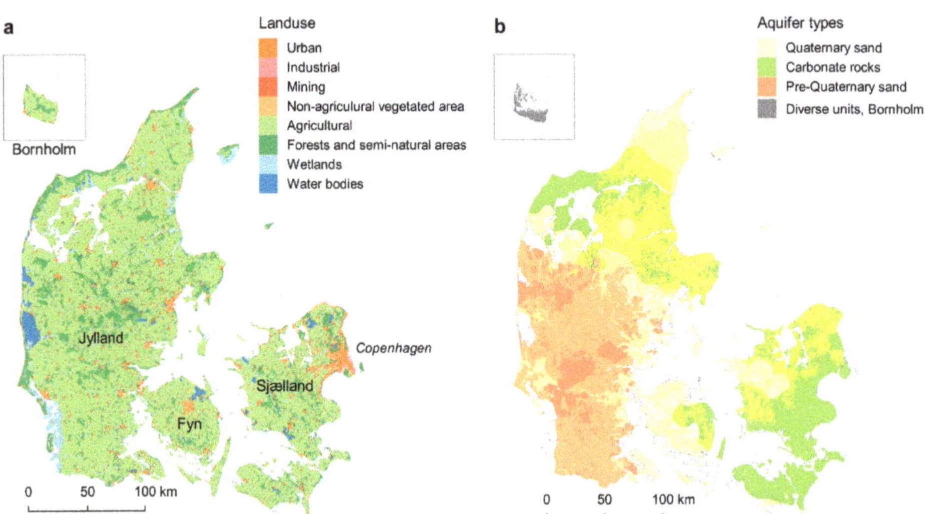

Figure 1. (**a**) Land-use map (CORINE 2012, downloaded from https://download.kortforsyningen.dk/content/corine-land-cover (accessed on 20 May 2020)). (**b**) Groundwater bodies included in this study; the Quaternary sand is shown with transparency above the other aquifer types, so the underlying carbonate and pre-Quaternary sand aquifers can also be seen.

2.2. TV and NBL Assessments in Denmark

The national TVs for trace metals in Denmark originate from the national drinking water legislation, which agrees with the EU Drinking Water Directive [13]. In accordance with Guidance Document No. 18 [3], higher TVs are established for Danish groundwater bodies, where the NBLs exceed the national TVs. The national TVs could also be lowered if the drinking water standards do not provide sufficient protection of groundwater-dependent terrestrial or associated aquatic ecosystems [4], but this approach has not yet been applied in Denmark. Table 1 shows the current national groundwater TVs, set by the Danish Environmental Protection Agency (Danish EPA), and the Danish drinking water standards (BEK nr 1070 af 28/10/2019).

Table 1. Drinking water quality standard and groundwater threshold values in Denmark.

	Unit	As	Cd	Cu	Cr	Ni	Zn
National threshold value	µg/L	5	0.5	100	25	10	100
National drinking water standard	µg/L	5	3	2000	50	20	3000

NBLs for trace metals have been derived in Denmark as part of the chemical status assessment of Danish groundwater bodies in the second and the ongoing third River Basin Management Plans, abbreviated further as MP2 (period 2015–2021) and MP3 (2021–2027); and as part of the research projects BRIDGE and HOVER. BRIDGE is an acronym for the project *"Background criteria for the identification of groundwater thresholds"*, which ran in the period January 2005–December 2006 (https://cordis.europa.eu/project/id/6538 (accessed on 29 April 2021)) and HOVER is an acronym of the ongoing project *"Hydrogeological processes and Geological settings **over** Europe controlling dissolved geogenic and anthropogenic elements in groundwater of relevance to human health and the status of dependent ecosystems"* in *the GeoERA program* (https://geoera.eu/ (accessed on 29 April 2021)).

In MP2, NBLs for As and Ni were derived as the 90th percentile of the mean concentrations at the sampling points for 2007–2013, or if not available, for 2000–2006 (see Supplementary Materials for more details).

The national NBLs for trace elements (Al, As, Pb, Cd, Hg, Ni, Pb, Cu, Cr, and Zn) is currently under revision for the MP3. This time, a larger focus is placed on the different hydrogeochemical conditions in Danish aquifers. Groundwater types are distinguished based on the redox, pH, and organic matter content (as non-volatile organic carbon, NVOC) [14]. Other differences from MP2 include aquifer type classification simplification and a different sampling period (2000–2018). NBLs were derived as the 90th percentile of the mean concentration calculated from the annual means in the period at pre-selected sampling points. Expert assessment deemed some of the calculated NBLs unreliable due to potential anthropogenic influences [14]. For example, the NBL for Ni in some aquifers were not used due to known elevated concentrations caused by oxidation and reduction of iron sulfides and manganese oxides, respectively, because of (1) initial lowering of the water table due to excessive abstraction and the oxidation of nickel containing pyrite followed by (2) rising water tables that submerge and reduce manganese oxides releasing adsorbed nickel [15–17].

BRIDGE project developed a tiered methodology for TV derivation [4]. The Danish case study focused on the TVs for N and P based on environmental objectives for dependent ecosystems at the Odense river basin, located on the isle of Fyn (Figure 1). Additionally, NBLs for other elements, including As, were calculated based on pre-selection criteria. A summary of these criteria commonly referred to in the NBL literature as the "BRIDGE method" is presented in SM.

A task in the ongoing HOVER project aims at proposing a common methodology to identify the main geological factors and hydrogeological processes controlling the distribution of NBLs of selected dissolved elements [18]. The tested method included identifying anthropogenic pressures and relating those to specific dissolved solutes (As, Cd, Cr, Cu, Ni, Zn, F, Cl, SO_4) in areas under agricultural, industrial, mining, and urban influences. A variety of applications are presented for different settings in Europe in [18].

3. Materials and Methods

3.1. Identifying Anthropogenically Influenced Water Sampling Points

To develop a road map for determining NBLs for trace metals, we apply, compare, and critically assess three methods for identifying and excluding anthropogenically influenced groundwater sampling points: (1) the *basis* level method relying on a pre-selection of sampling points only, (2) a land-use based method developed in HOVER [18], and (3) a modification of the BRIDGE method (Figure 2).

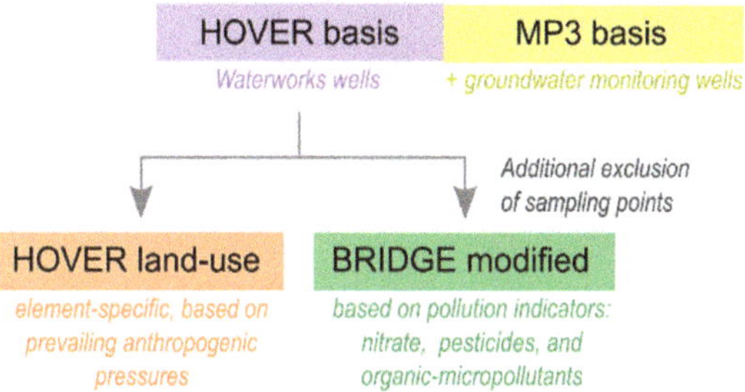

Figure 2. Compared methods and how they relate to each other, the *basis* method has two versions—from the HOVER project and from the third River Basin Management Plan (MP3). The dataset from *HOVER basis* is further used with *HOVER land-use* and *BRIDGE modified* methods.

The *basis level* method relies on a pre-selection of wells based on their primary usage. The chemical analyses of groundwater samples from these pre-selected wells form the dataset used further. Two versions of this method are compared to each other to assess the effect of both sampling pre-selection and different data handling. The *HOVER basis* version includes only waterworks wells used for drinking water production. While for

MP3 basis version (the official national NBL derivation part of MP3), wells of the national groundwater monitoring program are also included [14]. Even though such pre-selection is possible, there could still be sampling points influenced by anthropogenic activities. Thus, to further "clean" the dataset, we tested *HOVER land-use* and *BRIDGE modified* methods.

HOVER land-use method [18] relies on a land-use-based approach for identifying anthropogenic pressures and excluding sampling points potentially influenced by it. The dominating anthropogenic pressure for each sampling point was established based on the areal proportion of the land-use types in the vicinity of the well. For each heavy metal, the groups of sampling points with different anthropogenic pressure were compared, and the significantly different groups were excluded. The group comparison was performed based on Kruskal–Wallis test with a post-hoc Nemenyi test, and the statistical significance was assessed at the 95% confidence level [18]. The statistical testing is performed for each element separately, so different groundwater sampling points are excluded for different elements.

BRIDGE modified relies on a combination of pollution indicators, including NO_3, pesticides, and organic micropollutants. Nitrate and pesticides are used here as indicators for agricultural pollution, while the organic micropollutants serve as indicators for urban or industrial pollution. We consider a water sampling point to be polluted and exclude it from the dataset if at least one of the following conditions is true:

- $NO_3 > 10$ mg/L—a condition from the original BRIDGE method [4];
- At least one of the analyzed pesticides (metabolites, degradation, or transformation products) is exceeding the drinking water standard for individual pesticides (0.1 µg/L) or the sum of the quantified pesticides (0.5 µg/L);
- At least one of the organic micropollutants is exceeding the specific drinking water standards.

The *HOVER basis* dataset is used as a starting point for applying both the *HOVER land-use* and *BRIDGE modified*. Thus, when comparing the NBLs derived from the *HOVER land-use* and *BRIDGE modified* to those derived based on *HOVER basis*, we assess the effect of additional removal of potentially polluted sampling points. The assumption underlying our assessment is that this removal should result in lowered NBLs for the selected trace metals.

3.2. Trace Metals—Sources and Geochemical Controls

The main geogenic and anthropogenic sources and the relevant geochemical controls of the selected trace metals (As, Cd, Cr, Cu, Ni, and Zn) are summarized in Table A1 [5,19]. The main geochemical processes governing their concentrations in groundwater are dissolution of minerals or desorption from either Fe and Mn (oxy)hydroxides, clays, calcite surfaces, or organic matter (Table A1). Both industrial and urban areas may be sources of potential anthropogenic pollution, but intensive agriculture could also be a major contributor, as some of these elements are present in either pesticides or fertilizers (Table A1). The mobility of these elements in groundwater is directly or indirectly controlled by pH and redox conditions (Table A1); therefore, the groundwater pH and redox state at the sampling point should also be considered when calculating the NBLs for these elements.

We classified the sampling points into the following pH classes [18]:

- Acidic (pH < 7);
- Basic (pH > 7.5);
- Neutral ($7 \leq$ pH ≤ 7.5).

For the redox classification, we used an algorithm based on the O_2, NO_3 and Fe content [16]. The algorithm conditions are given in the brackets (more details in [16]):

- Oxic (A type, if $NO_3 > 1$ mg/L and Fe < 0.2 mg/L and $O_2 \geq 1$ mg/L);
- Anoxic, nitrate reducing (B type, if $NO_3 > 1$ mg/L and Fe < 0.2 mg/L and $O_2 < 1$ mg/L);
- Reduced (C and D types, if $NO_3 \leq 1$ mg/L and Fe ≥ 0.2 mg/L);
- Mixed (X and Y types, do not fulfil the conditions for A, B, C, and D types).

3.3. Data Sources and Processing

All chemical data originate from the open and freely accessible national well database Jupiter, hosted by the Geological Survey of Denmark and Greenland (GEUS). Groundwater sampling follows the procedures outlined in the national guidelines (e.g., https://www.geus.dk/media/8324/g02_proevetagning-okt12_uk.pdf (accessed on 29 April 2021)). Dataset-specific details are provided below.

3.3.1. Primary Chemical Dataset (HOVER Basis)

The raw chemical data are from the Danish groundwater monitoring program dataset extracted in July 2019 [20]. The general quality assurance is reported in [20], and the project-specific data pre-treatment is detailed in [18]. In brief, the data pre-treatment included various element-specific quality checks, treatment of all values below the limit of detection, and aggregation on sampling point level. The sampling points were limited to the waterworks wells used for the drinking water supply in Denmark. This differs from the *MP3 basis* dataset, in which also the groundwater monitoring wells were included. The other methodological differences in the data preparation for *MP3 basis* and *HOVER basis* methods are summarized in Table 2.

Table 2. Methodology comparison for NBL assessment based on the HOVER and MP3.

Method	HOVER Basis	MP3 Basis	Comparison
Data source	[20]		Same
Sampling points	Waterworks wells	Waterworks and monitoring wells	Overlap
Period	2009–2018 (incl.)	2000–2018 (incl.) [a]	Overlap
Limit of detection	<LOD = 1.5 × LOD		Same [b]
Aggregation (intake)	Median	Mean of annual means	Different
Aquifer types: - Geology - Location - pH - Redox - Organic matter	Carbonates, pre-Quaternary sand, Quaternary sand, Diverse - Acidic, neutral, basic Oxic, anoxic, reduced, mixed -	 Jylland, Sjælland, Fyn, Bornholm Low, high Oxic, anoxic high, low	Same Different Different Different Different
Representativeness [c]	20 (30)/50		Same
NBL computation	90th percentile		Same
Target spatial scale and use	Pan-European and non-regulatory	National and regulatory	Overlap

[a] trace elements data for this period, but only for sampling points that could be classified into pH, redox, NVOC classes based on data for 2009–2018; [b] MP3 had more stringent exclusion of data with high detection limits; [c] minimum number of sampling points for aquifer type to calculate NBL.

3.3.2. Complementary Data

The data necessary for classifying all water sampling points according to geology, pH, and redox types is also from [20]. The geology classification is derived from the established link between groundwater bodies and sampling points [12]. All sampling points included in this study were linked to a specific groundwater body ($n = 451$, Figure 1b); thus, it was possible to classify them in one of the main aquifer types (Figure S2). The same NO_3 data was used also for *BRIDGE modified*.

To apply *HOVER land-use*, we used Corine land cover (CLC-12) for Denmark (v.1 from October 2014 in 1: 100,000 scale with reference year 2012; https://download.kortforsyningen.dk/content/corine-land-cover (accessed on 20 May 2020)) (Figure 1a). A buffer of 1 km around each sampling point was used to determine the areal proportion of the different land-cover types. This approximates the catchment area of individual wells, as these are unknown at the national scale. Most likely, the actual catchment areas are not circular and differ in size. The prevailing anthropogenic pressure based on the dominating type (i.e., the type with the largest areal proportion within the 1 km buffer) includes:

- *Urban*—continuous and discontinuous urban fabric (CLC-12, Level 2 "urban fabric");
- *Industrial*—industrial or commercial units, road and rail networks, and associated land, port areas, and airports (CLC-12, Level 2 "industrial, commercial and transport units");
- *Agricultural*—non-irrigated arable land, fruit trees, and berry plantation, pastures, complex cultivation patterns, land principally occupied by agriculture with significant areas of natural vegetation (CLC-12, Level 1 "agricultural areas");
- *Mining*—mineral extraction sites, dump sites, and construction sites (CLC-12, Level 2 "mine, dump, and construction sites").

A sampling point was assigned the value *"natural or other"* if none of the listed above were dominating in 1 km buffer area, i.e., the dominating land-use type belonged to "forests and semi-natural areas", "wetlands", or "water bodies" (CLC-12 Level 1).

To apply *BRIDGE modified*, we also used two other aggregated datasets that were originally prepared (extracted, quality checked, cleaned, and aggregated) for the purposes of MP3. The first one contained data for five pesticides of interest: DEIA (desethyl-desisopropyl-atrazine, a degradation product of atrazine), BAM (2,6-dichlorbenzamide, a degradation product of dichlobenil and chlorthiamid), triazole, DPC (desphenyl chloridazon, a degradation product of chloridazon), and DMS (N,N-dimethylsulfamide, a degradation product of tolylfluanid and dichlofluanid); and, the maximum concentration of all analyzed pesticides and the sum of the pesticides for all sampling points with at least one pesticide analysis in the period 2013–2019. "Pesticide" term here also includes the metabolites, degradation, and transformation products of the pesticides. The aggregation on a sampling point for all MP3 projects was based on the mean calculated from the annual mean concentrations (2013–2019). This data was available for 12,688 sampling points, so we could classify almost all sampling points of the *BRIDGE basis* dataset ($n = 6221$, 97.4%) into (a) complying with, or (b) exceeding the drinking water quality criteria (0.1 µg/L for individual pesticides, 0.5 µg/L for the sum of pesticides). If there was exceedance for at least one of the parameters, then the water sampling point was assumed to be influenced by anthropogenic pollution.

The second aggregated dataset contained data for 50 different chemical compounds belonging to the groups' chlorinated solvents and degradation products, water-soluble solvents, phenolic compounds, MTBE (methyl tert-butyl ether), BTEXN (benzene, toluene, ethylbenzene, xylenes, and naphthalene) compounds, PFAS (per- and polyfluoroalkyl substances), cyanides, as well as the sum of PFAS and sum of chlorinated solvents and degradation products. Although there were data for 15,235 sampling points, only 48.2% ($n = 3079$) of the sampling points included in the *HOVER basis* dataset were covered. We classified these sampling points into (a) complying with or (b) exceeding the specific national standards set by the Danish EPA. If there was exceedance for at least one of the parameters, then the water sampling point was assumed to be influenced by anthropogenic pollution.

Additionally, we discuss the potential and limitations of applying other methods. We also use Cl and Na data [20] as in the original BRIDGE method [4]. Cl and SO_4 trends [21] were discussed as indicators for overexploitation of groundwater; and, the groundwater age based on tritium and chlorofluorocarbon CFC (11, 12, 113) [20] was discussed as an indicator for pre-industrial groundwater quality and pollution vulnerability.

3.4. Statistics and Software

TVs, and, respectively, NBLs, can be established at the national, river basin district, or at the level of the groundwater body, or groups of groundwater bodies [2]. Here we calculate the NBLs as the 90th percentile for aquifer types (i.e., groups of groundwater bodies of the same type). The 90th percentile is calculated with R package *stats* v.4.0.2 [22] ("quantile" function, type 7 [23]). A two-sided nonparametric confidence interval (CI) for the 90th percentile at the 95% confidence level was calculated with R package *EnvStats* v. 2.4.0 [24] ("eqnpar" function). CI could not be calculated for sub-sets with less than 30 sampling points; thus, we only report NBLs and their 95% CI for the classes with 30 or more sampling points. We construe no difference between the NBLs calculated based on the different methods if there is an overlap between the 95% CI (visualized as error bars). The distributions of heavy metal concentrations are presented graphically with empirical cumulative distribution plots. All calculations and graphs are completed in R v.4.0.2 [22] with the additional R packages: *stringr* v. 1.4.0 [25], *ggplot2* v. 3.3.2 [26], *tidyr* v. 1.1.2 [27], *dplyr* v. 1.0.2 [28], and *data.table* v. 1.13.0 [29]. All maps were prepared in QGIS [30], and the roadmap was created in Inkscape [31].

4. Results

4.1. Trace Metals in Danish Groundwater Used for Drinking Water Purposes

The concentration distribution for trace metals in Danish groundwater used for drinking water production, based on the *HOVER basis* dataset, is shown in Figure 3. The concentration levels on a national scale rank in the order from low to high: Cd < Cr < Cu < Ni < As < Zn. Figure 4 shows the spatial distribution of sampling points exceeding the national TVs for As and Ni (Table 1) for the main aquifer types. The rest of the trace metals do not exceed the TVs and have sparse spatial coverage, so they are not shown.

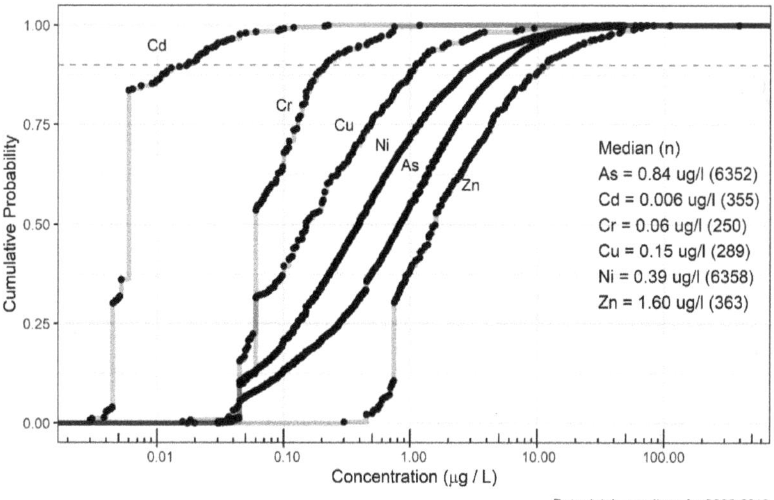

Figure 3. Empirical distribution for As, Cd, Cr, Cu, Ni, and Zn in Danish groundwater at waterworks wells used for drinking water production; each point is a median concentration for 2009–2018 at each sampling point.

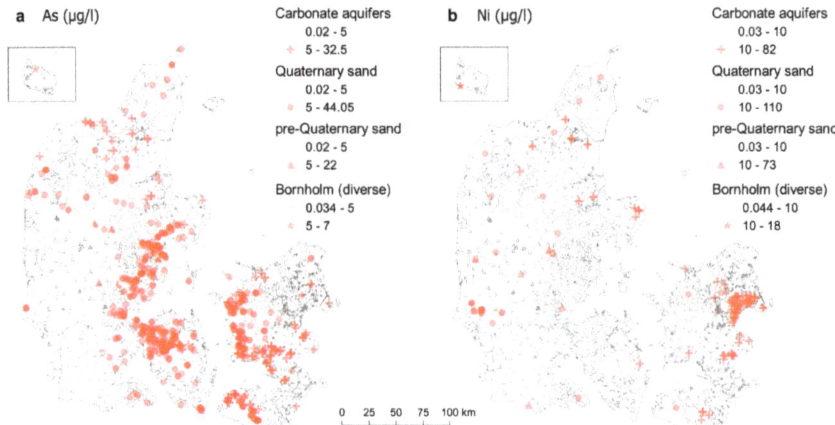

Figure 4. Spatial distribution of sampling points with As (**a**) and Ni (**b**) exceeding the national threshold values for As (5 µg/L) and Ni (10 µg/L).

4.2. Dataset Representativity

The percent sampling points in the *HOVER basis* dataset falling in different aquifer types, pH and redox classes, and prevailing anthropogenic pressure are presented in Table A2. The dataset is biased toward Quaternary sand aquifers (53–64%) with neutral pH (53–57%), reduced conditions (76–83%), and agricultural pressures (75–86%) (Table A2). The bias toward agriculturally dominated locations is because of the high percentage of arable land in Denmark (61%). The waterworks wells are usually located in areas with only minor point-source pollution, as the Danish drinking water supply relies on clean groundwater, which is treated only with conventional systems (aeration and sand-filtration). Thus, the low number of points with dominating industrial, mining, or urban influences could be explained by the pre-selection of sampling points, which here include only the waterworks wells.

There is an incomplete data coverage for pH, where 10–14% of the sampling points with trace metal data had no reported pH in the period (2009–2018). For redox, this percentage was <1%. The spatial distribution by pH and redox is shown in Figure 5. Empirical cumulative distribution plots stratified by aquifer, pH, redox, and prevailing anthropogenic pressure type can be seen in Figures S3–S6.

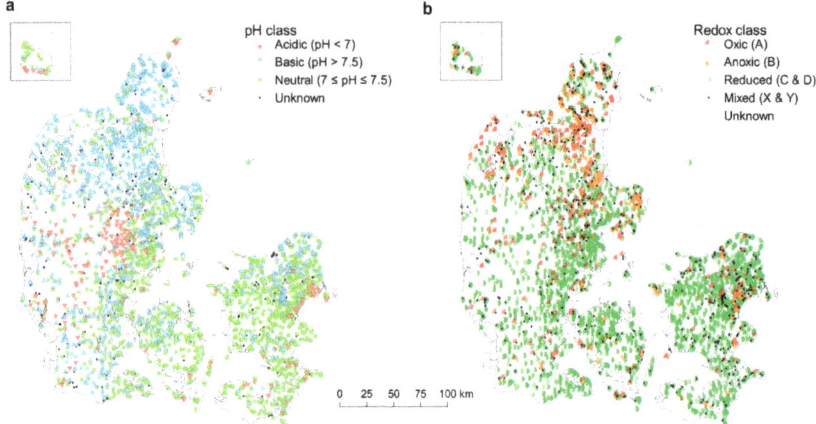

Figure 5. Spatial distribution of the groundwater sampling points based on their pH (**a**) and redox (**b**) class.

4.3. Excluding Sampling Points Due to Anthropogenic Influences

There was no additional assessment of pressures and exclusion of potentially polluted sampling points for the *HOVER basis* (and *MP3 basis)* method prior to NBL calculation. However, many sampling points were initially excluded as they did not belong to waterworks wells (or groundwater monitoring wells for *MP3 basis*). For example, polluted wells (e.g., decommissioned due to pollution), those for monitoring of polluted sites (point-source monitoring), and the remediation wells were excluded. To put this into context, the cleaned and aggregated trace elements dataset for the status assessment of the Danish groundwater bodies (part of MP3, sampling period 2013–2019) included 9343 sampling points belonging to (1) waterworks wells used for drinking water purposes (71.1%), (2) point-source pollution monitoring wells (6.3%), (3) wells part of the national groundwater monitoring or mapping program (21.6%), and (4) wells serving other purposes (1.1%). Thus, we could conclude that the pre-selection used in *HOVER basis* excluded about one-third of the sampling points with reliable data for trace metals in Denmark. Even though such pre-selection is made, it is possible that some of the sampling points in the *HOVER basis* dataset are influenced by anthropogenic activities. To further "clean" the dataset, we tested the *HOVER land-use* and *BRIDGE modified* methods.

Before calculating the NBLs for *HOVER land-use*, we excluded all sampling points with both *industrial* and *urban* prevailing pressure for As, only *industrial* for Cd and Cr, and only *urban* for Ni (Table A2 and Figure 6a) [18]. There was no statistically significant difference between any of the groups for Cu and Zn; thus, no sampling points were excluded for these elements [18]. The decision for excluding groups of sampling points specific to each element was based on the results of the Kruskal–Wallis test and post-hoc Nemenyi test. These statistical tests showed which of the groups of sampling points had a statistically significant difference in their elemental distributions at the 95% confidence level; thus, the different groups were excluded. In this statistical comparison, the *agricultural* pressure group was used as a background to which the distributions of the *urban* and *industrial* groups were compared. This was done because the *natural* group (without anthropogenic pressure) represented 1% or less of all sampling points (Table A2), all located near the coast (Figure S2b), thus not representative for the assessed aquifer types. At the same time, the group with agricultural pressure is the prevailing type throughout Denmark (Table A2). Thus, Mining was ignored due to the negligible representativity. Additionally, six water sampling points belonging to five wells were removed after manual assessment for potential pollution [18].

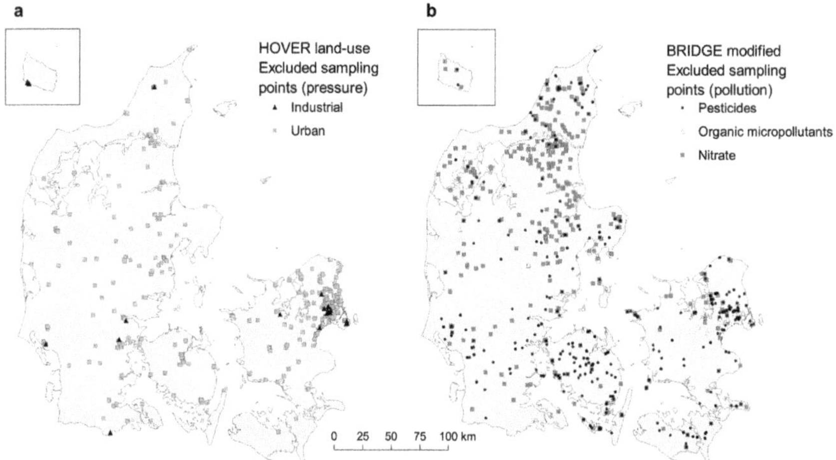

Figure 6. Location of sampling points that were excluded when calculating NBLs for the *HOVER land-use* (**a**) and the *BRIDGE modified* (**b**) methods.

Before calculating the NBLs for the *BRIDGE modified* method, we excluded 950 sampling points (14.9%) influenced by anthropogenic pollution (Figure 6b). They were identified based on the combination of pollution indicators: *pesticides (n = 448, 7.0%), organic micropollutants (n = 31, 0.5%),* and *nitrate (n = 552, 8.6%).*

Table 3 provides an overview of the number of sampling points in the *HOVER basis, HOVER land-use,* and *BRIDGE modified* methods after excluding the anthropogenically influenced points.

Table 3. Groundwater sampling points in the datasets after excluding the anthropogenically influenced points.

n	As	Cd	Cr	Cu	Ni	Zn
HOVER basis	6352	355	250	289	6358	363
HOVER land-use	5508	337	241	285	5558	359
BRIDGE modified	5410	297	208	239	5414	300
MP3 basis	5671	1666	913	1424	5672	1689

4.4. Comparison of NBLs Derived by the Different Methods

The NBLs of trace metals for the main aquifer types in Denmark, calculated based on the three methods, are compared in Figure 7. The only statistically significant difference (based on 95% CI) is for Ni in the carbonate aquifers and in the Quaternary sand, where *BRIDGE modified* results in lower values than *HOVER basis* (Figure 7). For all other types, the 95% CI is overlapping; thus, the differences are not significant. The NBLs for As in Quaternary sand exceed the national TV (5 µg/L) irrespective of the method.

We calculated NBLs considering redox and pH only for As and Ni (Figure 8) due to insufficient data for the rest of the elements. There is no significant difference between the NBLs obtained by the three methods for both Ni and As (Figure 8). The methods also agree that the NBLs for As are exceeding the national TV for carbonate aquifers with mixed redox conditions and neutral pH, and for Quaternary sand aquifers with reduced conditions and basic to neutral pH.

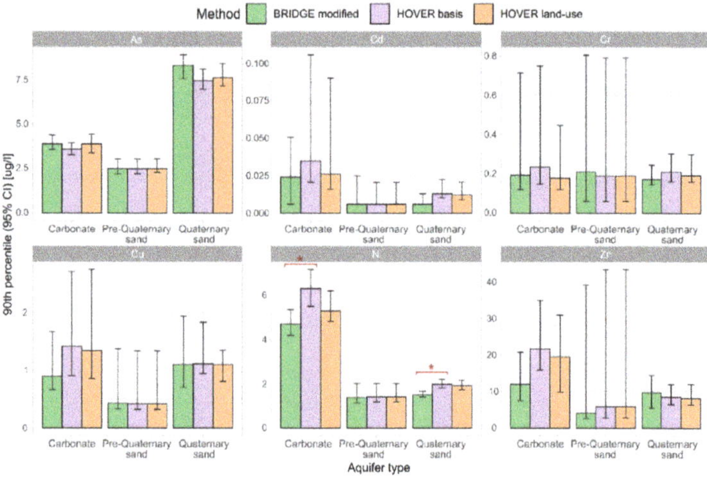

Figure 7. NBLs (y-axis, µg/L) for As, Cd, Cr, Cu, Ni, and Zn (see gray panel labels) based on the three methods. The symbol * indicates a statistically significant difference based on the 95% confidence intervals (CI); For values and number of sampling points, see Table S2.

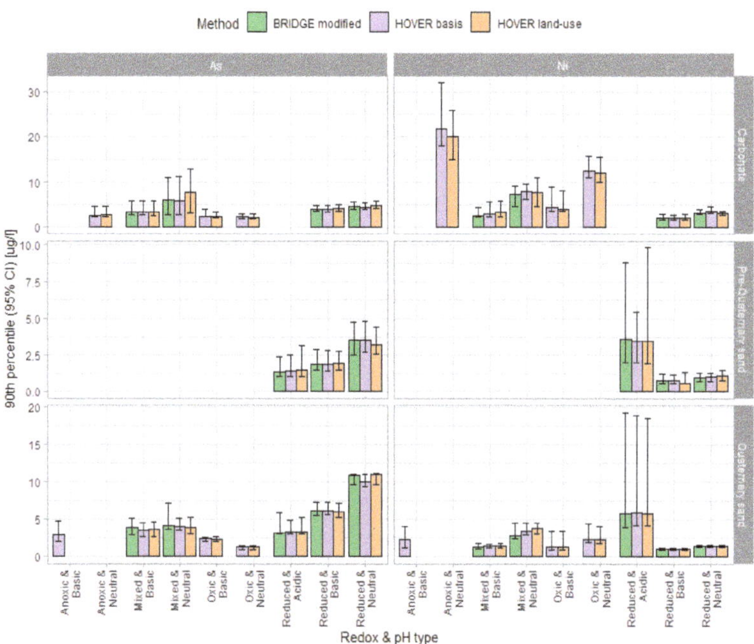

Figure 8. NBLs (y-axis, µg/L) for As and Ni (gray panel labels, top) considering the pH and redox (x-axis) of the aquifer types (gray panel labels, right). No significant differences are observed based on the 95% confidence interval (CI). Tables S3 and S4 provide values and number of samples.

HOVER basis and *HOVER land-use* agree that the NBLs for Ni exceed the national TV (10 µg/L) for carbonate aquifers with oxic/anoxic redox conditions and neutral pH. NBLs could not be derived based on the *BRIDGE modified* for this aquifer type as there were not enough sampling points. If the upper limit of the 95% CI is considered as well, there are few other classes with exceedance of the national TV (Tables S3 and S4).

The NBLs derived by the two versions of the *basis level* method (*HOVER rbasis* and *MP3 basis*) are compared in Figure 9. For this comparison, the location of the aquifer (Jylland, Sjælland, Fyn, Figure 1a) was also considered. This is how the aquifer types were defined in *MP3* [14], so to compare NBLs, we had to apply the same classification with the *HOVER basis* dataset. The following statistically significant differences in NBLs are observed:

- Quaternary sand aquifers on Jylland—Cd, Cr, Cu, Ni, and Zn;
- Quaternary sand aquifers on Fyn—Cu;
- Pre-quaternary sand aquifers on Jylland—Cd, Ni.

The two methods agree that the NBLs for As are exceeding the national TV for Quaternary sand, irrespective of the location (Fyn > Sjælland > Jylland). The NBL for Ni in carbonate aquifers on Sjælland is also elevated in comparison to those on Jylland and Fyn, but it does not exceed the national TV (10 µg/L). The upper confidence limit for *MP3 basis* is, however, exceeding it.

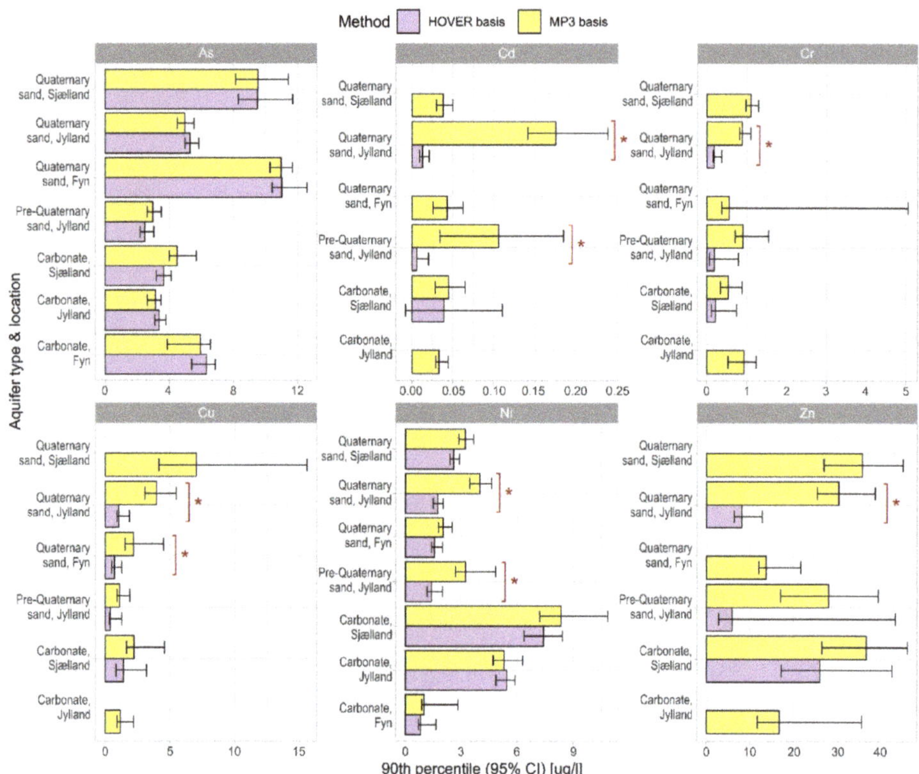

Figure 9. NBLs (x-axis, µg/L) for As, Cd, Cr, Cu, Ni, and Zn (see gray panel labels), calculated based on *HOVER basis* and *MP3 basis* methods considering the aquifer type and location (y-axis). The symbol * indicates a statistically significant difference based on the 95% confidence intervals (CI). See also Table S5 for the numbers and Figure 1 for geographical reference.

5. Discussion

5.1. Comparative Analysis of the Tested Methods

Table 4 summarizes our experience with the three methods for excluding anthropogenically influenced sampling points. The summary is organized into three categories: Requirements, advantages, and disadvantages.

The NBLs resulting from these three methods (Figures 7 and 8) did not differ significantly at the 95% CI. This could indicate that as a whole *HOVER basis* dataset was not affected by pollution in a significant way. The only exception was for Ni in the carbonate and Quaternary sand aquifers where *BRIDGE modified* resulted in lower NBLs than *HOVER basis*. These lower NBLs for Ni could arise because many of the excluded NO_3-containing sampling points (>10 mg/L) also had high Ni concentrations. Figure 8 shows that the highest NBLs for Ni for both *HOVER basis* and *land-use* are found in anoxic carbonate aquifers with neutral pH; however, no NBLs could be computed for *BRIDGE modified* due to the additional exclusion of sampling points with NO_3 > 10 mg/L.

Table 4. Comparative analysis for *HOVER basis*, *HOVER land-use*, and *BRIDGE modified*.

HOVER basis	Requirements	Availability of information about the sampling purpose, enabling exclusion of sampling points used for monitoring of polluted sites (as a minimum). Meta-data for most sampling points in Denmark is available in the Jupiter database.
	Advantages	Low data and labor intensity
	Disadvantages	The anthropogenic pressures are not assessed directly. Data from polluted yet active waterworks wells may be present in the data set. The data set is not representative for all groundwater types, only for those favored for drinking water abstraction and supply.
HOVER land-use	Requirements	Mapping prevailing anthropogenic pressures in the catchment of the well (recharge zone) in GIS software.
	Advantages	Moderately data and labor-intensive, CORINE land cover can be downloaded freely from https://land.copernicus.eu/pan-european/corine-land-cover (accessed on 29 April 2021).
	Disadvantages	Anthropogenic pressure in the catchment does not necessarily result in groundwater pollution. Other factors are not considered. The catchments (or groundwater recharge zones) are unknown for all wells at the national scale. The approximation of a 1 km buffer around the well may under- or overrepresent the actual area. No delineation between intensive/extensive/organic agriculture is included. All anthropogenic pressures were given equal weight, and only their areal proportions mattered when assigning prevailing pressure to each well. The proximity to roads was not included in the analysis, even though storm runoff may contribute to heavy metal loads. The method can only be applied partially if there are no representative sampling points without anthropogenic pressures (prevailing natural areas).
BRIDGE modified	Requirements	Availability of groundwater quality data for other chemical compounds indicating anthropogenic pressure from agricultural activities (e.g., nitrate, pesticides) or urban/industrial activities (e.g., organic micropollutants).
	Advantages	A more holistic assessment of potential pollution as opposed to basing the analysis on a single trace element at a time.
	Disadvantages	Very data and labor-intensive if it is done on a national scale. The NO_3 condition limits the assessment to aquifer types with reduced conditions mostly. The sampling points representing shallow, oxidized groundwater below agricultural land with $NO_3 > 10$ mg/L are excluded from the dataset, which in the Danish conditions means that NBLs for the shallow oxic and anoxic aquifers cannot be derived by this method, as those water types are mostly affected by diffuse pollution. However, any method that provides NBLs for such water types must be carefully analyzed and tested with independent data. In addition, this method is not particularly suitable for screening against industrial or mining pollution when only heavy metals are released into the groundwater, as other pollution indicators are used here.

The primary source of elevated Ni concentrations in Danish groundwater has been linked to the release of Ni during pyrite oxidation due to lowering of the water table due to abstraction or changes in the barometric pressure causing barometric pumping in the vicinity of the well [15–17]. Some of the Ni is then demobilized due to sorption on carbonate sediments, with rates dependent on the relative clay content of the sediment [16,17]. When the groundwater level is re-established, this secondary pool of Ni is also mobilized due to ion exchange with Ca-containing groundwater [17]. It was also shown that the elevated groundwater concentrations of Ni in eastern Sjælland (Figure 4b) were due to Ni mobilization over short distances (<500 m) rather than due to a regional groundwater transport issue [17]. When assessing the groundwater bodies' status in MP3, the NBLs exceeding the national TVs were used instead of the national TV (i.e., a new TV was set to be equal to the NBL). However, because of this documented process of anthropogenic influence, the NBLs for Sjælland were not used when the NBLs of Ni were established as part of the MP3.

The second comparison was between the two *basis level* methods, which included only pre-selection of sampling points (*HOVER basis* and *MP3 basis*). The observed differences in NBLs between the *HOVER basis* and *MP3 basis* can be attributed to the cumulative effect of the different methodological specifics (Table 2). To isolate the effect of the different period and aggregation method (median vs. mean of the annual means, MAM), we compared the two datasets, including only the sampling points present in both datasets (see Figure S7). The representative concentrations at sampling points based on the MAM aggregation are overall higher than the aggregation based on a median (Figure S7). The difference is negligible for As and Ni, but for Cd, Cr, Cu, and Zn, it is substantial (Figure S7).

The wells' pre-selection did not affect the NBLs for As, but there are significant differences for all other elements (Figure 9). It is possible that the observed significant differences could be because the *MP3 basis* dataset also includes water sampling points representing the shallower groundwater bodies. When the depth of the sampling points included in *MP3 basis* and *HOVER basis* datasets are compared, the median top/bottom of the abstraction screens are at 34/46 m below terrain for *MP3 basis*, while they are at 39/53 m below terrain for *HOVER basis*. However, further studies are needed to evaluate the effect of potential leaching of trace metals from the agricultural topsoil to the shallow groundwater resources.

The relative size of anthropogenic sources of trace metals to Danish soils was estimated in the early 90s based on a nationwide dataset of their content in the top-25 cm (regular grid n = 393, 1992) [32]. The levels of Ni, Zn, and Cr in soil were attributed to mainly natural sources (high correlation with soil texture and small difference between land-use), while anthropogenic sources were influencing the levels of Cd, Cu, and As [32]. The soil monitoring campaign was repeated in 2014 [8] and uncovered a significant increase in both Cu (36%) and Zn (41%) concentrations in the topsoil. Their primary sources on arable land in Denmark come from the use of organic fertilizers such as manure, slurry, and sewage sludge [8]. Cu and Zn can reach high concentrations in manure and slurry due to their use as growth-promoting additives in livestock feed and their use in the prevention of diarrhea associated with *E. coli* [8].

Selecting only the drinking water wells (as in *HOVER basis*) limits the representativity of the dataset to only groundwater types abstracted for drinking water purposes. Next to limiting representativity, it also limited the data availability for Cd, Cr, Cu, and Zn (Table 3) and, respectively, the ability to derive NBLs for different hydrogeochemical conditions (Figure 8) or different aquifer locations (Figure 9). Considering these limitations, we can conclude that including the wells from the national groundwater monitoring network, as in *MP3 basis*, improves the data availability and representativity of the NBLs (with respect to the shallow groundwaters), but also results in higher NBL for most of the studied trace metals.

5.2. Other Possibilities for Assessing Anthropogenic Influences

The original *BRIDGE method* [4] also included a condition for identifying hydrothermal, brackish/saline groundwaters, based on the concentrations of Na and Cl ions ($[Na^+]$ + $[Cl^-]$ > 1000 mg/L). Hydrothermal waters are not characteristic for Denmark, and in addition to that, groundwater used for drinking water purposes usually has low salinity, as it should comply with the national and EU drinking water standards for Cl (250 mg/L) and Na (175 mg/L) (BEK nr 1070 af 28/10/2019). There was Cl and Na data for 99.7% of all sampling points included in the *HOVER basis* dataset. The sum of Na and Cl exceeded 1000 mg/L for only two of the sampling points, but both of those had low median As (0.04 µg/L and 0.045 µg/L) and Ni (0.3 µg/L and 0.045 µg/L). Thus, we decided that this condition is not relevant for our study, so we kept only the NO_3 condition of the original BRIDGE method [2] and modified the method by adding other pollution indicators (pesticides and organic micropollutants).

Another possibility for assessing the anthropogenic influence due to unsustainable pumping practices, characteristic for some urbanized areas in Denmark, is performing Cl and SO_4 trend analysis. Increasing Cl trends are indicative for saltwater intrusion (due to sea-water intrusion or up-coning of deeper saline groundwater) and potential lowering of the groundwater table. Linear trend analysis for Cl and SO_4 was performed for 92 groundwater bodies at risk of bad quantitative state for MP3 [21]. However, there was only enough data (min 8 y of data at sampling points in 1988–2016) for 26% of these groundwater bodies, most of which are located on Sjælland and Fyn [21]. If we consider the sampling points included in the *HOVER basis* dataset, Cl and SO_4 trends [21] are available for 297 or 253, respectively. Of those, 43.8% had significant ($p < 0.05$) increasing Cl trends, and 32.8% significantly increasing SO_4 trends. As there were no high Cl concentrations, the effect of marine conditions is minor even with increasing Cl trends. Using Cl and SO_4 trends is limited by data availability and requires in-depth assessment at the groundwater body level, which is beyond the scope of this study. However, understanding the impacts of groundwater abstraction is relevant, especially in areas with urban anthropogenic pressures where long-term trends in different water quality parameters could be used as indicators for unsustainable aquifer exploitation [33].

Groundwater age could be another potential indicator, where the older ("pre-industrial") groundwaters [34] would be representing the baseline groundwater quality. An example of such a study in New Zealand can be found in [35]. Unfortunately, groundwater age (CFC-based) could be determined only for 69 of the sampling points of the *HOVER basis* dataset. Only a few waterworks in Denmark have dated the groundwater of their abstraction wells, and even fewer have reported it to Jupiter. In addition, the groundwater abstracted from long filters can be a mixture of very different ages [36], and the CFC in most waterworks wells can be expected to be partly degraded due to the reduced conditions [37]. Tritium data was also available for only a few of the sampling points ($n = 58$). If combined, there were in total 125 sampling points with either CFC or tritium data; thus, due to this low data coverage, we deemed this method as inappropriate for the current study. In future studies, if the pre-selection of wells was extended to the groundwater monitoring wells (as in MP3) or all wells that have been dated, this method should also be explored.

5.3. Implications and Recommendations

According to EU policies and guidelines [1–3,13], NBLs derived for groundwater bodies should be used by the national authorities of EU member states to derive and establish TVs based on criteria values for legitimate groundwater uses (e.g., drinking water) and the environment to protect human health and the ecological status of dependent terrestrial and associated aquatic ecosystems. Hence, the selected methods for the derivation of NBLs have important implications for the protection of human health, ecosystems, and biodiversity across Europe. Based on the NBL analyses presented in this study for Denmark—a country with intensive agriculture and widespread anthropogenic pressures—we prepared the following roadmap for method selection when determining NBLs for trace elements (Figure 10). This roadmap is applicable for NBL derivation at the national scale in countries with varying amounts of data, and it can be adjusted based on the local hydrogeological and hydrogeochemical conditions. Our assessment was based on aquifer types (groups of groundwater bodies) representing carbonate aquifers, Quaternary and pre-Quaternary sand aquifers, and their pH and redox conditions. However, the definitions of aquifer types should be adjusted to suit best the local conditions and data availability. The pre-selection of sampling points based on the primary use of the well or the sampling purpose is an important step in assuring that anthropogenically influenced points are removed from the dataset. However, this "cleaning" of the dataset should be performed with attention to the representativity and data availability, as well.

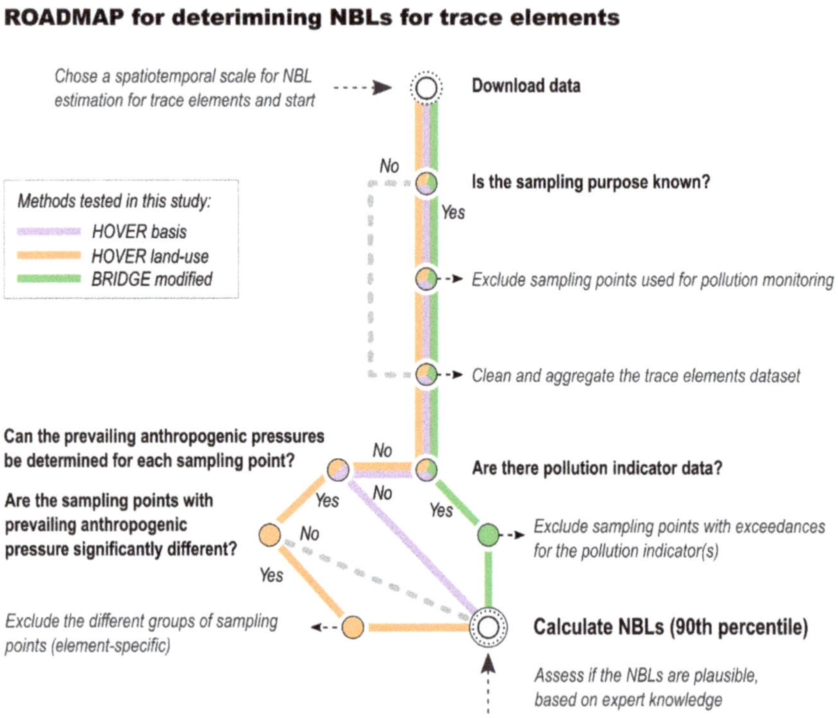

ROADMAP for deterimining NBLs for trace elements

Figure 10. Roadmap for determining natural background levels (NBLs) for trace elements showing the different steps in the data analysis based on data availability; the 3 methods for excluding anthropogenic influences that were tested in this study are also mapped.

6. Conclusions

We developed a roadmap for deriving NBLs for trace metals (As, Cd, Cr, Cu, Ni, Zn), in the context of intensive and widespread agriculture and extensive groundwater pumping for drinking water supply in urban areas. This work contributes to the need for harmonization in the NBL derivation by EU member states for the purposes of assessing the chemical status of groundwater bodies stipulated by the WFD and GWD. It provides a systematic way of selecting an appropriate method or combination of methods to assure that NBLs are calculated based on groundwater data representing no or only very minor anthropogenic alterations to undisturbed conditions.

We applied and compared three different methods for excluding anthropogenically influenced sampling points: *HOVER basis*, *HOVER land-use*, and *BRIDGE modified*. Denmark was used as an example of a country with widespread agricultural pressures (diffuse pollution). We found that the *HOVER basis* provided already a relatively "clean" dataset; thus, the two other methods that removed additional sampling points potentially affected by anthropogenic pollution (*HOVER land-use* and *BRIDGE modified*) did not result in significantly different NBLs, except for Ni, for which *BRIDGE modified* performed best (resulted in lower NBL). Data availability limited the derivation of NBLs, accounting also for the redox and pH conditions, except for As and Ni.

Furthermore, we critically assessed these three methods, i.e., we discussed the specific data requirements, the advantages, and the disadvantages of the individual methods. This critical assessment generalizes the outcomes of our study and will hopefully help other researchers or water managers when setting NBLs for trace metals in groundwater at the national scale. We demonstrated how combining several methods and using several types of data may compensate for the individual limitations of the methods. Since the methods have different data availability requirements, the roadmap accounts for this too. Our results showed that the simplest of the three methods performed well in almost all cases, stressing the importance of excluding known polluted water sampling points. In the Danish case, this was possible because the sampling purpose and the well use are known. Thus, we recommend, if this information is available, to use it for initial pre-screening of the datasets.

Supplementary Materials: The following are available online at https://www.mdpi.com/article/10.3390/w13091267/s1, one file including supplementary text, and the following figures, and tables: Figure S1: Map with landscape types and the pre-quaternary stratigraphic succession in Denmark, Figure S2: Maps with sampling points classified based on aquifer type and prevailing anthropogenic pressure, Figure S3: ECDFs for the trace elements stratified by aquifer type, Figure S4: ECDFs for the trace elements stratified by pH class, Figure S5: ECDFs for the trace elements stratified by redox class, Figure S6: ECDFs for the trace elements stratified by prevailing anthropogenic pressure, Figure S7: Comparison between HOVER basis and MP3 basis datasets, Table S1: Excluded analyses during quality control, Table S2: NBLs for the trace elements in the main aquifer types based on BRIDGE modified, HOVER basis and HOVER land-use, Table S3: NBLs for As in different aquifer types based on redox, pH, and geology for the 3 methods, Table S4: NBLs for Ni in different aquifer types based on redox, pH, and geology for the 3 methods, Table S5: NBLs for trace elements for different aquifer types and locations, based on HOVER basis and MP3 basis.

Author Contributions: Conceptualization, D.D.V., L.T., V.E. and K.H.; methodology, D.D.V.; validation, J.S. and L.T.; data curation, D.D.V.; writing—original draft preparation, D.D.V.; visualization, D.D.V.; supervision, L.T., V.E., B.H. and K.H.; project administration and funding acquisition, K.H. All authors participated in the writing, review, and editing of the manuscript. All authors have read and agreed to the published version of the manuscript.

Funding: This project has received funding from the European Research Council (ERC) under the European Union's Horizon 2020 research and innovation program (grant agreement n° 731166) and by the Innovation Fund Denmark, funding agreement number 8055-00073B.

Institutional Review Board Statement: Not applicable.

Informed Consent Statement: Not applicable.

Data Availability Statement: The raw data used in this study is publicly available in the Danish national well database (Jupiter). This data can be found here: https://eng.geus.dk/products-services-facilities/data-and-maps/national-well-database-jupiter (accessed on 29 April 2021).

Acknowledgments: This work was carried out under work package 3 of the GeoERA project HOVER; we would like to thank all our colleagues from the European geological surveys participating in this work package, especially for everyone involved in Task 3-3. The *HOVER land-use* method presented in this paper is an adaptation of the original method proposed by Eline Malcuit and Laurence Gourcy (BRGM) [18]. We also thank the Danish EPA for including us in the River Basin Management Plan work on assessing the status of groundwater bodies for trace elements, which allowed us to focus our attention on the practical side of NBLs derivation.

Conflicts of Interest: The authors declare no conflict of interest. The funders had no role in the design of the study; in the collection, analyses, or interpretation of data; in the writing of the manuscript; or in the decision to publish the results.

Appendix A

General guideline for establishing TVs, according to Annex II.A of GWD [2]:

" . . .

(1)   the determination of TVs should be based on:

- *the extent of interactions between groundwater and associated aquatic and dependent terrestrial ecosystems;*
- *the interference with actual or potential legitimate uses or functions of groundwater;*
- *all pollutants which characterise bodies of groundwater as being at risk . . .*

- *hydrogeological characteristics including information on background levels and water balance;*
(2) *the determination of [TVs] should also take into account the origins of the pollutants, their possible natural occurrence, their toxicology and dispersion tendency, their persistence and their bioaccumulation potential; . . .*
(3) *wherever elevated background levels of substances or ions or their indicators occur due to natural hydrogeological reasons, these background levels in the relevant body of groundwater shall be taken into account when establishing threshold values".*

Table A1. Geogenic and anthropogenic sources and geochemical controls (summarized from [5,19]).

Arsenic (As)	Geogenic	Sulfide minerals (e.g., pyrite, arsenopyrite, arsenian pyrite); feldspars; phosphate minerals [1]; sorbs to clays, Fe oxyhydroxides, and OM;
	Anthropogenic	Pesticides; pig and poultry farming; combustion processes; ore roasting;
	Controls	pH and redox dependent; reductive dissolution and desorption (sulfide minerals); oxidation reactions (iron oxides)
Cadmium (Cd)	Geogenic	Sphalerite [2]; micas, amphiboles; phosphorite; due to affinity to OM, enrichment in coal and peat; sorbs to calcite surfaces, clay minerals, and OM
	Anthropogenic	Fertilizers; sewage sludge; traffic (wear of tires); incinerators; coal combustion; metal smelters; iron and steel mills; electroplating
	Controls	pH and redox dependent; soluble in oxidizing conditions at pH < 8; co-precipitates with Fe and Mn hydroxides
Chromium (Cr)	Geogenic	Ferromagnesian minerals (e.g., olivine, pyroxene, amphibole); micas; garnets; enriched in mafic and ultramafic rocks, shales, and other argillaceous rocks; sorbs to clays, Fe and Mn oxyhydroxides, and OM
	Anthropogenic	Tanning and wood impregnation; steel industry;
	Controls	pH and redox dependant; mobile under acidic oxidizing conditions and forms inorganic and organic complexes
Copper (Cu)	Geogenic	Sulfide minerals (e.g., chalcopyrite); accessory in many common minerals (e.g., micas and amphiboles); strong sorption to OM, Fe, and Mn oxyhydroxides;
	Anthropogenic	Farm effluents and sewage sludge [3]; wide range of industrial and urban uses (e.g., roofing, pipework, plumbing, and water components; electrical industry);
	Controls	pH and redox dependant; highest mobility under acidic and oxidizing conditions; forms inorganic and organic complexes; co-precipitates with Fe and Mn hydroxides
Nickel (Ni)	Geogenic	Ni-minerals; accessory in sulfide minerals (e.g., pyrite, chalkopyrite) and other common minerals (e.g., micas and amphiboles); closely associated with Cr and Co; sorbs to Fe and Mn oxides, clay edges, calcite
	Anthropogenic	Phosphate fertilizers ("contaminant" along with Zn, Cr, and Cd); industrial and urban pollution (alloys, batteries, magnets, plating, pigments); landfill leachates
	Controls	pH and redox dependant [4]; highly mobile under acidic and reducing conditions; in near-neutral waters, it may form carbonate complexes
Zinc (Zn)	Geogenic	Sphalerite; range of Zn-carbonates (e.g., smithsonite) and oxides; can be present as a trace constituent in calcite; in clays, it may be in secondary oxide and silicate minerals; sorbs to oxide and oxyhydroxide minerals
	Anthropogenic	Used as anticorrosion coating of steel, in alloys, pipework, plumbing, and water components; pigment in paint; in rubber products
	Controls	pH and redox dependant [5]; highest mobility under acidic and oxidizing conditions; mobile also in circum-neutral and alkaline conditions

[1] possible substitution of P^{5+} with As^{5+} [19]; [2] Cd could be replacing 0.5% and up to 1.5% of the Zn [19]; [3] Cu may be administered as a supplement to farm animals [19]; [4] Ni is not redox-sensitive, but it is affected by redox processes as the redox-sensitive elements Mn and Fe are linked with its mobilization in water; [5] Zn has only one oxidation state, so insensitive to redox, but its ability to be adsorbed to metal oxyhydroxides may have indirect control.

Table A2. Groundwater sampling points in each class; the dominating classes are shown in bold.

	As	Cd	Cu	Cr	Ni	Zn
HOVER basis dataset (n)	6352	355	289	250	6358	363
Aquifer type (%)						
• Carbonate	35	23	24	23	35	24
• **Quaternary sand**	**53**	**57**	**64**	**64**	**53**	**56**
• Pre-Quaternary sand	10	16	12	12	10	15
• Bornholm (various)	1	5	-	-	1	5
pH class (%)						
• Acidic	5	5	4	3	5	6
• Basic	27	26	27	30	27	27
• **Neutral**	**57**	**57**	**56**	**53**	**57**	**55**
• Unknown	10	12	12	14	10	12
Redox class (%)						
• Oxic	8	5	5	4	8	4
• Anoxic	4	3	2	3	4	3
• **Reduced**	**76**	**81**	**82**	**82**	**76**	**82**
• Mixed	13	11	10	11	13	10
• Unknown	<1	-	-	-	<1	<1
Prevailing pressure (%)						
• **Agricultural**	**86**	**81**	**75**	**75**	**86**	**77**
• Industrial	1	5	2	3	1	5
• Urban	13	15	22	22	13	17
• Mining	<1	-	-	-	<1	-
• No pressure (natural)	1	-	<1	-	1	<1

References

1. European Commission. Directive 2000/60/EC of the European Parliament and of the Council, of 23 October 2000, establishing a framework for community action in the field of water policy. *Off. J. Eur. Communities* **2000**, *327*, 1–73.
2. European Commission. Directive 2006/118/EC of the European Parliament and of the Council of 12 December 2006 on the Protection of Groundwater against Pollution and Deterioration. *Off. J. Eur. Union* **2006**, *372*, 19–31.
3. European Commission. *Guidance on Groundwater Status and Trend Assessment (Guidance Document No. 18 of the Common Implementation Strategy for the Water Framework Directive (2000/60/EC)*; Technical Report; Office for Official Publications of the European Communities: Luxembourg, 2009; ISBN 978-92-79-11374-1.
4. Hinsby, K.; Condesso de Melo, M.T.; Dahl, M. European case studies supporting the derivation of natural background levels and groundwater threshold values for the protection of dependent ecosystems and human health. *Sci. Total Environ.* **2008**, *401*, 1–20. [CrossRef] [PubMed]
5. Edmunds, W.M.; Shand, P. (Eds.) *Natural Groundwater Quality*; Blackwell Pub: Malden, MA, USA, 2008; ISBN 978-1-4051-5675-2.
6. Scheidleder, A. *Groundwater Threshold Values: In-Depth Assessment of the Differences in Groundwater Threshold Values Established by Member States*; Umweltbundesamt (Environment Agency Austria): Vienna, Austria, 2012; p. 57.
7. EC; CIS Working Group Groundwater. *Threshold Values—Initial Analysis of 2015 Questionnaire Responses*; European Commission: Brussels, Belgium, 2015; p. 46.
8. Jensen, J.; Larsen, M.M.; Bak, J. National monitoring study in denmark finds increased and critical levels of copper and zinc in arable soils fertilized with pig slurry. *Environ. Pollut.* **2016**, *214*, 334–340. [CrossRef] [PubMed]
9. Koch, J.; Stisen, S.; Refsgaard, J.C.; Ernstsen, V.; Jakobsen, P.R.; Højberg, A.L. Modeling depth of the redox interface at high resolution at national scale using random forest and residual gaussian simulation. *Water Resour. Res.* **2019**, *55*, 1451–1469. [CrossRef]
10. Frei, R.; Frei, K.M.; Kristiansen, S.M.; Jessen, S.; Schullehner, J.; Hansen, B. The link between surface water and groundwater-based drinking water—Strontium isotope spatial distribution patterns and their relationships to Danish sediments. *Appl. Geochem.* **2020**, *121*, 104698. [CrossRef]
11. Sandersen, P.B.E.; Jørgensen, F. Buried tunnel valleys in Denmark and their impact on the geological architecture of the subsurface. *GEUS Bull.* **2017**, *38*, 13–16. [CrossRef]
12. Troldborg, L. *Afgrænsning af de Danske Grundvandsforekomster: Ny Afgrænsning og Delkarakterisering Samt Fagligt Grundlag for Udpegning af Drikkevandsforekomster*; GEUS Rapport 2020/1; Geological Survey of Denmark and Greenland (GEUS): Copenhagen, Denmark, 2020; p. 82. (In Danish)
13. European Commission. Council Directive 98/83/EC of 3 November 1998 on the Quality of Water Intended for Human Consumption. *Off. J. Eur. Union* **2015**, *330*, 32–54. Available online: http://data.europa.eu/eli/dir/1998/83/2015-10-27 (accessed on 29 April 2021).

14. Ernstsen, V.; Mortensen, M.H.; Voutchkova, D.D.; Thorling, L. *Udvikling af Metode til Vurdering af Grundvandsforekomsters Kemiske Tilstand for Udvalgte Uorganiske Sporstoffer og Salte*; GEUS Rapport 2021/19; Geological Survey of Denmark and Greenland (GEUS): Copenhagen, Denmark, 2020. (In Danish)
15. Larsen, F.; Postma, D. Nickel mobilization in a groundwater well field: Release by pyrite oxidation and desorption from manganese oxides. *Environ. Sci. Technol.* **1997**, *31*, 2589–2595. [CrossRef]
16. Jensen, T.F.; Larsen, F.; Kjøller, C.; Larsen, J.W. *Nikkelfrigivelse ved Pyritoxidation Forårsaget af Barometerånding—Pumpning*; Arbejdsrapport 5; Miljøstyrelsen: Odense, Denmark, 2003; p. 131. (In Danish)
17. Kjøller, C.; Jessen, S.; Larsen, F.; Postma, D.; Jakobsen, R. *Binding af Nikkel til og Frigivelse fra Naturlige Kalksedimenter*; Arbejdsrapport 8; Miljøstyrelsen: Odense, Denmark, 2006; p. 144. (In Danish)
18. Lions, J.; Malcuit, E.; Gourcy, L.; Voutchkova, D.; Hansen, B.; Schullehner, J.; Forcada, E.G.; Olmedo, J.G.; Elster, D.; Camps, V.; et al. *Proposing a Common Methodology to Calculate the Natural Concentration of Dissolved Elements Based on Lithological/Geological Water Families Taking into Account Possible Anthropogenic Influences*; Report type Deliverable of HOVER Project No. D 3-3; 2021; p. 206.
19. Reimann, C.; Birke, M. (Eds.) *Geochemistry of European Bottled Water*; Borntraeger Science Publishers: Stuttgart, Germany, 2010; ISBN 978-3-443-01067-6.
20. Thorling, L.; Ditlefsen, C.; Ernstsen, V.; Hansen, B.; Johnsen, A.R.; Troldborg, L. *Grundvandovervågning. Status og Udvikling 1989–2018*; Teknisk Rapport; Geological Survey of Denmark and Greenland (GEUS): Copenhagen, Denmark, 2019; p. 132. (In Danish)
21. Henriksen, H.J.; Voutchkova, D.; Troldborg, L.; Ondracek, M.; Schullehner, J.; Hansen, B. *National Vandressource Model. Beregning af Udnyttelsesgrader, Afsænkning og Vandløbspåvirkning Med DK Model 2019*; GEUS Rapport 2019/32; Geological Survey of Denmark and Greenland (GEUS): Copenhagen, Denmark, 2019; p. 84. (In Danish)
22. R Core Team. *R: A Language and Environment for Statistical Computing*; R Foundation for Statistical Computing: Vienna, Austria, 2020.
23. Hyndman, R.J.; Fan, Y. Sample quantiles in statistical packages. *Am. Stat.* **1996**, *50*, 361. [CrossRef]
24. Millard, S. *EnvStats: An R Package for Environmental Statistics*; R Package; Springer: New York, NY, USA, 2013; ISBN 978-1-4614-8455-4.
25. Wickham, H. *Stringr: Simple, Consistent Wrappers for Common String Operations*; R Package. 2019. Available online: https://CRAN.R-project.org/package=stringr (accessed on 5 January 2021).
26. Wickham, H. *Ggplot2: Elegant Graphics for Data Analysis*; R Package; Springer: New York, NY, USA, 2016; ISBN 978-3-319-24277-4.
27. Wickham, H. *Tidyr: Tidy Messy Data*; R Package. 2020. Available online: https://CRAN.R-project.org/package=tidyr (accessed on 5 January 2021).
28. Wickham, H.; François, R.; Henry, L.; Müller, K. *Dplyr: A Grammar of Data Manipulation*; R Package. 2020. Available online: https://CRAN.R-project.org/package=dplyr (accessed on 5 January 2021).
29. Dowle, M.; Srinivasan, A. *Data.Table: Extension of 'Data.Frame'*; R Package. 2020. Available online: https://CRAN.R-project.org/package=data.table (accessed on 5 January 2021).
30. QGIS Development Team. *QGIS Geographic Information System*; Long Term Release; QGIS Association. 2021. Available online: https://qgis.org/en/site/ (accessed on 5 January 2021).
31. Inkscape's Contributors. *Inkscape—A Free and Open Source Vector Graphics Editor*; The Inkscape Project. 2021. Available online: https://inkscape.org/ (accessed on 5 January 2021).
32. Bak, J.; Jensen, J.; Larsen, M.M.; Pritzl, G.; Scott-Fordsmand, J. A heavy metal monitoring-programme in Denmark. *Sci. Total Environ.* **1997**, *207*, 179–186. [CrossRef]
33. Gejl, R.N.; Rygaard, M.; Henriksen, H.J.; Rasmussen, J.; Bjerg, P.L. Understanding the impacts of groundwater abstraction through long-term trends in water quality. *Water Res.* **2019**, *156*, 241–251. [CrossRef] [PubMed]
34. Hinsby, K.; Edmunds, W.M.; Loosli, H.H.; Manzano, M.; Condesso De Melo, M.T.; Barbecot, F. The modern water interface: Recognition, protection and development—advance of modern waters in european aquifer systems. *Geol. Soc. Lond. Spec. Publ.* **2001**, *189*, 271–288. [CrossRef]
35. Morgenstern, U.; Daughney, C.J. Groundwater age for identification of baseline groundwater quality and impacts of land-use intensification—The national groundwater monitoring programme of New Zealand. *J. Hydrol.* **2012**, *456–457*, 79–93. [CrossRef]
36. Jakobsen, R.; Hinsby, K.; Aamand, J.; van der Keur, P.; Kidmose, J.; Purtschert, R.; Jurgens, B.; Sültenfuss, J.; Albers, C.N. History and sources of co-occurring pesticides in an abstraction well unraveled by age distributions of depth-specific groundwater samples. *Environ. Sci. Technol.* **2020**, *54*, 158–165. [CrossRef] [PubMed]
37. Hinsby, K.; Højberg, A.L.; Engesgaard, P.; Jensen, K.H.; Larsen, F.; Plummer, L.N.; Busenberg, E. Transport and degradation of chlorofluorocarbons (CFCs) in the pyritic rabis creek aquifer, Denmark. *Water Resour. Res.* **2007**, *43*. [CrossRef]

Article

A Broad-Scale Method for Estimating Natural Background Levels of Dissolved Components in Groundwater Based on Lithology and Anthropogenic Pressure

Julie Lions [1,*], Nicolas Devau [1], Daniel Elster [2], Denitza D. Voutchkova [3], Birgitte Hansen [3], Jörg Schullehner [3,4], Tanja Petrović Pantić [5], Katarina Atanasković Samolov [5], Victor Camps [6], Georgina Arnó [6], Ignasi Herms [6], Nina Rman [7], Sonja Cerar [7], Juan Grima [8], Elena Giménez-Forcada [8], Juan Antonio Luque-Espinar [8], Eline Malcuit [1] and Laurence Gourcy [1]

[1] Brgm, Geological Survey of France, 45100 Orléans, France; n.devau@brgm.fr (N.D.); e.malcuit@brgm.fr (E.M.); l.gourcy@gmail.fr (L.G.)
[2] Geological Survey of Austria, A-1030 Vienna, Austria; daniel.elster@geologie.ac.at
[3] Geological Survey of Denmark and Greenland (GEUS), K 1350 Copenhaguen, Denmark; dv@geus.dk (D.D.V.); bgh@geus.dk (B.H.); jsc@geus.dk (J.S.)
[4] Department of Public Health, Research Unit for Environment, Work and Health, Aarhus University, C 8000 Aarhus, Denmark
[5] Geological Survey of Serbia, 11000 Belgrade, Serbia; tanjapetrovic.hg@gmail.com (T.P.P.); k.samolov@gmail.com (K.A.S.)
[6] Institut Cartogràfic i Geològic de Catalunya (ICGC), Parc de Montjuïc, 08038 Barcelona, Spain; victor.camps@icgc.cat (V.C.); ignasi.herms@icgc.cat (I.H.); georgina.arno@icgc.cat (G.A.)
[7] Geological Survey of Slovenia (GeoZS), 1000 Ljubljana, Slovenia; Nina.Rman@GEO-ZS.SI (N.R.); sonja.cerar@geo-zs.si (S.C.)
[8] Instituto Geológico y Minero de España (IGME), 28003 Madrid, Spain; j.grima@igme.es (J.G.); e.gimenez@igme.es (E.G.-F.); ja.luque@igme.es (J.A.L.-E.)
[*] Correspondence: j.lions@brgm.fr

Citation: Lions, J.; Devau, N.; Elster, D.; Voutchkova, D.D.; Hansen, B.; Schullehner, J.; Petrović Pantić, T.; Samolov, K.A.; Camps, V.; Arnó, G.; et al. A Broad-Scale Method for Estimating Natural Background Levels of Dissolved Components in Groundwater Based on Lithology and Anthropogenic Pressure. *Water* 2021, 13, 1531. https://doi.org/10.3390/w13111531

Academic Editors: Elisabetta Preziosi, Marco Rotiroti, M. Teresa Condesso de Melo and Klaus Hinsby

Received: 30 March 2021
Accepted: 25 May 2021
Published: 29 May 2021

Publisher's Note: MDPI stays neutral with regard to jurisdictional claims in published maps and institutional affiliations.

Abstract: The Water Framework Directive (WFD) requires EU member states to assess the chemical status of groundwater bodies, a status defined according to threshold values for harmful elements and based on/the natural background level (NBL). The NBL is defined as the expected value of the concentration of elements naturally present in the environment. The aim of this study is to propose a methodology that will be broadly applicable to a wide range of conditions at the regional and national scale. Using a statistical approach, the methodology seeks to determine NBLs for SO_4, As, Cd, Cr, Cu, Ni, Zn, and F based on the lithology of aquifers from which groundwater monitoring data were collected. The methodology was applied in six EU countries to demonstrate validity for a wide range of European regions. An average concentration was calculated for each parameter and chosen water point and linked to a lithology. Based on the dataset created, significant differences between lithologies and pressure categories (urban, agricultural, industrial, and mining) were tested using a nonparametric test. For each parameter, 90th percentiles were calculated to provide an estimation of the maximum natural concentrations possible for each lithology.

Keywords: natural background level; groundwater; water quality; anthropogenic pressure; trace element; groundwater monitoring

1. Introduction

The Water Framework Directive 2000/60/EC (WFD) [1] requires EU member states to assess the chemical status of groundwater bodies, a status defined by threshold values that should take into account concentrations of elements that may be both naturally present in aquatic environments and discharged by human activities. Member states must therefore characterize the natural background levels (NBLs) for each component and groundwater body while recognizing that a wide range of approaches have been developed. The NBL is defined as the concentration of a component in a groundwater body that corresponds to zero, or only a very minor anthropogenic alteration or undisturbed conditions [2]. A natural concentration is defined as an element that results entirely from a natural source, whether geological, biological, or atmospheric, under conditions not disturbed by human activities. In addition to natural, a semi-natural origin is also possible, when elements derive naturally from the soil matrix or from the aquifer, but are mobilized by a change of conditions following human intervention, e.g., aquifer dewatering, an accident involving acid or a plume that contains salt and/or water rich in organic matter [1,3].

Since 2000, the WFD has promoted the establishment of river basin monitoring networks, resulting in a large amount of available data. These data, if well-selected, can contribute to the identification of the expected trace element concentrations in groundwater and a determination of the NBLs. However, the methodology needs to be tailored and harmonized among river basins by taking into account aquifer conditions and anthropogenic inputs. Moreover, groundwater chemistry in a given aquifer provides an important basis for defining the nature and extent of current pollution, and aids in the identification of past contributions and necessary corrective measures.

A key objective of the WFD is the achievement of "good ecological and chemical status" for surface water bodies based on environmental quality standards (EQS) [4] and a "good quantitative and chemical status" for groundwater bodies based on quality standards or threshold values (mainly based on standards listed in the Drinking Water Directive 2020/2184) [5]. According to the Groundwater Directive 2006/118/EC (GWD) [2] requirements, (1) assessment of flow direction and (2) exchange rates between groundwater bodies and associated surface water are crucial aspects of river basin management plans [6]. Table 1 illustrates the standard groundwater and drinking water values. For several elements (i.e., Zn and Cu), a significant difference exists between groundwater threshold values that are usually applied and the EQS applied to surface water [4,7]. This fact underlines the importance of defining NBLs, and is possible when groundwater supplies are associated surface water (e.g., streams, lakes, wetlands, etc.), and may affect chemically-dependent surface water status. In addition, it is useful to know natural concentration ranges and average values to guide, support, and justify the chemical status assessment of surface water and groundwater bodies.

Table 1. Drinking water (DW) standards [5] and groundwater threshold value (GW-TV).

Unit (µg/L)	DW Standards	GW-TV		
		Austria, France, Spain, Slovenia	Denmark	Serbia [c]
As	10 (5) [a]	10	5	10
Cd	5	5	0.5	3
Cr	50	50	25	50
Cu	2000	2000	100	2000
Ni	20	20	10	20
Zn	5000 [b]	not defined (5000–France)	100	3000
F	1500	1500	1500	1200
SO$_4$	250,000	250,000	250,000	

[a] Danish legislation for drinking water supply, LBK nr 118, 22 February 2018, and Ministerial Order BEK nr 1070, 28 October 2019; [b] French decree of 11 January 2007; [c] Official Gazette of FRY volume 42/98 and 44/99 and Official Gazette of RS. Regulation on the hygienic acceptability of potable water.

Previous EU studies focused on WFD procedures for estimating the chemical status of groundwater and consideration of NBLs. The BRIDGE (Background Criteria for the Identification of Groundwater Thresholds) approach is based on the generalization that aquifers with similar petrographic properties have similar water composition under analogous hydrodynamic and hydrologic conditions [8]. The BRIDGE approach uses the BRIDGE aquifer typology to distinguish between NBLs [9,10].

BRIDGE methods consist of using only sampling points unaffected by anthropogenic factors (natural areas, groundwater with low NO_3 concentrations, etc.) [11]. As a result, this approach limits NBL determination to specific areas. Lack of data and analytical method limitations (reporting limit being too high) made it difficult to determine NBLs for most trace elements. For this reason, first estimations were based primarily on bibliographic review [12,13].

Groundwater chemistry variation depends on a number of factors, including rainfall composition, aquifer lithology, groundwater flow pathways, and residence time [14]. Accordingly, each European country developed its own studies to meet its obligation to determine NBLs; several key European studies have been published [15–23]. More broadly, NBLs have been determined for the purposes of groundwater and health risk assessment and drinking water quality management [24–26]. Based on the numerous data, statistical methods have been applied, e.g., frequency distributions, probability plots, box plots, and histograms. Concentrations that deviate from the basic distribution (between the 10th and 90th percentiles) are generally excluded from the derivation of background concentrations. Masciale et al. [27] used Huber's non-parametric test to identify and eliminate anomalous data before NBL calculation, also introducing a confidence level for NBL values. Recent studies have also considered model-based statistical approaches, such as iterative 2-δ techniques, maximum normalized residual tests, and geostatistical multi-model approaches [26,28–31]. Furthermore, an online tool evaluation of natural background levels (eNaBLe) was developed in Italy in 2020 [32].

The purpose of this study is to propose and test a methodology common to multiple countries, taking into account national specificities and available data related to groundwater, lithology, and anthropogenic pressure. Prior to this work, previous studies had been conducted at different scales and in specific areas (high mountain karst aquifers, specific geologic units) using data provided by groundwater monitoring networks [33–38]. Each study has adapted the BRIDGE approach methodology in different ways. In some cases, data from wells with only a median nitrate concentration below 10 mg/L were used (e.g., Slovenia, France, Austria). In others, such as Slovenia, sites with up to 20% anthropogenic impact in recharge areas were considered [39]. In this study, a new methodology aims to work at a large scale (river basin or national scale) and focus on the components presented in Table 1.

However, this exercise faces two primary challenges in particular, because the data used are provided by monitoring under the WFD framework, which was not primarily designed to characterize NBLs. The challenges are:

- Monitoring network sampling points that are selected to be representative for a groundwater body, but may not be representative for the NBL calculation.
- Laboratory reporting limits (LRLs) that may not be in the most appropriate ranges for NBL calculation, because they were initially designed to verify compliance with drinking water standards.

2. Methodology

2.1. Background

2.1.1. Reference Concentrations

Several approaches can be used to estimate the ranges of natural concentrations for water management plans, either by determining the average value and standard deviation or by describing low and high values after eliminating outliers. Spatial heterogeneity and temporal variability of groundwater composition make it difficult to determine a single value. Heterogeneity is scale-dependent, which leads to the consideration of a concentration range or a confidence limit adapted to each scale of work [40]. Indeed, upper limits for groundwater NBLs may exceed environmental or drinking water quality standards, particularly for elements such as As and F [41,42].

2.1.2. Censored Data

A primary challenge in using statistical approaches to process geochemical datasets is the presence of values that are below the limit of quantification (LOQ)/detection limit (LOD), also known as left-censored values. In some datasets, more than 50% of the values may be left-censored. Sophisticated methods require, for example, work on the data distribution law (distributional methods) or an extrapolation from >LOQ data to <LOQ data (robust methods), and are therefore less direct than the substitution of the value by a constant [43]. The data substitution method also depends on the amount of data below the LOQ. For example, to estimate the median, Antweiler (2015) [44] confirmed that, for a low-censoring dataset, LOQ/2 can be used. In addition, when 50% of the data are censored, no technique gives a reliable distribution estimate. Other more robust approaches for the imputation of left-censored values take multivariate approaches into account [45].

Analytical methods have evolved significantly in recent years, and the quality of the results has improved significantly, reducing uncertainties, detection limits, and/or quantification limits. For these reasons, the temporal evolution of the methods and laboratory reporting limits needs to be reviewed before selecting the dataset (see Section 2.3).

2.1.3. Uncertainties

Before interpreting water quality data, it is important to consider the uncertainty associated with the measurement. Traditionally, only analytical uncertainties have been applied to water quality interpretation [46]. However, these uncertainties represent only part of the uncertainty related to interpreting environmental measurements. Water sample collection is also part of the measurement acquisition process. Ghestem (2009) [47] evaluated the overall uncertainty in groundwater measurement. Overall variations were estimated at 5–10% for most substances, including the major components (SO_4, NO_3, Cl) and a range of 10 to 35% for trace elements. For As, Cu, Cr, Ni, and Zn (>5 µg/L), the uncertainties are between 15 and 20%.

2.2. Proposed Methodology

Based on the WFD context and data limitations, the following methodology (Figure 1) shows the steps for determining NBLs based on groundwater network data and a consideration of the potential anthropogenic effects.

2.3. Groundwater Quality Dataset

The method involves processing data stored in national/regional groundwater monitoring databases (Table 2), which usually contain large datasets. The first step (step 1, Figure 1) is an inventory of the available and usable data; this step involves the selection of relevant parameters for NBL calculation and information on the lithological and hydrogeological context of each sampling point. Additional chemical parameters are needed to determine the hydrochemical groundwater conditions, such as pH, redox potential, NO_3, and dissolved O_2 and Fe. Thermal and mineral water are excluded from this assessment. After considering data availability and laboratory reporting limits (LOD/LOQ), step 2 is an examination of the time period for which the data are considered (reference period) for each dataset (step 2).

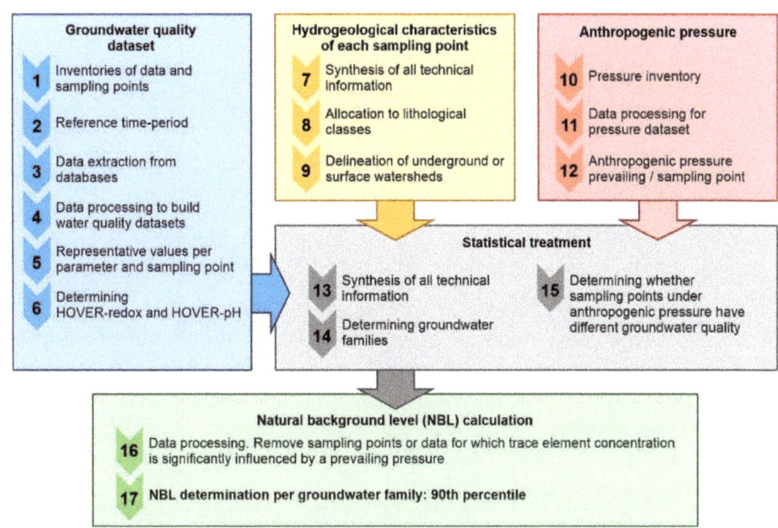

Figure 1. Flowchart for NBL calculation applied in this study.

Table 2. Dataset descriptions and methodologies applied by area. Simplified lithology refers to the lithology classes defined in this study [1].

Area	Reference Period	Sampling Points/Analyses	Land Use and Prevailing Pressure (Methodology)	Lithologies (Methodology and Classes)	LOQ Treatment (Methodology)	Data Source
Loire- Bretagne basin, France (155,000 km^2)	2009–2020	4200/78559	Watershed approach based on "TAUDEM" module, CLC 2012 [48] completed by inventories (agricultural uses: Agreste; industrial location: ICPE, BASIAS, BASOL; mining inventory: SIG Mines)	Sampling point link to BDLisa (1:50 000) simplified lithology: sedimentary:(carbonate, gravel, sand), crystalline, metamorphic, and volcanic	Substituted value with LOQ/2—high LOQ removed.	ADES
Nationwide Denmark (42,933 km^2)	2009–2018	6388/125106	Buffer of 1 km, pressures extracted from CLC 2012 [48]	Sampling point link to the groundwater bodies delineated based on the DK model [49]. simplified lithology: sedimentary:(carbonate, sand)	Values < LOD substituted with LOQ/2, where LOQ = 3*LOD. High LOD removed	JUPITER
Nationwide, Austria (83,879 km^2)	2010–2020	2024/604353	Point data (10 m buffer) pressures extracted from CLC 2018 [50]	Derived from national hydrogeological map (1:500.000) [51] simplified lithology: sedimentary: gravel, metamorphic	Values < LOD or LOQ substituted with LOD/2 or LOQ/2	WISA
Internal basins of Catalonia (Spain) (32,108 km^2)	1994–2018 (long period due to low number of data available)	1336/8316	Buffer of 1 km, pressures extracted from CLC 2018 [50]	Aquifer map produced by the ACA (1: 50.000); simplified lithology: crystalline rocks and sedimentary: carbonate.	"Log-Ratio EM Algorithm" method with the lrEM function (zCompositions v1.3.4) [45] or the Wilcoxon transformation method applied by calculating < LOD/2.	SDIM-ACA

Table 2. *Cont.*

Area	Reference Period	Sampling Points/Analyses	Land Use and Prevailing Pressure (Methodology)	Lithologies (Methodology and Classes)	LOQ Treatment (Methodology)	Data Source
Mountain Fruška Gora (Serbia)	2006–2017	33/132	Drastic vulnerability map, pressures extracted from CLC 2006 [52,53]	Simplified lithology: sedimentary: sand (quaternary deposits).	Values < LOD substituted with LOD/2	Studies
Nationwide, Slovenia (20,271 km^2)	2016	203/2848	Water body pressure and land use from the National Water Management Plan 2016–2021	Lithostratigraphy according to Basic Geological Map (1:100.000 and 1:250.000) [54] and points linked to 21 groundwater bodies. Simplified lithology: sedimentary:(carbonate, sands)	Values < LOQ substituted with LOQ/2	ARSO
Spain, Duero River Basin (78,859 km^2)	2010–2020	465/6457	Point data (1 km buffer) pressures extracted from CLC 2018 [50]	Simplified lithology: metamorphic and sedimentary: others (Cenozoic)	Values < LOQ substituted with LOQ/2	IGME

[1] Abbreviations: CLC—Corine land cover map (https://land.copernicus.eu/pan-european/corine-land-cover (accessed on 20 March 2021); LOQ—limit of quantification, LOD—limit of detection.

Dataset construction and formatting are based on the extraction of data and metadata (e.g., sampling date, unit, analytical method, LOD/LOQ) from national or regional databases (step 3) (Table 2), followed by the use of a set of controls and corrections prior to any treatment (step 4). Controls are applied to check systemic errors that may be observed in large databases (unit mg/L instead of µg/L), duplicated values (multiple uploads), and anomalous data (pH, temperature); when an error can be corrected (generic error), it is possible to substitute the value. Otherwise, the sample is deleted.

A large proportion of trace elements were censored at various LRLs. The censoring level was adjusted (Table 2), thus providing a lower LRL for each element. This is a simple way to deal with non-detected data, but it is a rather arbitrary approach depending on the analytical method and laboratory. In the case of the LOQ exceeding expected NBL ranges, it is more appropriate to remove the censored value with such a high LOQ from the dataset to avoid bias [36]. This removal may require using an iterative approach.

The number of analyses and parameters varies for each dataset. The objective is to have one value for each parameter per sampling point (step 5), so that each sampling point has the same representativeness in the dataset. When several analyses (>3) were available for water points, we chose to calculate a median for these datasets so as to assign a single value to each sampling point. The median value is preferred because it is a more robust value than the mean as a result of being less affected by outliers. When several LOQ levels were reported for one sampling point, we applied the median LOQ.

Summary statistics, which include the minimum, maximum, and common percentiles, were computed for all parameters. Particular attention must be paid to the number of analyses available before conducting this first statistical analysis. In addition, redox and pH types were defined for each sampling point (step 6) according to the following classes (Tables 3 and 4): pH and redox states that control mineral solubility and trace metal mobility, often through sorption/desorption processes for pH and the redox process for redox potential. If no direct measurements are available, dissolved O_2, NO_3, SO_4, and Fe can be used to estimate the groundwater redox state based on the definition in Table 4.

2.4. Hydrogeological Characteristics of Sampling Points

For each sampling point, hydrogeological information (aquifer, lithology, depth) are linked to the data (step 7). In addition (step 8), lithological information related to water wells aids in the definition of local geology at the sampling point and in controlling consistency with the assigned groundwater body. For a common approach among all investigated aquifers, major categories were defined on the basis of eight major simplified lithologies (sedimentary: sand, gravel, carbonates, clay/marls, others; volcanic rocks; crystalline bedrock; and metamorphic rocks). This global approach makes it possible to define NBLs for simplified lithologies.

Table 3. Redox and pH types applied.

pH Type (n = 3)	Redox Types (n = 3)
Acidic (pH < 7)	Oxic or anoxic (A, B redox types)
Neutral (pH $\in$ [7, 7.5])	Weakly or strongly reduced (C, D redox types)
Basic (pH > 7.5)	Mixed (X redox type)

Table 4. Definition of redox types (modified from [55]).

REDOX TYPE	Redox Condition	NO_3- mg/L	Fe mg/L	O_2 mg/L	SO_4 mg/L
A	Oxic water	>1	<0.2	$\geq$1	-
B	Nitrate-reducing anoxic water	>1	<0.2	<1	-
C	Weakly reduced water	$\leq$1	$\geq$0.2	-	$\geq$20
D	Strongly reduced	$\leq$1	$\geq$0.2	-	<20

Finally, each sampling point is linked to its groundwater catchment or associated surface watershed (step 9). This information is necessary for the assessment of anthropogenic pressure at each sampling point. For superficial aquifers, the link is quite important, whereas, for deep aquifers that are disconnected from the surface, the link is not critical. This link is also an illustration of groundwater vulnerability for the following steps.

2.5. Anthropogenic Pressure

Evaluating anthropogenic pressure and relating activities to specific dissolved components would make it possible to determine the expected NBLs in the area under agricultural, industrial (including mining), and urban conditions. The accuracy of this approach would depend on the information regarding anthropogenic activities and the relationship between the activities and dissolved components released. The pressure inventory (step 10) is provided for anthropogenic pressure, such as diffuse pollution (agriculture, urban), point source pollution (industrial), and mining areas. Anthropogenic pressure is determined on the basis of the Corine Land Cover (CLC) inventory and the following classification: urban pressure; industrial pressure; agricultural pressure; and mining activities. This inventory is complete when databases related to industrial sites and service activities, as well as a register of pollution discharges, are available (Table 2).

All information for each water point and anthropogenic pressure is summarized using geographical information systems (GISs) (step 11). This facilitates a mapping approach to check areas with abnormal concentrations or exceedances of DWS. Due to the heterogeneity of the available data and anthropogenic conditions among countries, different GIS approaches have been tailored to determine the prevailing pressure at each sampling point (step 12). This approach can be performed at the groundwater body scale, by lithology, or for the entire dataset. The approach for determining the pressure area around a water point will vary depending on hydrogeological and watershed characteristics. The goal of this step is to identify elements affected by anthropogenic activity (after a statistical test to evaluate the impact on concentration distribution) and to discard data for which an anthropogenic pressure is recognized.

2.6. Statistical Treatments

Using one average value for each water point, descriptive statistics (step 13) are calculated for each parameter:

- Median—Quartile 1 and 3—10th percentile and 90th percentile
- Percentage of quantified data by group in order to assess the weight of unquantified (censored) values in the calculation

The description of the distribution is illustrated by boxplots or a cumulative frequency plot. Cumulative probability curves may be preferred to facilitate data comparison, because they are useful for evaluating different data populations. Spatial distributions were highlighted by mapping with GIS tools. Outliers potentially illustrate water with anthropogenic contamination or a high geochemical background, which are to be distinguished from the NBLs defined in this study.

Discriminant function analysis (DFA) (step 14) is applied to find patterns within the dataset and classify them. DFA is useful for determining whether water points can be grouped into one water family for NBL determination or if subgroups are more appropriate.

Distribution comparison and non-parametric tests (step 15) may be applied to each parameter to determine if sampling points have the same concentrations under anthropogenic pressure or to compare different classes of sampling points using simplified lithology or geochemical type (acidic or basic water, redox conditions) so as to determine whether lithology or geochemical parameters may affect NBLs. This step makes it possible to identify water families for which NBLs need to be determined. Assumptions of data normality and homogeneity of variances between types within a domain are not always met. In this situation, a parametric test is less powerful than non-parametric alternatives. In a dataset containing groups with large sample numbers (>100), a parametric approach, such as ANOVA, may still be suitable. Nevertheless, groups with a limited amount of data (five to 20) are included in the datasets in this study. For this reason, non-parametric one-way tests, such as Mann–Whitney and Kruskal–Wallis tests, were applied to test the null hypothesis assuming identical distribution for different groups [43]. Results from the non-parametric tests were analyzed with the understanding that they were less powerful than parametric tests. Non-parametric statistics do not require distributional assumptions, just mutual independence of samples. The significance of differences is established from $p < 0.05$. In the event that the Kruskal–Wallis test was significant, that is, the null hypothesis could be clearly rejected, a post hoc analysis (Nemenyi test) was performed to determine which groups were identified in the dataset.

2.7. NBL Calculation

Based on the results of step 15, sampling points identified as being affected by anthropogenic pressure were removed from each parameter data subset (step 16).

NBL determination (step 17) consists of a calculation of natural maximum concentrations representative of a water family. A water family is a lithology type or groundwater body with a dominant lithology. NBLs are calculated for SO_4, As, Cd, Cr, Cu, Ni, Zn, and F within each lithology. When geochemical parameters, such as pH and redox, are identified as affecting the data distribution (based on the Kruskal–Wallis test result, steps 14–15), the NBL is determined for subgroups. For each water family, the 90th percentile is determined. This percentile is usually applied as a cutoff to define natural concentrations by lithology without any anthropogenic effects or geogenic anomaly; the 90th percentile defines the concentration at which 90% of sampling points associated with a simplified lithology are of a lower concentration. The concentration range between the 10th and 90th percentiles represents the element's "expected" concentration [11]. For this reason, the 90th percentile was selected for NBL determination in this study.

3. Results of Methodology Application to Regional Studies

The NBL study was conducted in six (6) countries; a summary of groundwater chemistry data is shown in the following tables and detailed in the Lions et al. report [56].

3.1. Groundwater Quality Dataset Construction (Steps 1–9)

Each dataset was based on the available analysis with consideration of the specificities and limits of each dataset (Tables S1 to S7). In several cases, the LRL varies significantly from year to year and between laboratories. For example, LRLs for As can range from 0.1 to 10 µg/L in France, whereas, in Denmark, they range from 0.01 to 1 µg/L. For the highest LOQs, which may be higher than NBLs, only data with a maximum LOQ were considered. LOQ cutoff values were optimized according to data distribution, because data with a large amount of high LOQs (i.e., higher than NBLs) disturb the natural concentration distribution [36]. In the internal basins of Catalonia, assuming that a censored value may imply decreased data quality, only parameters (As, Cu, Zn, SO_4) with a LOD/LOQ below 33% were considered in the analysis. By considering a large amount of data < LOQs, 90th percentiles may be identical to the LOQ/2 or <LOQ; in this case, the NBL is indicated as <LOQ, and no concentration is calculated.

3.2. Anthropogenic Pressure (Steps 10–15)

Various GIS approaches have been used to determine the pressure at each sampling point.

3.2.1. Watershed (Loire-Bretagne)

In the Loire-Bretagne river basin, the methodology consists of linking the prevailing pressure to watershed units at each sampling point (Table S8). Watershed units are determined using the TauDEM (Terrain Analysis Using Digital Elevation Models) toolbox plugin combined with the Digital Terrain Model (DEM). This DEM (BD ALTI) has a spatial resolution of 25 m [57]. The territory was divided into 36,453 basins and sub-basins. Upstream and downstream basins were identified for each basin. Sampling points and pollution sources, such as discharge points, industrial sites, mining areas, non-point sources, or diffuse pollution per municipality and CLC [48], were linked to each watershed.

3.2.2. Individual Analysis

In Slovenia, for groundwater bodies that were declared to be of poor quality, all sampling points were listed as having anthropogenic impacts. Surrounding sites were investigated where potential pollution sources or urban areas were identified. This approach was used because the number of sampling points was limited. This data management led to the removal of 24 sampling points from the original dataset of 203 and one sampling point as an outlier for As, Cd, Cr, and Fe.

3.2.3. Buffer around the Sampling Point

For each sampling point, the areal proportion of CLC inventory types was calculated for the area surrounding sampling points (Table 2, Tables S9 and S10). Based on the large number of wells studied in Denmark (Figure 2) and Catalonia (>1000), it was decided that a one (1) km buffer would be an adequate proxy for well catchment zones, because actual well catchment areas are unknown at these national scales (Figure 3). In Austria, CLC [50] information was extracted as point data (10 m buffer), because most sampling points are found in gravel aquifers with complex catchment areas.

3.2.4. DRASTIC Method

The possibility of contaminants infiltrating from the surface in the Fruška Gora national park area was studied using the DRASTIC method [58], which was developed to evaluate groundwater vulnerability and make vulnerability maps [52,53]. This study noted that, in the central part of Fruška Gora (ridge), the vulnerability level was low (75–100) to very low (<75). Down the Fruška Gora slope, the vulnerability level increased; thus, ridges in the mountains, which are densely populated, are the most vulnerable (150–185). By combining DRASTIC with CLC, anthropogenic influences (discontinuous urban fabric, industrial units, mineral extraction sites, agriculture areas) can be seen/are clearly observable in areas of high vulnerability. This approach can be useful for an improved determination of NBLs, because prevailing pressure depends on all DRASTIC factors and hydrogeological settings.

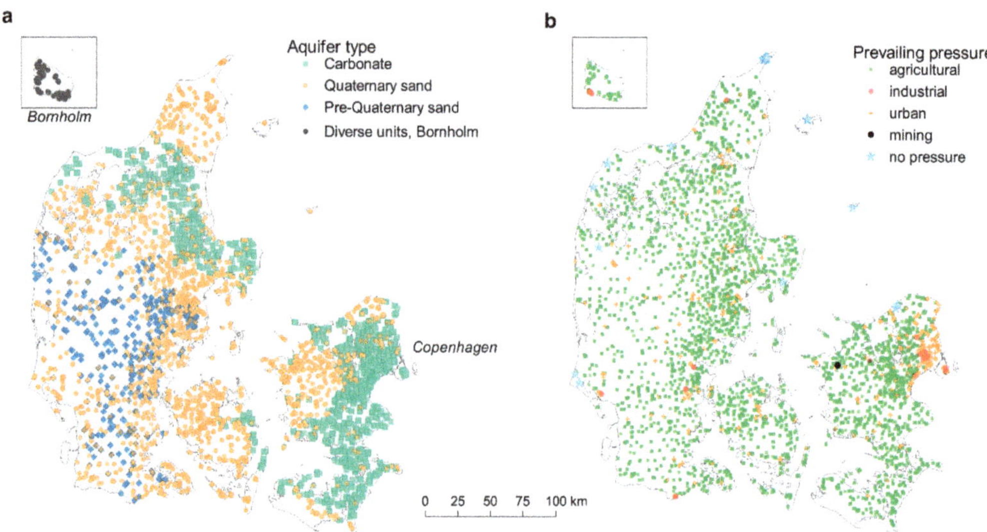

Figure 2. Sampling points in Denmark classified on the basis of (**a**) aquifer type and (**b**) prevailing pressures.

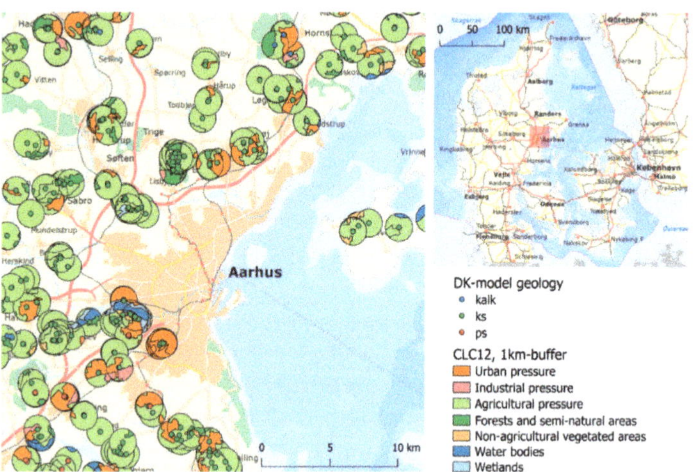

Figure 3. Example of 1 km buffer around sampling points. Abbreviations in DK model geology: kalk—carbonate aquifers; ks—quaternary sand, ps—pre-quaternary sand (figures from [56]).

3.3. Geochemistry (pH, Redox, Lithology) (Steps 13–15)

Data analysis addresses the geological composition of aquifers and geochemical parameters (pH, redox), which are among the main factors that affect trace element occurrence. Some examples are presented here; for additional details, see Lions et al. [56].

Lithology is the primary factor that controls water composition, including pH, conductivity, major ions, and trace elements. In Austria, major cations Na, K, Ca, and Mg were considered as quantitative variables to characterize aquifer lithology using a confusion matrix (Table S11). DFA (specifically, linear discriminant analysis) showed that, for Denmark, based on predictor variables (pH, SO_4, As, Ni, F, Cl, O_2, NO_3, Fe), it was possible to discriminate easily between the two lithology classes sedimentary: carbonates and sedimentary: sand (accuracy and 95% confidence interval: 81.7% (79.3–83.9%)). Therefore, these lithology classes were used for determining Danish NBLs. For Slovenia, the Mann–Whitney U test results showed significant differences between NBLs of carbonates and sands at the 95% confidence level ($\alpha = 0.05$). This was true for all elements except Cd, due to its low concentration (0.01 µg/L).

In Loire Bretagne, the Kruskal–Wallis test and Nemenyi post hoc test showed that trace elements belong to different groups that overlap when the lithology is not well-constrained, such as for clays and marls or gravels (Figure 4). Results show that volcanic rocks belong to an individual group for components such as Ni, Zn, F, or SO_4. Crystalline rocks are in a group of their own for elements such as As (including the group sedimentary–others, which contains crystalline and sedimentary formations) and Cr, whereas Ni is individualized for metamorphic rocks and sand. The distribution of an element across lithology classes was tested to justify the NBL definition for each lithology. In addition, descriptive statistics showed that crystalline and metamorphic rocks were primarily represented by acidic water (>90% with a pH < 7). This fact is directly related to the water–rock interaction in these lithologies. Conversely, carbonates and clays/marls were represented by neutral water (7–7.5) (>83%), whereas basic water was present at 12% of the sampling points in carbonates. Water in gravels can be either acidic or neutral, whereas sand and volcanic aquifers are represented in each family (acidic, basic, neutral). Redox conditions will depend more on hydrogeological settings (depth, confined aquifer) than on lithology.

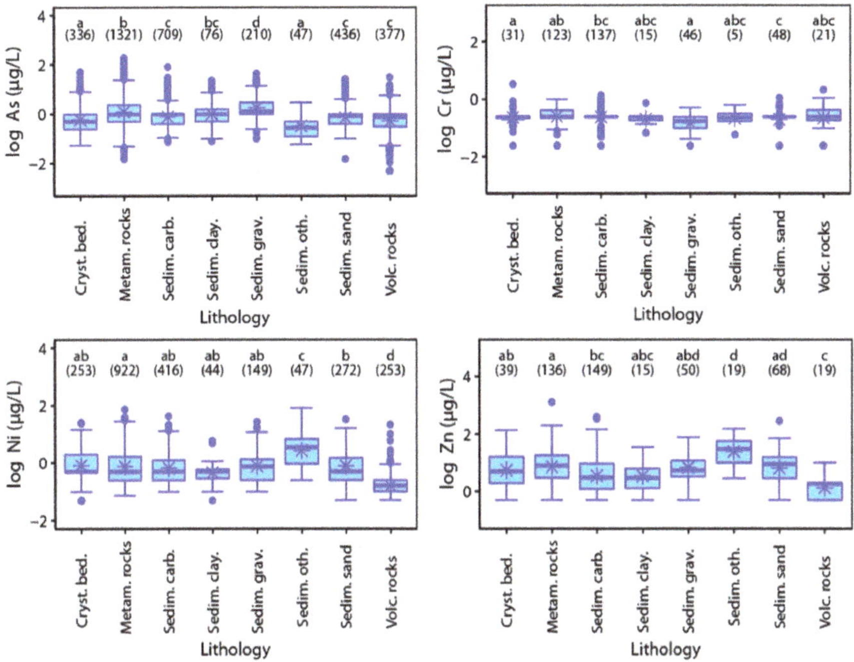

Figure 4. Results for the Kruskal–Wallis test and the Nemenyi post hoc test for lithology families in the Loire-Bretagne area (number of samples in brackets). The categories sharing letters (above the boxplot) are not significantly different at the 95% confidence level based on the post hoc Nemenyi test [56].

The Kruskal–Wallis test was used to evaluate the effect of pH and redox on trace element concentrations for each element (Figure S1). Reduced water predominates in Denmark, whereas oxic water is predominant in Loire-Bretagne. However, As is affected by the redox class, as illustrated in Figure 5 for sedimentary aquifers, whereas pH is not discriminant.

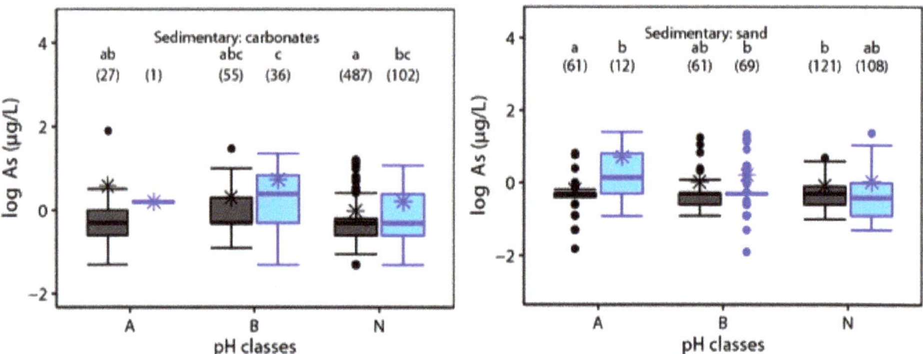

Figure 5. Results for Kruskal–Wallis test and Nemenyi post hoc test for lithology families according to pH and redox classes in the Loire-Bretagne area (number of samples in brackets). The categories sharing letters (above the boxplot) are not significantly different at the 95% confidence level based on the post hoc Nemenyi test [56].

3.4. NBLs (Steps 16–17)

The following NBLs were determined for each dataset based on aquifer lithology. Concentrations that deviated due to anthropogenic factors were excluded from the dataset prior to NBL derivation according to the conclusions of each dataset. Only families with more than five (5) sampling points are presented in Tables 5 and 6. For Denmark, the minimum number of sampling points for a lithological group was set at 20, but when the pH and redox were used, this threshold was increased to 50.

The resulting NBLs considering pH and redox conditions were determined for As, Ni, SO_4, and F (Table S12). For the Loire-Bretagne area, the results showed that Ni and Zn were clearly affected by acidic conditions with higher concentrations. Reduced conditions led to high concentrations of Fe, Mn, and As. For this dataset, NBLs were determined for As based on redox only and for Ni and Cu based on redox and pH. For carbonates, only neutral and basic waters were considered. NBLs for Zn in sandy aquifers were considered based on pH (Table S13).

Table 5. NBLs for sedimentary aquifers. DK—Denmark, SR—Serbia, SI—Slovenia, FR—Loire-Bretagne, Cat—Internal Basins Catalonia, A—Austria, (n, number of sampling points). Values higher than the respective threshold value are in bold.

	DK		SR		SI		FR		DK		Cat		SI		FR		A		FR	
	n	NBL	n	NBL	n	NBL	n	NBL	n	NBL	n	NBL	n	NBL	n	NBL	n	NBL	n	NBL
As (µg/L)	3639	**6.8**	24	**29**		<0.5	343	1.46	1830	3.9	124	3.5		<0.5	554	3	1168	1.1	33	5
Cd (µg/L)	259	0.01	24	<1	9	0.01	221	<0.5	77	0.03	–	–	51	<0.5	723	<0.5	135	0.1	723	<0.5
Cr (µg/L)	186	0.2	13	20.3	9	2.4	5	<0.45	55	0.18	–	–	51	0.5	17	<0.4	673	0.6	<5	–
Cu (µg/L)	218	1	15	4.6	9	1.3	39	4.4	67	1.3	229	8	51	0.6	131	3.1	126	2.2	9	1.4
Ni (µg/L)	3661	1.8	14	6.3	9	3.3	231	2.3	1856	5.3	–	–	51	0.5	375	2.8	135	0.5	27	7.8
Zn (µg/L)	271	8.1	19	41	9	380	38	12.3	85	20	235	99	51	4.5	133	24	–	–	10	65
F (mg/L)	4046	0.4	24	0.3		<0.1	388	0.9	2213	1.2	–	–		<0.1	673	0.3	–	–	51	0.2
SO$_4$ (mg/L)	3642	87	24	35.8	9	16	394	69.9	1834	89	8	60	47	8.2	686	44	–	–	52	56

Table 6. NBLs for sedimentary: others (SI—quaternary sand; DRB—cenozoic), crystalline, metamorphic, and volcanic aquifers: SI—Slovenia, DRB—Duero River Basin (Spain), FR—Loire-Bretagne, Cat—Internal Basins Catalonia, A—Austria (*n*, number of sampling points).

	Sedimentary: Others				Crystalline Rocks				Metamorphic Rocks						Volcanic Rocks	
	SI		DRB		Cat		FR		DRB		A		FR		FR	
	n	NBL	*n*	NBL	*n*	NBL	*n*	NBL	*n*	NBL	*n*	NBL	*n*	NBL	*n*	NBL
As (μg/L)	–	–	202	8.1	90	4.2	336	3.6	16	3.3	53	3.4	1321	6.7	373	2.4
Cd (μg/L)	22	0.02	203	0.025	–	–	380	<0.5	16	<0.05	35	0.04	1437	<0.5	378	<0.5
Cr (μg/L)	22	2	–	–	–	–	31	<0.5	–	–	50	0.5	123	<0.5	21	1
Cu (μg/L)	22	0.7	203	2.7	237	10.8	25	9.6	16	< 5	34	0.5	95	11	13	<1
Ni (μg/L)	22	15	–	–	–	–	211	5.1	–	–	35	0.5	741	5	139	0.3
Zn (μg/L)	22	27	–	–	234	44.8	40	23.8	–	–	–	–	139	40.4	19	<5
F (mg/L)	–	–	197	0.95	–	–	318	0.1	16	0.27	–	–	1146	0.1	310	0.2
SO$_4$ (mg/L)	22	15.6	203	100.3	43	33.8	383	24	16	30.9	–	–	1469	18	392	7.3

4. Discussion

4.1. Lithology

For the elements in this study, lithology is one of the primary factors that controls NBLs. The results show that using simplified aquifer lithologies is appropriate for comparing large datasets from different countries. Further subdivisions of lithology groups would lead in many cases to very low numbers of data per lithology. The NBL variations shown in Tables 5 and 6 can be partly explained by the broad spectrum of elements found. NBLs can also show multiple and different geological settings (e.g., paleo-environment) that explain the possible variations of a parameter across several countries. However, the approach of this study highlights valuable primary characteristics of lithology groups, including NBL ranges for specific elements.

A discussion of lithology groups by country may be helpful in clarifying NBL variations that can be linked to lithology.

For Denmark, it is important to distinguish between different depositional ages for the sedimentary: sand class. For example, the DFA analysis showed a difference between quaternary sands (mostly of glacial origin) and pre-quaternary sands when pH, SO$_4$, As, Ni, F, Cl, O$_2$, NO$_3$, and Fe are used as predictors [56].

We correctly identified some national specificity. In Denmark and Serbia, the NBL of As in sandy aquifers exceeds threshold values. In Denmark, groundwater resources have often geogenic As with high concentrations [59]. The Serbian dataset refers to a location north of Fruška Gora (Vojvodina), where groundwater usually contains elevated As [60–62].

The sedimentary: others class includes a wide range of area-specific lithologies, usually described as having distinct characteristics linked to local geology (specific environment, mineralization). Therefore, it is not a well-defined lithology, and NBL determination will reflect each regional lithology. In Loire-Bretagne, the sedimentary: others class includes aquifers composed of a mixture of sedimentary and other lithologies, such as volcanic strata or weathered crystalline or metamorphic rocks. In the Duero River Basin, there is an undifferentiated sequence of interbedded sedimentary sand, gravel, and clay. In such multilayered aquifer faces, NBLs are calculated for all sedimentary units. For this reason, all of them were aggregated into a single category (cenozoic). In Slovenia, the sedimentary: others class points that sample quaternary alluvial aquifers are mixtures of sand and gravel. Results for the clays/marls class are not presented here, because they were found only in France and parts of Austria [56].

Furthermore, regarding the crystalline or metamorphic classes, all formations cannot be merged into only one lithological class, because each geological domain (e.g., variscan, hercynian) has specific petrologies and mineralization that may affect the NBLs differently. This approach also excludes the contribution of local mineralization, faults, and mineral deposits. This study provides several standard NBLs for crystalline and metamorphic rocks, but further studies are needed to characterize local NBLs.

4.2. Effects of Physico-Chemical Parameters

Lithology and water chemistry control pH buffering in groundwater: acidic water in crystalline aquifers and neutral to basic water in carbonate aquifers. These factors will directly control the geochemical equilibrium for trace element release and resulting NBLs. In addition, redox potential may be directly controlled by depth or confined/unconfined conditions. In this case, redox potential is an indicator of groundwater conditions, which is correctly demonstrated by reduced water speciation (low NO_3, low Fe, low dissolved O_2) in deep or confined aquifers. In the Loire-Bretagne area and Denmark, F concentrations differ according to redox classes. For this element, redox is not directly involved in F mobilization, because it is not sensitive to redox processes; however, F is higher in confined aquifers and deep groundwater.

4.3. Anthropogenic Effects

This land use-based method for assessing anthropogenic effects aids in the identification of statistical differences between sampling points subject to prevailing agricultural pressure compared to those affected by industrial and/or urban pressures. Sampling point distribution according to prevailing pressures is presented in [56] and Table S8–S10. Regarding statistical significance among various lithologies and anthropogenic impacts, it is possible to conclude that prevailing pressures may affect trace element concentrations in several cases.

In the Duero River Basin (Spain), most sampling points are mainly affected by agricultural pressure. In other words, within a 1 km buffer around each sampling point, most of the land is used for agricultural purposes. The pre-selection of sampling locations to include only drinking water supply wells may explain the absence of points affected by industry, mining, or urban activities.

In the Danish case, less than 1% of the sampling points were devoid of any anthropogenic pressure (i.e., are natural). All are located near the coast, and are not considered to be representative of undisturbed conditions. Depending on the components considered, between 74.8 and 85.8% of the sampling points show effects of agricultural pressure. This finding limits the application of the method, because it means that a statistical test for comparing elemental distributions cannot be used against the natural group. As a result, the normal state is assumed to be dominant agricultural pressure. In this case, only sampling point groups for which industrial or urban pressure predominates can be compared to the agricultural group. The shortcoming of this method is that it fails to distinguish between polluted and potentially polluted sampling points, but excludes the entire group with the specific dominating pressure. As a consequence, an entire data group is excluded, and some NBLs increase because sampling points with low concentrations (potentially the NBL) were excluded.

In Slovenia, a previous study [38] used only sampling points with a minimum probability of being affected by anthropogenic activities, and therefore determined NBLs for NO_3 (3.8 mg/L) and SO_4 (6.9 mg/L). Serianz et al. (2020) [34] assigned 6.0 mg/L for NO_3 after removing all analyses exceeding 10 mg/L of nitrate; SO_4 NBLs were 17.0 mg/L (by the probability method) and 26.0 mg/L (by the pre-selection method). In this study, the natural group NBLs are 15.5 mg/L for NO_3 and 14.0 mg/L for SO_4, but as much as 41.5 mg/L and 25.0 mg/L for the anthropogenic group, respectively. This result indicates that many points in the Slovenian national monitoring network used for groundwater body assessment are noticeably affected by anthropogenic pressures, so trace element concentrations should be evaluated with care. As a result, only 87 sites without obvious anthropogenic impact were used for the NBL calculation. Thus, it is important to carefully assess the statistical test results (step 15, Figure 1), which can serve as a first screening followed by expert assessment before data removal. In addition, a more detailed assessment of anthropogenic pressure and exclusion of polluted sites can be done for a target area within a specific groundwater body or aquifer.

The results of each prevailing pressure assessment (Table 7) could not be generalized to other groundwater bodies, because they are directly linked to the distribution and nature of anthropic activities. However, anthropogenic impacts have been statically observed for each trace components (As, Cd, Cr, Cu, Ni, Zn, F, SO_4) under the following pressures: agriculture (As, Cd, Cu, Ni, Zn), urban and industrial (all), and mining (As, Cu). Other trends may exist, but they have not been observed in the investigated datasets.

Table 7. Differences statistically observed for anthropogenically influenced values.

	SI	FR	Cat	DRB	A
As		Significant differences between natural state and anthropogenic influences (agriculture, urban, industrial, mining)	Significant differences between natural state and anthropogenic influences (mining)		No significant differences between anthropogenic influences and natural state
Cd	No significant differences between anthropogenic influences and natural state				Significant differences between natural state and anthropogenic influences (agricultural, industrial, and urban)
Cr	Significant differences between anthropogenic influences and natural state				Significant differences between natural state and anthropogenic influences (industrial and urban)
Cu	Significant differences between anthropogenic influences and natural state	Significant differences between natural state and anthropogenic influences (mining)	Significant differences between natural state and anthropogenic influences (mining)		Significant differences between natural state and anthropogenic influences (agricultural, industrial, and urban)
Ni	Significant differences between anthropogenic influences and natural state	Significant differences between natural state and anthropogenic influences (agriculture)			Significant differences between natural state and anthropogenic influences (agricultural, industrial, and urban)
Zn	Significant differences between anthropogenic influences and natural state	Significant differences between natural state and anthropogenic influences (agriculture, urban, industrial)	No significant differences between anthropogenic influences and natural state		

Table 7. *Cont.*

	SI	FR	Cat	DRB	A
F		Significant differences between natural state and anthropogenic influences (urban)		Significant differences between anthropogenic influences and natural state	
SO$_4$	Significant differences between anthropogenic influences and natural state	Significant differences between natural state and anthropogenic influences (urban)	Significant differences between natural state and anthropogenic influences (agriculture, urban, industrial, mining)	Significant differences between anthropogenic influences and natural state	Significant differences between anthropogenic influences and natural state

4.4. NBLs

Discriminant analyses of lithology classes (qualitative variables) and selected major and minor elements (quantitative variables) were performed to characterize the most prevalent water families in each dataset. While the median is representative of the average concentration, the 90th percentile gives a high range for natural concentrations [63]. The 90th percentile makes it possible to estimate the possible concentrations of natural origin in each group. Values defined by higher percentiles, for example, 95 or 97.7%, would be also appropriate as a reference for maximum natural concentrations, but higher percentiles require larger datasets and well-selected sampling points for higher confidence intervals [64]. When a large variability is observed, NBLs could be much higher than the average concentration in groundwater. Therefore, NBLs are not average concentrations, but concentrations in a higher range. These values are helpful to emphasize anomalous concentrations in groundwater. The NBLs obtained from these seven case studies are well within the range of NBLs determined by international studies [65].

In this study, we were able to apply the proposed method and establish NBL values that are within the expected range compared to previous studies [34–38,40,66]. In the case of the Duero River Basin, NBLs are compared to threshold values provided by local authorities. In the case of SO$_4$, the NBL obtained when considering the cenozoic aquifer is about 241 mg/L (anthropogenically affected value), while the value without the anthropogenically affected data is about 100 mg/L.

For groundwater management, these NBLs help to highlight groundwater bodies where high concentrations need further investigation, so as to identify the origin of the anomaly. An anomalous concentration in a groundwater body may be a result of groundwater contamination, but may also come from a natural occurrence in a specific aquifer or in a particular geological context, such as mineralization or thermal waters [67–69]. Even so, the NBLs established in this study should not be applied directly to management and planning purposes at the groundwater body scale, because the data processing was conducted at a different scale (regional, national). In addition, in the case of high geological heterogeneity, such as in the Catalonian aquifers, NBL estimation needs more criteria to define water families for groundwater management implementation.

5. Conclusions

The methodology developed in this study was demonstrated in multiple countries and groundwater monitoring contexts. The use of the methodology made it possible to distinguish water families on the basis of lithology and geochemical conditions (pH, redox), for which different NBLs are expected. The proposed method can be used for the broad contexts investigated in this study; it provides coherent results. Therefore, it can be used as a step to explore a large dataset from groundwater monitoring networks. The methodology is an overall lithological approach; its results can be extrapolated to aquifers of comparable lithology by excluding local specificity links, for example, to the presence of mineralization, mining (metals and metalloids), or evaporites (gypsum). For each aquifer, this first estimation is open to the addition of new qualitative and hydrogeological data to refine the preliminary NBLs and the area of application.

The statistical tests applied depend on the amount of available data and the number of hydrogeological and lithological families. This approach may be limited for some hydrogeological entities (e.g., heterogeneous aquifers) depending on the number of available sampling points and characterization of hydrogeological settings (depth, confined/unconfined, pumped layer). NBLs determined for families characterized by only a few water points or that are not representative of the groundwater body should be viewed with caution. NBLs may be refined on the basis of additional data, the definition of lithology or groundwater entities, and the lithologic and petrographic description of the sampled aquifer. Hydrogeological settings for each groundwater need to be carefully addressed.

Determination of NBLs in alluvial aquifers is challenging because of obvious contamination and presence of anthropic activities. This caution is also relevant for unconsolidated and unprotected shallow aquifers, such as the aeolian or glacio-fluvial aquifers in Denmark [70]. In general, these vulnerable aquifers are usually impacted by nitrates or/and organic pollutants according to prevailing pressures [71]. Conversely, confined aquifers are naturally protected from surface infiltration and, as a result, anthropogenic contamination is less common.

Using monitoring network data and comparing measured concentrations to calculated NBLs, it is possible to identify groundwater bodies for additional investigation to identify the origin of anomalous concentrations or anthropogenic inputs. Delineation of these entities and the interpretation of exceedances are beyond the scope of this study, because they require local investigation. Specific geogenic features (mineralization, evaporites, thermal waters) require a different methodology from the lithological NBL as defined in this study. The results of each prevailing pressure assessment cannot be generalized to other groundwater bodies, because trace element contaminations are directly linked to the type of anthropic activity; only general trends for each element can be highlighted.

This work constitutes a first step in the definition of NBLs. At this scale, data analytical methods might also be accompanied by process-oriented analyses and supported by expert background knowledge at the local scale. This methodology will depend strongly on the selection of relevant water points and the use of sampling and analytical methods suitable for reducing uncertainty in NBL characterization. To provide better estimates of trace element NBLs in the future, monitoring networks should be improved or additional sites besides the national monitoring network should be used with the selection of water points less subject to anthropogenic impacts to provide additional information regarding trace element concentrations. In addition, targeted analytical methods should be used to obtain lower LOQs and thereby provide more useful datasets.

Supplementary Materials: The following are available online at https://www.mdpi.com/article/10.3390/w13111531/s1, Table S1. Denmark: The number of chemical analyses for each element (n), account of LODs in the dataset, range of LODs, the number and percentage of analyses < LOD, the max substitute value, and the number of analyses with the max substituted. Table S2. Serbia: The number of chemical analyses for each element (n), account of LODs in the dataset, the percentage of analyses < LOD, the max substitute value, and max LOD substituted. Table S3. Slovenia: The number of chemical analyses for each element, account of LOQs in the dataset (two were reported in 2016), range of LOQs, the number and percentage of analyses < LOQ, the max substitute value, and the number of analyses with the max substitute. Table S4. Loire-Bretagne, France: The number of chemical analyses for each element (n), account of value > LOQs in the dataset, range of LOQs (low, high), percentage of analyses < LOQ, percentage with high LOQ, and the maximum value substituted. Table S5. Internal Basins, Catalonia: The number of analyses for each element (n) and number and percentage of analyses < LOD, range of LOD, and maximum value imputed with each method. Table S6. Austria: Dataset overview, ranked by percentage of values below LOD or LOQ. Table S7. Spain, DRB: The number and percentage of chemical analyses below the limit of quantification (LOQ), range of LOQs in the dataset, and min and max substituted value. Table S8. Loire-Bretagne, France: The number of water sampling points with data for each element and each group of prevailing pressure. Table S9. Denmark: The number of water sampling points with data for each element and each group of prevailing pressure and results from the Kruskal–Wallis rank sum test. Table S10. Spain, DRB: The number of water sampling points with data for each element and each group of prevailing pressure and results from the Kruskal–Wallis rank sum test. Table S11. Austria: The confusion matrix for lithology. Table S12. Denmark: Natural background levels (NBLs) for sedimentary aquifers combined with pH and redox conditions. Table S13. Loire-Bretagne, France: Natural background levels (NBLs) for the sedimentary aquifers combined with pH and redox conditions.

Author Contributions: Conceptualization, L.G., D.E.; methodology, N.D., E.M. and J.L.; data curation, J.L., N.D., V.C., T.P.P., N.R., D.E., J.S., S.C. and D.D.V.; writing—original draft preparation, J.L.; writing—review, G.A., I.H., T.P.P., K.A.S., N.R., J.S., J.G., E.G.-F., J.A.L.-E., D.D.V. and B.H.; funding acquisition, L.G. All authors have read and agreed to the published version of the manuscript.

Funding: This research was funded by the European Union's Horizon 2020 research and innovation program (GeoERA Groundwater HOVER) under grant agreement number 731166. N. Rman participation was supported by the Slovenian Research Agency, research program P1-0020 Groundwaters and Geochemistry.

Data Availability Statement: Final Report for HOVER WP3.3 is available here: https://repository.europe-geology.eu/egdidocs/hover/hover_33_statistical_data_treatment_backgroundlev.pdf. For France, monitoring data (ADES) are available https://ades.eaufrance.fr/ (accessed on 12 February 2021) and information for aquifer are available on https://bdlisa.eaufrance.fr/ (accessed on 15 September 2020). For Denmark, database JUPITER is available at https://eng.geus.dk/products-services-facilities/data-and-maps/national-well-database-jupiter (accessed on 3 July 2019). Austrian data are available on the Austrian groundwater monitoring network (WISA): Wasserinformationssystem; Gewässerzustandsüberwachungsverordnung, BGBl. II Nr. 479/06 i.d.g.F. Catalonian data from SDIM-ACA are available on http://aca-web.gencat.cat/sdim21/ (accessed on 25 September 2020). Data from the Slovenian Environmental Agency are available on https://www.arso.gov.si/vode/podatki/ (accessed on 15 July 2020) and land use on www.mkgp.gov and https://rkg.gov.si/vstop/ (accessed on access on 21 September 2020). Data from Duero River Basin (Spain) are available on https://www.chduero.es/web/guest/red-control-estado-quimico (accessed on 30 October 2020) and land use on http://centrodedescargas.cnig.es/CentroDescargas/catalogo.do?Serie=SIOSE (accessed on 29 December 2020).

Conflicts of Interest: The authors declare no conflict of interest.

References

1. *Directive 2000/60/EC of the European Parliament and of the Council of 23 October 2000 Establishing a Framework for Community Action in the Field of Water Policy*; European Union: Brussels, Belgium, 2000.
2. *Directive 2006/118/EC of the European Parliament and of the Council of 12 December 2006 on the Protection of Groundwater against Pollution and Deterioration*; European Union: Brussels, Belgium, 2006.
3. Shand, P.; Edmunds, W.M. The Baseline Inorganic Chemistry of European Groundwaters. In *Natural Groundwater Quality*; Blackwell Publishing: Oxford, UK, 2009; pp. 22–58.
4. *Directive 2008/105/EC of the European Parliament and of the Council of 16 December 2008 on Environmental Quality Standards in the Field of Water Policy*; European Union: Brussels, Belgium, 2008.
5. *Directive 2020/2184 of the European Parliament and of the Council of 16 December 2020 on the Quality of Water Intended for Human Consumption*; European Union: Brussels, Belgium, 2020.
6. Ballesteros-Navarro, B.J.; Díaz-Losada, E.; Domínguez-Sánchez, J.A.; Grima-Olmedo, J. Methodological proposal for conceptualization and classification of interactions between groundwater and surface water. *Water Policy* **2019**, *21*, 623–642. [CrossRef]
7. *Directive 2013/39/UE of the European Parliament and of the Council of 12 August 2013 Amending Directives 2000/60/EC and 2008/105/EC as Regards Priority Substances in the Field of Water Policy*; European Union: Brussels, Belgium, 2013.
8. Appelo, C.; Postma, D. *Geochemistry, Groundwater and Pollution*; Balkema Publishers: Amsterdam, The Netherlands, 2005.
9. Blum, A.; Pauwels, H.; Wenland, F.; Grifioen, J. Background levels under the Water Framework Directive. In *Groundwater Monitoring (Water Quality Measurement Series)*; John Wiley and Sons: Wiltshire, UK, 2009; pp. 145–153.
10. Wendland, F.; Blum, A.; Coetsiers, M.; Gorova, R.; Griffioen, J.; Grima, J.; Hinsby, K.; Kunkel, R.; Marandi, A.; Melo, T.; et al. European aquifer typology: A practical framework for an overview of major groundwater composition at European scale. *Environ. Earth Sci.* **2007**, *55*, 77–85. [CrossRef]

11. Hinsby, K.; de Melo, M.T.C.; Dahl, M. European case studies supporting the derivation of natural background levels and groundwater threshold values for the protection of dependent ecosystems and human health. *Sci. Total. Environ.* **2008**, *401*, 1–20. [CrossRef]

12. Brenot, A.; Gourcy, L.; Mascré, C. *Identification des Zones à Risque de Fond Géochimique Élevé en Éléments Traces Dans les Cours D'eau et les Eaux Souterraines*; BRGM: Orléans, France, 2007.

13. Hobiger, G.; Klein, P.; Denk, J.; Grösel, K.; Heger, H.; Kohaut, S.; Kollmann, W.F.H.; Lampl, H.; Lipiarsky, P. *Österreichweite Abschätzung von Regionalisierten, Hydro-Chemischen Hintergrundgehalten in Oberflächennahen Grundwasserkörpern auf der Basis Geochemischer und Wasserchem-Ischer Analysedaten zur Umsetzung der Wasserrahmenrichtlinie (Geogene Hintergrundgehalte Oberflächennaher Grundwasserkörper)*; Geologische Bundesanstalt: Wien, Austria, 2004.

14. Edmunds, W.; Shand, P.; Hart, P.; Ward, R. The natural (baseline) quality of groundwater: A UK pilot study. *Sci. Total Environ.* **2003**, *310*, 25–35. [CrossRef]

15. Shand, P.; Edmunds, W.M.; Lawrence, A.R.; Smedley, P.L.; Burke, S. *The Natural (Baseline) Quality of Groundwater in England and Wales*; British Geological Survey: Nottingham, UK, 2007.

16. Wendland, F.; Berthold, G.; Blum, A.; Elsass, P.; Fritsche, J.-G.; Kunkel, R.; Wolter, R. Derivation of natural background levels and threshold values for groundwater bodies in the Upper Rhine Valley (France, Switzerland and Germany). *Desalination* **2008**, *226*, 160–168. [CrossRef]

17. Molinari, A.; Guadagnini, L.; Marcaccio, M.; Guadagnini, A. Natural background levels and threshold values of chemical species in three large-scale groundwater bodies in Northern Italy. *Sci. Total Environ.* **2012**, *425*, 9–19. [CrossRef] [PubMed]

18. Coetsiers, M.; Blaser, P.; Martens, K.; Walraevens, K. Natural background levels and threshold values for groundwater in fluvial Pleistocene and Tertiary marine aquifers in Flanders, Belgium. *Environ. Earth Sci.* **2008**, *57*, 1155–1168. [CrossRef]

19. Urresti-Estala, B.; Carrasco-Cantos, F.; Vadillo-Pérez, I.; Jiménez-Gavilán, P. Determination of background levels on water quality of groundwater bodies: A methodological proposal applied to a Mediterranean River basin (Guadalhorce River, Málaga, southern Spain). *J. Environ. Manag.* **2013**, *117*, 121–130. [CrossRef]

20. Sellerino, M.; Forte, G.; Ducci, D. Identification of the natural background levels in the Phlaegrean fields groundwater body (Southern Italy). *J. Geochem. Explor.* **2019**, *200*, 181–192. [CrossRef]

21. Busico, G.; Cuoco, E.; Kazakis, N.; Colombani, N.; Mastrocicco, M.; Tedesco, D.; Voudouris, K. Multivariate statistical analysis to characterize/discriminate between anthropogenic and geogenic trace elements occurrence in the Campania Plain, Southern Italy. *Environ. Pollut.* **2018**, *234*, 260–269. [CrossRef] [PubMed]

22. Rotiroti, M.; Di Mauro, B.; Fumagalli, L.; Bonomi, T. COMPSEC, a new tool to derive natural background levels by the component separation approach: Application in two different hydrogeological contexts in northern Italy. *J. Geochem. Explor.* **2015**, *158*, 44–54. [CrossRef]

23. Nisi, B.; Buccianti, A.; Raco, B.; Battaglini, R. Analysis of complex regional databases and their support in the identification of background/baseline compositional facies in groundwater investigation: Developments and application examples. *J. Geochem. Explor.* **2016**, *164*, 3–17. [CrossRef]

24. Morgenstern, U.; Daughney, C.J. Groundwater age for identification of baseline groundwater quality and impacts of land-use intensification—The National Groundwater Monitoring Programme of New Zealand. *J. Hydrol.* **2012**, *456–457*, 79–93. [CrossRef]

25. Missimer, T.; Teaf, C.M.; Beeson, W.T.; Maliva, R.G.; Woolschlager, J.; Covert, D.J. Natural Background and Anthropogenic Arsenic Enrichment in Florida Soils, Surface Water, and Groundwater: A Review with a Discussion on Public Health Risk. *Int. J. Environ. Res. Public Health* **2018**, *15*, 2278. [CrossRef]

26. Li, J.; Shi, Z.; Liu, M.; Wang, G.; Liu, F.; Wang, Y. Identifying anthropogenic sources of groundwater contamination by natural background levels and stable isotope application in Pinggu basin, China. *J. Hydrol.* **2021**, *596*, 126092. [CrossRef]

27. Masciale, R.; Amalfitano, S.; Frollini, E.; Ghergo, S.; Melita, M.; Parrone, D.; Preziosi, E.; Vurro, M.; Zoppini, A.; Passarella, G. Assessing Natural Background Levels in the Groundwater Bodies of the Apulia Region (Southern Italy). *Water* **2021**, *13*, 958. [CrossRef]

28. Guadagnini, L.; Menafoglio, A.; Sanchez-Vila, X. Probabilistic assessment of spatial heterogeneity of natural background concentrations in large-scale groundwater bodies through Functional Geostatistics. *Sci. Total Environ.* **2020**, *740*, 140139. [CrossRef]

29. Kim, K.-H.; Yun, S.-T.; Kim, H.-K.; Kim, J.-W. Determination of natural backgrounds and thresholds of nitrate in South Korean groundwater using model-based statistical approaches. *J. Geochem. Explor.* **2015**, *148*, 196–205. [CrossRef]

30. Gao, Y.; Qian, H.; Huo, C.; Chen, J.; Wang, H. Assessing natural background levels in shallow groundwater in a large semiarid drainage Basin. *J. Hydrol.* **2020**, *584*, 124638. [CrossRef]

31. Molinari, A.; Guadagnini, L.; Marcaccio, M. Geostatistical multimodel approach for the assessment of the spatial distribution of natural background concentrations in large-scale groundwater bodies. *Water Res.* **2019**, *149*, 522–532. [CrossRef]

32. Parrone, D.; Frollini, E.; Preziosi, E.; Ghergo, S. eNaBLe, an On-Line Tool to Evaluate Natural Background Levels in Groundwater Bodies. *Water* **2020**, *13*, 74. [CrossRef]

33. Herms, I.; Jódar, J.; Soler, A.; Lambán, L.; Custodio, E.; Núñez, J.; Arnó, G.; Ortego, M.; Parcerisa, D.; Jorge, J. Evaluation of natural background levels of high mountain karst aquifers in complex hydrogeological settings. A Gaussian mixture model approach in the Port del Comte (SE, Pyrenees) case study. *Sci. Total Environ.* **2021**, *756*, 143864. [CrossRef] [PubMed]

34. Serianz, L.; Cerar, S.; Šraj, M. Hydrogeochemical characterization and determination of natural background levels (NBL) in groundwater within the main lithological units in Slovenia. *Environ. Earth Sci.* **2020**, *79*. [CrossRef]

35. Brielmann, H.; Legerer, P.; Schubert, G.; Wemhöner, U.; Philippitsch, R.; Humer, F.; Zieritz, I.; Rosmann, T.; Schartner, C.; Scheidleder, A.; et al. *Hydrochemie und Hydrogeologie der Österreichischen Grundwässer und Deren Natürliche Metall- und Nährstoffgehalte (Update Geohint 2018)*; Bundesministerium für Nachhaltigkeit und Tourismus: Wien, Austria, 2018.

36. Lions, J.; Mauffert, A.; Devau, N. *Evaluation des Concentrations de Rérérence des Fonds Hydrogéochimiques des Eaux Souterraines par Lithologie D'aquifères*; BRGM: Orléans, France, 2016.

37. Voutchkova, D.; Ernstsen, V.; Schullehner, J.; Hinsby, K.; Thorling, L.; Hansen, B. Roadmap for Determining Natural Background Levels of Trace Metals in Groundwater. *Water* **2021**, *13*, 1267. [CrossRef]

38. Cruz, J.V.; Pacheco, D.; Cymbron, R.; Mendes, S. Monitoring of the groundwater chemical status in the Azores archipelago (Portugal) in the context of the EU water framework directive. *Environ. Earth Sci.* **2010**, *61*, 173–186. [CrossRef]

39. Mezga, K. Natural Hydrogeochemical Background and Dynamics of Groundwater in Slovenia. Ph.D. Thesis, University of Nova Gorica, Nova Gorica, Slovenia, 2014.

40. Grima, J.; Luque-Espinar, J.A.; Mejía, J.A.; Rodríguez, R. Methodological approach for the analysis of groundwater quality in the framework of the Groundwater Directive. *Environ. Earth Sci.* **2015**, *74*, 4039–4051. [CrossRef]

41. Preziosi, E.; Giuliano, G.; Vivona, R. Natural background levels and threshold values derivation for naturally As, V and F rich groundwater bodies: A methodological case study in Central Italy. *Environ. Earth Sci.* **2010**, *61*, 885–897. [CrossRef]

42. Alcaine, A.A.; Schulz, C.; Bundschuh, J.; Jacks, G.; Thunvik, R.; Gustafsson, J.-P.; Mörth, C.-M.; Sracek, O.; Ahmad, A.; Bhattacharya, P. Hydrogeochemical controls on the mobility of arsenic, fluoride and other geogenic co-contaminants in the shallow aquifers of northeastern La Pampa Province in Argentina. *Sci. Total Environ.* **2020**, *715*, 136671. [CrossRef] [PubMed]

43. Helsel, D.R.; Hirsch, R.M. Techniques of Water-Resources Investigations of the United States Geological Survey. Hydrologic Analysis and Interpretation: Statistical Methods in Water Resources. *Environ. Sci. Pollut. Res. Int.* **2002**, *14*, 297–307.

44. Antweiler, R.C. Evaluation of Statistical Treatments of Left-Censored Environmental Data Using Coincident Uncensored Data Sets. II. Group Comparisons. *Environ. Sci. Technol.* **2015**, *49*, 13439–13446. [CrossRef]

45. Palarea-Albaladejo, J.; Martín-Fernández, J.A. zCompositions—R package for multivariate imputation of left-censored data under a compositional approach. *Chemom. Intell. Lab. Syst.* **2015**, *143*, 85–96. [CrossRef]

46. Guigues, N.; Lepot, B.; Desenfant, M.; Durocher, J. Estimation of the measurement uncertainty, including the contribution arising from sampling, of water quality parameters in surface waters of the Loire-Bretagne river basin, France. *Accredit. Qual. Assur.* **2020**, *25*, 281–292. [CrossRef]

47. Ghestem, J.P. Incertitudes Liées à L'échantillonnage: Exemples D'estimation sur eau de Surface et eau Souterraine, Aquaref. Available online: https://www.aquaref.fr/ (accessed on 20 March 2021).

48. Corine Land Cover. Available online: https://land.copernicus.eu/pan-european/corine-land-cover/clc-2012 (accessed on 20 March 2021).

49. Troldborg, L. *Afgrænsning af de Danske Grundvandsforekomster: Ny Afgrænsning og Delkarakterisering Samt Fagligt Grundlag for Udpegning af Drikkevandsforekomster*; GEUS Rapport 2020/1; Geological Survey of Denmark and Greenland (GEUS): Copenhagen, Denmark, 2020; p. 82. (In Danish)

50. Corine Land Cover. Available online: https://land.copernicus.eu/pan-european/corine-land-cover/clc2018 (accessed on 20 March 2021).

51. Schubert, G.; Lampl, H.; Shadlau, S.; Wurm, M.; Pavlik, W.; Pestal, G.; Bayer, I.; Freiler, M.; Schild, A.; Stöckl, W. *Hydroge-ologische Karte von Österreich 1:500,000*; Geologische Bundesanstalt: Wien, Austria, 2003.

52. Pantić, T.P.; Veljković, Ž.; Tomić, M.; Samolov, K. Hydrogeology and vulnerability of groundwater to polluting in the area of national park "Fruška Gora". *VODOPRIVREDA* **2017**, *49*, 288–290.

53. Veljković, Ž.; Pantić, T.P.; Tomić, M.; Mandić, M.; Samolov, K. Vulnerability of groundwater pollutants in the area of national park "Fruška Gora". In Proceedings of the XXIV International Conference Eco-Ist, Vrnjačka Banja, Serbia, 12–15 June 2016; pp. 464–470.

54. Buser, S. *Geological Map of Slovenia in the Scale of 1:250,000*; GeoZS: Ljubljana, Slovenia, 2010.

55. Hansen, B.; Thorling, L. *Kemisk Grundvandskortlægning (Geo-Vejledning No. 2018/2)*; GEUS: Copenhagen, Denmark, 2018.

56. Lions, J.; Malcuit, E.; Gourcy, L.; Voutchkova, D.; Hansen, B.; Schullehner, J.; Forcada, E.G.; Olmedo, J.G.; Elster, D.; Camps, V.; et al. HOVER Deliverable 3.3—Data Set of the Results of the Statistical Data Treatment Allowing the Preparation of the Raw Elements for the Tasks 4 and 5 i.e., Concentrations of Elements of Natural Origin Per Typologies; Orléans, France. 2021. Available online: https://repository.europe-geology.eu/egdidocs/hover/hover_33_statistical_data_treatment_backgroundlev.pdf (accessed on 28 May 2021).

57. Devau, N.; Lions, J.; Schomburgk, S.; Bertin, C.; Blanc, P.; Mathurin, F.; Thinon-Larminach, M.; Lucassou, F.; Le Guern, C.; Tourlière, B.; et al. *Etude par Approche Globale des Fonds Hydrogéochimiques des Eaux Souterraines Sur Le Bassin Loire-Bretagne*; BRGM: Orléans, France, 2017.

58. Aller, L.; Bennet, T.; Lehr, J.; Petty, R.; Hackett, G. *DRASTIC: A Standardized System for Evaluating Groundwater Pollution Potential Using Hydrogeologic Settings*; U.S. EPA: Chicago, IL, USA, 1985.

59. Ramsay, L.; Petersen, M.M.; Hansen, B.; Schullehner, J.; van der Wens, P.; Voutchkova, D.; Kristiansen, S.M. Drinking Water Criteria for Arsenic in High-Income, Low-Dose Countries: The Effect of Legislation on Public Health. *Environ. Sci. Technol.* **2021**, *55*, 3483–3493. [CrossRef]

60. Dalmacija, B.; Bečelić-Tomin, M.; Krčmar, D.; Lazić, N. Water. In *Environment in Autonomous Province of Vojvodina*; Provincial Secretariat for Urbanism, Construction and Environmental Protection: Novi Sad, Serbia, 2011; pp. 94–135.

61. Jovanović, D.; Jakovljević, B.; Rašić-Milutinović, Z.; Paunović, K.; Peković, G.; Knežević, T. Arsenic occurrence in drinking water supply systems in ten municipalities in Vojvodina Region, Serbia. *Environ. Res.* **2011**, *111*, 315–318. [CrossRef] [PubMed]

62. Petrović, T.; Zlokolica-Mandić, M.; Veljković, N.; Papić, P.; Poznanović, M.; Stojković, J.; Magazinović, S. Macro- and micro-elements in bottled and tap waters of Serbia. *Hem. Ind.* **2012**, *66*, 107–122. [CrossRef]

63. Griffioen, J.; Passier, H.F.; Klein, J. Comparison of Selection Methods to Deduce Natural Background Levels for Groundwater Units. *Environ. Sci. Technol.* **2008**, *42*, 4863–4869. [CrossRef] [PubMed]

64. Edmunds, W.M.; Shand, P. *Natural Groundwater Quality*; Blackwell Publishing: Oxford, UK, 2008.

65. Tedd, K.; Coxon, C.; Misstear, B.; Daly, D.; Craig, M.; Mannix, A.; Williams, T.H. *Research 183: Assessing and Developing Natural Background Levels for Chemical Parameters in Irish Groundwater*; Environmental Protection Agency: Dublin, Ireland, 2017.

66. Agència Catalana de l'Aigua (ACA). *Pla de Gestió del Districte de la Conca Fluvial de Catalunya: Període 2016–2021*; Agència Catalana de l'Aigua: Barcelona, Spain, 2015.

67. Binda, G.; Pozzi, A.; Livio, F.; Piasini, P.; Zhang, C. Anomalously high concentration of Ni as sulphide phase in sediment and in water of a mountain catchment with serpentinite bedrock. *J. Geochem. Explor.* **2018**, *190*, 58–68. [CrossRef]

68. Hossain, M.; Patra, P.K. Hydrogeochemical characterisation and health hazards of fluoride enriched groundwater in diverse aquifer types. *Environ. Pollut.* **2020**, *258*, 113646. [CrossRef] [PubMed]

69. Cinti, D.; Poncia, P.; Brusca, L.; Tassi, F.; Quattrocchi, F.; Vaselli, O. Spatial distribution of arsenic, uranium and vanadium in the volcanic-sedimentary aquifers of the Vicano–Cimino Volcanic District (Central Italy). *J. Geochem. Explor.* **2015**, *152*, 123–133. [CrossRef]

70. Hansen, B.; Sonnenborg, T.O.; Møller, I.; Bernth, J.D.; Høyer, A.-S.; Rasmussen, P.; Sandersen, P.B.; Jørgensen, F. Nitrate vulnerability assessment of aquifers. *Environ. Earth Sci.* **2016**, *75*, 999. [CrossRef]

71. Lopez, B.; Ollivier, P.; Togola, A.; Baran, N.; Ghestem, J.-P. Screening of French groundwater for regulated and emerging contaminants. *Sci. Total Environ.* **2015**, *518–519*, 562–573. [CrossRef] [PubMed]

Article

Deriving Natural Background Levels of Arsenic at the Meso-Scale Using Site-Specific Datasets: An Unorthodox Method

Maria Filippini [1], Chiara Zanotti [2], Tullia Bonomi [2], Vito G. Sacchetti [1], Alessandro Amorosi [1], Enrico Dinelli [1] and Marco Rotiroti [2,*]

[1] Department of Biological, Geological and Environmental Sciences, Alma Mater Studiorum, University of Bologna, Piazza di Porta S. Donato, 1, 40126 Bologna, Italy; maria.filippini3@unibo.it (M.F.); vito.sacchetti@studio.unibo.it (V.G.S.); alessandro.amorosi@unibo.it (A.A.); enrico.dinelli@unibo.it (E.D.)

[2] Department of Earth and Environmental Sciences, University of Milano-Bicocca, Piazza della Scienza 1, 20126 Milan, Italy; chiara.zanotti@unimib.it (C.Z.); tullia.bonomi@unimib.it (T.B.)

* Correspondence: marco.rotiroti@unimib.it

Abstract: Arsenic is found in groundwater above regulatory limits in many countries and its origin is often from natural sources, making the definition of Natural Background Levels (NBLs) crucial. NBL is commonly assessed based on either dedicated small-scale monitoring campaigns or large-scale national/regional groundwater monitoring networks that may not grab local-scale heterogeneities. An alternative method is represented by site-specific monitoring networks in contaminated/polluted sites under remediation. As a main drawback, groundwater quality at these sites is affected by human activities. This paper explores the potential for groundwater data from an assemblage of site-specific datasets of contaminated/polluted sites to define NBLs of arsenic (As) at the meso-scale (order of 1000 km^2). Common procedures for the assessment of human influence cannot be applied to this type of dataset due to limited data homogeneity. Thus, an "unorthodox" method is applied involving the definition of a consistent working dataset followed by a statistical identification and critical analysis of the outliers. The study was conducted in a highly anthropized area (Ferrara, N Italy), where As concentrations often exceed national threshold limits in a shallow aquifer. The results show that site-specific datasets, if properly pre-treated, are an effective alternative for the derivation of NBLs when regional monitoring networks fail to catch local-scale variability.

Keywords: natural background levels; arsenic; groundwater quality; sites under remediation; site-specific data; Ferrara

Citation: Filippini, M.; Zanotti, C.; Bonomi, T.; Sacchetti, V.G.; Amorosi, A.; Dinelli, E.; Rotiroti, M. Deriving Natural Background Levels of Arsenic at the Meso-Scale Using Site-Specific Datasets: An Unorthodox Method. *Water* **2021**, *13*, 452. https://doi.org/10.3390/w13040452

Academic Editor: Fernando António Leal Pacheco

Received: 13 January 2021
Accepted: 5 February 2021
Published: 9 February 2021

Publisher's Note: MDPI stays neutral with regard to jurisdictional claims in published maps and institutional affiliations.

1. Introduction

Heavy metals and metalloids affect the quality of groundwater in many parts of the world [1] and are by far the most abundant group of contaminants and pollutants affecting European groundwaters [2]. Among them, arsenic is a well-known threat to human health and ecosystems due to its toxicity and carcinogenicity. Arsenic is found in groundwater above local regulatory limits in many countries [3]. Its origin is often from natural sources, such as arsenic-bearing minerals occurring in sediments and rocks, with release to groundwater driven by certain geochemical conditions that favor the As mobilization [4]. For this reason, the assessment of the Natural Background Level (NBL) of arsenic in groundwater is crucial, especially in urbanized and industrialized areas, where natural arsenic pollution should be distinguished from contamination caused or triggered by human activities in order to set proper remediation goals [5]. Moreover, the definition of NBLs supports the correct management of groundwater resources by highlighting potential issues related to chronic human exposure to naturally occurring arsenic.

Essential for the definition of NBLs, besides methods and protocols, is the availability of a set of hydrogeochemical data representative of the pristine groundwater composition, including the concentration of the pollutant/contaminant of interest and major ions, as well as the physicochemical parameters of water. The higher the quality, quantity, and homogeneous distribution of the data, the more accurate the NBL definition and the disentanglement of processes that cause pollutant release in groundwater [6–8]. The dataset is most often acquired with dedicated monitoring campaigns in pre-existing or new boreholes and springs (feasible for relatively small areas in the order of 100 km^2) [9–13] or from groundwater quality monitoring networks implemented at the national level [14–18] or, as for Italy, at the scale of administrative regions [8,19–24]. In many European countries, national or regional groundwater quality networks were developed in the last decades to fulfill the requirement of the European Groundwater Directive (2006/118/EC) for a systematic assessment of the chemical status of groundwater bodies. Some authors recently pointed out that the definition of arsenic NBLs at the scale of the groundwater body may provide unreliable results, due to local-scale geological and geochemical variability influencing As release to groundwater [25,26]. To overcome this limitation, new approaches have been proposed, representing an improvement of the two commonly used methodologies (i.e., the preselection and component separation methods) defined by the EU BRIDGE research project (Background cRiteria for the IDentification of Groundwater thrEsholds), that provided guidelines aimed at harmonizing the methods for estimating NBLs at the European level [27]. The preselection approach involves the use of indicator chemical species, such as NO_3, NH_4, Cl, and K [15,28,29], to identify samples with most likely anthropogenic influences [14,19,30–34]. The component separation approach involves the subdivision of the working dataset into normally and log-normally distributed populations, considering the latter as representative of the natural background [8,20,35–37]. The new approaches mentioned above involve the combination of the preselection or component separation methods with geostatistical tools that take into account the actual distribution of the contaminant of concern and its correlation with other environmental parameters (e.g., indicator kriging [21,23,38–40] or object-oriented statistics [24]). These approaches allow for a spatial enhancement of the high-quality information provided by national or regional monitoring networks by producing maps of NBLs or associated probability of exceedance, instead of assigning a single background value or a range of values for the whole groundwater body. However, the regional-scale monitoring networks upon which the geostatistical approaches are based remain unable to grab the potential spatial complexity caused by local-scale natural heterogeneities, since they can only provide point data with relatively large mesh (i.e., a few km).

A pervasive and extensive source of information on groundwater chemical composition is related to contaminated/polluted sites under remediation. These sites are commonly found in urbanized and industrialized areas and are generally managed through site-specific monitoring networks made of densely spaced piezometers that are typically monitored over a timespan of years to decades [41,42]. Notwithstanding the large amount of monitoring data associated with contaminated and polluted sites, an obvious drawback towards the use of these data for the definition of NBLs is that groundwater quality at these sites is most likely affected by human activities to some degree. Some authors recently put effort into distinguishing between natural (or, in a broad sense, background) and anthropogenic arsenic and other compounds in the aquifers below urban landfills [43,44] or tried to identify unimpacted monitoring locations within the network of a large industrial complex to assess background concentrations [45]. These authors performed thorough analyses at the scale of individual contaminated sites using very well-informed databases. To the best of our knowledge, the potential for an aggregation of monitoring networks of sites under remediation to assess NBLs at a meso-scale (i.e., areas in the order of 1000 km^2) has never been explored in the literature. Given the large availability of this type of data in urbanized and industrialized settings, their potential is worth investigating at least for assessing NBLs of compounds of likely natural origin, being infrequently associated with common human activities, such as arsenic [4].

In the case of groundwater quality data collected from an assemblage of site-specific monitoring networks whose original scope was different from NBL definition, data pre-processing is of pivotal importance [8]. The first issue that should be dealt with is the variability of sample collection and analysis methods, since each site-specific network is managed independently, possibly following different strategies. This variability necessarily reflects on concentration data that should be critically evaluated in order to identify a working dataset as consistent as possible. The second major issue is the identification of possible anthropic impacts on NBLs, involving the contaminant of concern. The most common strategies recently used to distinguish anthropic from natural contributions (i.e., preselection and component separation) hardly apply to assemblages of site-specific datasets for the following reasons: (1) the preselection approach barely operates because the monitoring of sites under remediation generally involves the sole analysis of specific contaminants and pollutants, excluding some major ions which are frequently used as indicator species (e.g., NO_3, NH_4, Cl, and K); (2) the component separation can be inappropriate because different sites could be subjected to different geochemical processes, leading to different types of data distribution and generating, on the whole, a multi-modal distribution, for which the assumptions "lognormal component = natural background dataset" and "normal component = human-influenced dataset" hardly apply. In addition, natural background populations are often not lognormal [8,46,47]. With the inapplicability of such standardized procedures, careful statistical identification and critical analysis of the outliers of the compound of concern for NBLs can be a reasonable way to eliminate the most likely anthropic inputs [48].

This paper aims at assessing the potential for publicly available groundwater quality data from sites under remediation to define NBLs of As at the meso-scale (i.e., in the order of 1000 km^2). Since standardized procedures, such as the preselection and component separation cannot be applied to a dataset formed by an assemblage of site-specific datasets, an "unorthodox" method is here applied, involving statistical identification and critical analysis of the outliers, based on the conceptual model elaborated for the investigated system. This study was performed in a highly anthropized area (Ferrara, in the Po Plain, northern Italy), where As concentrations exceeding the national threshold limits (10 µg/L [49]) are often detected in the shallowest aquifer of a complex multilayered system. This shallow aquifer is made of alluvial or coastal deposits locally enriched in peat, which is well known to drive As release in groundwater in the Po Plain [50,51] and worldwide [3]. Thus, a potential issue of surficial natural pollution subsists in the area, that must be addressed carefully. In the study area, the shallow aquifer is monitored by an institutional regional network dedicated to groundwater quality monitoring, which consists of only 18 monitoring points. The lack of an adequate number of monitoring points from the regional monitoring network motivated the use of an assemblage of site-specific monitoring networks (leading to a total number of 980 monitoring points). This approach can be potentially applied to other anthropized areas worldwide, where many site-specific monitoring networks exist and the regional monitoring is not able to grab the existing hydrogeochemical heterogeneity.

2. Materials and Methods

2.1. Conceptual Model of the Investigated System

The investigated area is located in the southeastern sector of the lower Po River Plain and corresponds to the administrative Province of Ferrara (Emilia Romagna Region, northern Italy). It covers an area of about 2600 km^2 bounded by the Po River to the north, the Adriatic Sea to the east, and the Reno River to the south (Figure 1).

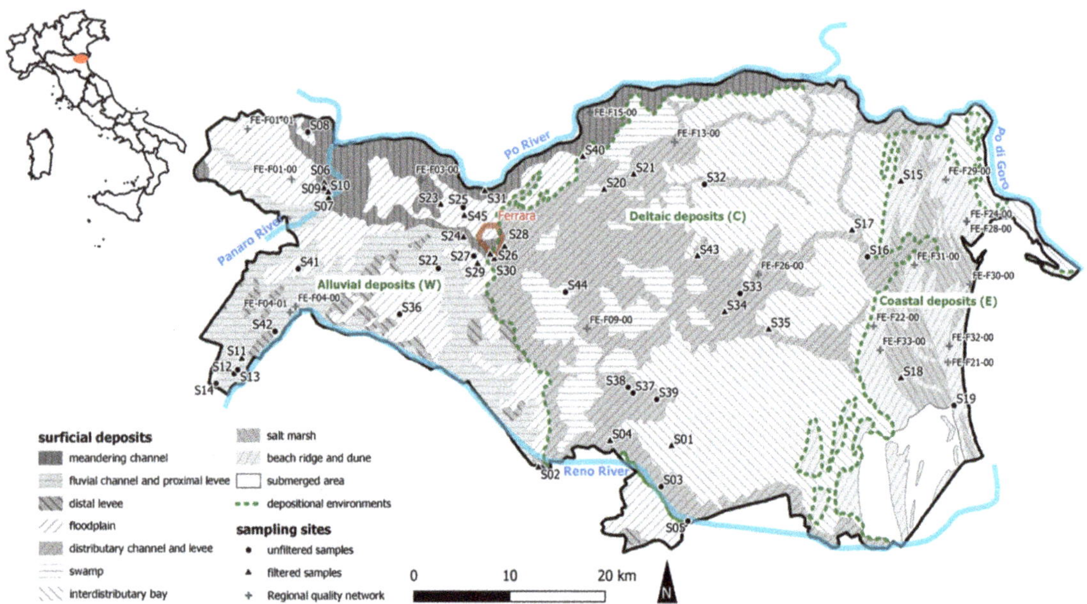

Figure 1. Surficial deposits in the administrative province of Ferrara (modified from the Geological Map of the Emilia Romagna plain 1:250,000) and sampling sites. In particular, sites under remediation are labeled as S*xx* whereas the sampling points of the regional monitoring network are labeled as FE-F*xx-xx*.

The subsurface stratigraphy of the study area consists of sub-horizontal alternations of coarse-grained (sand) and fine-grained (silt and clay) sediment bodies, tens of m thick, down to around 200 m below ground surface (bgs) that form the Pliocene-Pleistocene infill of the Po Plain foreland basin [52]. From a hydrogeological standpoint, such configuration corresponds to a number of vertically stacked aquifer-aquitard systems [53]. Our research is focused on the shallowest 10 m bgs, hosting heterogeneous Holocene deposits that constitute the shallowest aquifer system, known as "A0" [54]. Above the Pleistocene–Holocene boundary, the A0 system is made up of poorly interconnected sandy ribbons and lenses (aquifer) encased into a silty or clayey matrix (aquitard) [55]. Facies associations in the A0 aquifer system are part of three distinct depositional systems (Figure 1). The western sector is occupied by alluvial plain deposits supplied by the Po River and other Northern Apennine river systems south of Ferrara. Several facies associations have been recognized along a detailed cross-section cutting this sector [56]: fluvial-channel sands, crevasse splay and levee sand-silt alternations, and poorly-drained to well-drained floodplain silts and clays (thorough descriptions of facies associations are provided in the cited literature). The central sector corresponds to a delta plain depositional system mainly fed by the Po River. It is characterized by thick, laterally extensive mud-prone facies associations, including swamp, salt marsh, and lagoon/bay clays and silts. Distributary-channel sands and related overbank deposits form isolated lens-shaped bodies [57,58]. A peculiar feature of this sector is the abundance of organic matter, especially in fine-grained swamp and salt marsh facies showing frequent peat intercalations, several dm thick. Vegetal material, such as plant remains, plant debris, or roots, is typically associated with peat layers. The easternmost sector is a coastal plain facing the Adriatic Sea that consists almost exclusively of coastal sands, interpreted as beach-ridge or delta front facies associations [55]. Unlike the other two sectors, fine-grained deposits in the shallow subsurface of the coastal area are highly subordinate and correspond to thin, surficial paludal deposits that accumulated in topographic lows between individual sand ridges.

The aquifer is unconfined to semiconfined depending on the stratigraphic architecture and thickness of the encasing fine-grained deposits. Horizontal groundwater flow directions, hydraulic conductivity, and hydraulic gradients are extremely variable due to aquifer heterogeneity and complex sand-body geometry. The hydraulic head is close to the ground surface (max depth of about 1 m bgs) [59,60] and is generally dominant with respect to the deeper confined aquifers. Recharge of A0 is almost exclusively vertical, by rainfall or irrigation [61]. These features make the aquifer vulnerable to surface sources of pollution.

As for the most part of the Po River Plain, the Ferrara area is highly impacted by farming, with 70% of cultivated lands, and by industrial and urban activities occupying 8% of the territory [62–64], causing, in many cases, the deterioration of shallow groundwater quality. Among many organic and inorganic pollutants, arsenic is often detected in the A0 groundwater at concentrations exceeding the Italian regulatory limit of 10 µg/L. The study area hosts 45 sites under remediation supervised by local public authorities, i.e., the former Environmental Office of the administrative Province of Ferrara and the Regional Agency for Prevention, Environment, and Energy of Emilia-Romagna (ARPAE). These sites are listed in the Supplementary Material (SM) and their locations are shown in Figure 1. The sites are impacted by a number of different anthropic activities, such as gas stations (14 sites), urban (six sites), and inert (three sites) waste landfills, sugar factories (five sites), one large petrochemical complex, and other smaller industrial plants or various activities (16 sites) causing groundwater contamination. To the best of our knowledge, no direct anthropogenic inputs of As to soils and groundwater were registered in these sites. Thus, As occurrence is likely related to mobilization from sediments due to the reductive dissolution of Fe and Mn oxyhydroxides driven by the oxidation of organic matter (OM). The main source of OM in this aquifer is buried peat, as previously observed in other sectors of the Po Plain [65–67] and in similar fluvial systems around the world [3], however, additional anthropogenic sources of OM (e.g., hydrocarbons or landfill leachate spills) cannot be excluded. More specifically, some spills/leaks of hydrocarbons were reported for a sugar factory (site S02) and most likely occurred in the petrochemical complex (site S45).

Eh values between +150 and −385 mV, with a prevalence of negative values around –200 mV, have been previously detected in A0 all around the investigated area [68–72], confirming the occurrence of reducing conditions. The occurrence of the peat-rich facies in the central sector is expected to favor much higher arsenic concentrations in groundwater with respect to the western and eastern sectors.

2.2. Available Dataset

Hydrochemical data of the 45 sites were collected from a public registry of sites under remediation. The registry was managed by the former Environmental Office of the administrative Province of Ferrara until 2015 and is currently handled by ARPAE. Each site is equipped with 1 to 463 piezometers screened in the A0 aquifer, with variable screen depths ranging within 1.5 and 10 m bgs. For this study, arsenic concentrations were retrieved from 980 piezometers distributed in the 45 sites, over a variable timespan between 1997 and 2015, for a total of 4305 As values. Concentration analyses were performed either by private labs or by the public ARPAE labs, following a variety of sampling and analytical protocols. In a few instances (126 samples), reference samples were counter-analyzed by two different labs. Although details on sampling and analytical protocols were often unavailable, samples have been labeled as "filtered at 45 µm" or "unfiltered" when the information was retrievable. In some cases (363 samples), duplicate filtered and unfiltered samples were analyzed by the same lab. Data on major water ions and the specific contaminants affecting the sites were often incomplete or unavailable. When feasible, the location of the piezometer with respect to the source of contamination was defined based on groundwater flow directions provided by site-specific reports. Piezometers unequivocally located up-gradient to the source or labeled as "blanks" by site investigators were considered not affected by the site-specific anthropic contamination. When the location with respect to the source of contamination was unknown, piezometers were labeled as "internal", i.e., within the boundaries of the site as defined for remediation purposes, or "external". Internal piezometers were considered plausibly affected by the site-specific contamination due to likely proximity to the source of contamination.

Concerning the regional monitoring network, managed by ARPAE and aimed at supporting the periodic determination of the chemical status of groundwater bodies, As concentrations were available from the 18 monitoring points tapping aquifer A0 (locations in Figure 1) for the period 2010–2018, with a maximum of 20 values per point (four samples in 2010 and two samples per year from 2011 to 2018; the data are presented in the SM).

2.3. Data Pre-Processing and Calculation of Natural Background Levels of Arsenic

Since the entire database of As concentrations in groundwater was derived from the registry of sites under remediation, where some indirect anthropogenic influences on As concentrations (i.e., hydrocarbons and/or landfill leachate spills/leaks) cannot be excluded, the data were pre-processed prior to NBL calculation, following a slightly unconventional (unorthodox) procedure aimed at minimizing data shortcomings caused by the variability of sampling and analytical techniques and/or possible anthropic interferences. This procedure is described step-by-step in the following subsections and involves: (1) treatment of non-detects, (2) data quality assessment and preparation of the working dataset, (3) identification and treatment of outliers, and (4) calculation of NBL. The common procedure of considering a charge balance error below 10% as the criterium for evaluating data quality [27] could not be applied in the present study since the full set of major ions analyses were not available in the site-specific datasets. Instead, data quality was evaluated through the analysis of duplicate filtered/unfiltered samples and counter-analyzed (public and private labs) samples (i.e., the only information available for a significant number of samples), as described in detail in Section 2.3.2.

2.3.1. Treatment of Non-Detects

Non-detects (i.e., samples with an As concentration lower than the limit of detection—0—LOD) are 14.8% of the total As dataset. The literature suggests assigning a value of LOD/2 to the non-detects when these are a small proportion (<15%) of the total number of observations, without the need for further analyses [73]. The use of different sampling and analytical protocols in the considered dataset produced nine different LODs between 0.01 and 5 µg/L. To avoid a fabricated spatial and temporal variability of As concentrations, the lowest LOD that had a significant recurrence in the database (>5% of the non-detects) was selected, corresponding to 0.01 µg/L (23.1% of total non-detects), and its half value (0.005 µg/L) was assigned to all the non-detects.

2.3.2. Data Quality Assessment and Preparation of the Working Dataset

Different sampling strategies and sample pre-treatment and preservation may induce significant variability in As concentration measures, much larger than that expected from the application of different analytical methods, as long as the analyses follow standard protocols validated under national accreditation bodies [74]. In particular, field filtration of samples was reported to influence the measured concentration of several trace elements in water [75]. Filtered samples generally cause less variability of groundwater As concentration in a well compared to unfiltered samples [76].

The dataset of this study was split into three different categories based on field filtration of the sample: filtered, unfiltered, and unknown. In the last category, the information on field filtering was unavailable and the associated As data could not be considered for further observations on data quality. Duplicate filtered/unfiltered samples were employed to compare the first two categories. The alignment with respect to the 1:1 line of the two types of data was checked and the nonparametric Mann–Whitney U test [77] was applied to assess whether the data of the two groups were ascribable to the same statistical population. Counter-analyzed samples were used to assess the variability within each of the two categories. Scatter plots were employed to check whether the couples of counter-analyzed data were affected by analytical variability.

The aim of such comparisons was to assess data quality in order to identify a consistent, reliable, and robust working dataset of As concentration to be considered in the further steps.

2.3.3. Identification and Treatment of Outliers

The identification of anthropogenic inputs that may have affected the composition of shallow groundwater is crucial in the current study given the human-impacted nature of the considered database. The most widely used approaches at the EU level for the identification of such impacts are pre-selection and component separation [8,27]. These approaches could not be applied in the present case study for the reasons discussed in Section 1.

Therefore, potential impacts deriving from human activities were assessed here through a careful analysis of temporal and spatial outliers of As concentration informed by the knowledge on the conceptual model described in Section 2.1. The idea behind this choice is that impulsive anthropogenic influences may generate one or more outliers within the time series of a monitoring point (temporal outliers), whereas constantly impacted monitoring points may constitute an outlier with respect to the surrounding unaffected monitoring points (spatial outliers).

As a first step, a temporal analysis of the outliers was performed at each monitoring point. The low number of As data available for each piezometer (max six) did not allow for the statistical identification of the outliers. Instead, the interquartile range (IQR) approach was employed to analyze the temporal series [78]. The upper outliers (C) were defined as follows [79]:

$$C > Q3 + 1.5 \times IQR \tag{1}$$

where IQR corresponds to the difference between the 75th percentile (Q3) and 25th percentile (Q1) of the series. The lower outliers (i.e., < Q1—1.5 × IQR) were not considered, since they are not of interest in the definition of NBLs, being uninformative of possible anthropic impacts. After the exclusion of temporal outliers, each series was tested for the occurrence of a temporal trend using the nonparametric Mann–Kendall test [80,81], since the occurrence of such trends could be a further indication of anthropic influences [82]. Eventually, the outliers were critically analyzed on the basis of the site conceptual model, with a focus on the geology and location of the monitoring points with respect to the source of contamination and groundwater flow, to discern between natural or anthropic anomalies. The latter were excluded from the database prior to NBL calculation. As a general rule, upper outliers with arsenic concentrations $\leq$5 µg/L, i.e., half of the national regulatory limit, were maintained in the working database, since they were considered unaffected by relevant anthropogenic influences [83].

A similar procedure was followed for the identification of spatial outliers. In this case, a representative As concentration was considered for each piezometer corresponding to the median value of the temporal series [25] and the outliers were searched within each site, i.e., the IQR was calculated from the median values of each piezometer at a site. The use of the median as a representative value of a temporal series is suggested in several protocols for the estimation of NBLs [27,82], although it may lead to an underestimation of NBLs when the peak values of the series are unequivocally related to natural processes [83]. However, the median value was considered as a precautionary choice in the current research because the presence of anthropogenic influences on the extreme As values of some piezometers cannot be definitively excluded by the sole analysis of outliers.

2.3.4. NBL Calculation

For the definition of the NBLs of As in the A0 groundwater, the pre-processed database was split into three sub-populations corresponding to the three depositional systems identified in the conceptual model of Section 2.1. Indeed, the literature for the Po Plain area shows evidence of a strong control of the local geology (i.e., presence/absence of natural buried organic matter) on natural As release processes operating in the shallow aquifers [65]. The distinct groups of data were first checked with the Mann–Whitney U test to assess whether they could be statistically ascribed to different populations. Then, the NBL was calculated as the 90th percentile of the distribution of each distinct population in agreement with the BRIDGE protocol [27]. The choice of the 90th percentile (instead of higher suggested percentiles, e.g., 95th [84] or 97.7th [27]) aimed at mitigating the intrinsic uncertainty associated with possible undetected anthropic influences in the peak values of the distribution.

2.3.5. Comparison with NBLs Calculated from the Regional Monitoring Network

As a final step, the NBLs calculated with the methodology presented above, i.e., from the assemblage of site-specific datasets, were compared with the NBLs calculated using the data from the 18 points of the regional groundwater quality monitoring network of the same study area (see Section 2.2). Assuming, reasonably, that the regional monitoring points are located out of contaminated/polluted sites (in order to determine a chemical status representative of the whole groundwater body), the NBL was calculated as the 90th percentile of the median values calculated for each monitoring point omitting preliminary analyses on possible anthropic influences. The choice of the 90th percentile, rather than higher percentiles (e.g., 95th or 97th), is motivated by two reasons: (1) to have comparable NBL results (the same percentile used) between the regional dataset and the site-specific dataset, since the 90th percentile was used for the latter; (2) to adopt a precautionary approach: although we can reasonably assume that the regional monitoring points are unaffected by anthropogenic influences, we cannot definitely exclude it.

3. Results and Discussion

3.1. Identification of a Higher Quality Working Dataset

Statistical parameters for the total As dataset and for the sub-datasets made on the basis of the filtration procedure, i.e., filtered or unfiltered (see Section 2.3.2), are shown in Table 1 (the complete database is presented in the SM). Arsenic concentrations measured from unfiltered samples are generally higher than those from filtered samples, with mean values of 34.3 and 18.7 µg/L, respectively, and similar medians of 3.8 and 3.6 µg/L. The standard deviation is much higher for the unfiltered samples (304.7, compared to 55.7 of filtered data), suggesting higher variability within the unfiltered group. The scatter plot comparing filtered and unfiltered analyses from double sampling (Figure 2) shows a rather good alignment for many samples but, at the same time, a poor correlation for a significant number of samples, with unfiltered samples often providing higher concentrations. A reason for that may be the occurrence of solid particulates in unfiltered samples, comprising Fe-oxides [85] on which some As may be adsorbed.

Table 1. Statistics for the total, filtered, and unfiltered As datasets.

	TOT	Filtered	Unfiltered
max (µg/L)	9626.00	594.00	9626.00
min (µg/L)	0.005 [1]	0.005 [1]	0.005 [1]
mean (µg/L)	26.04	18.70	34.28
median (µg/L)	2.80	3.80	3.60
st. dev. (µg/L)	245.00	55.74	304.71
75 perc. (µg/L)	10.70	12.70	13.30
25 perc. (µg/L)	0.50	0.70	1.00
n. of observations	4305	815	2698
n. of non-detects	966	135	425

[1] LOD/2.

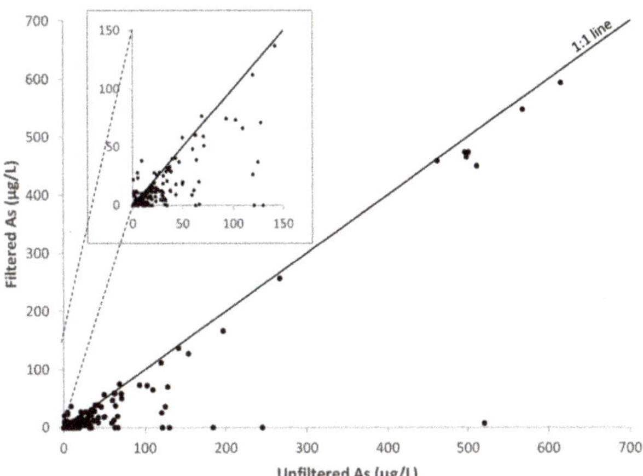

Figure 2. Scatter plot of filtered vs unfiltered As concentrations from the same samples.

The counter-analyzed As concentrations (Figure 3) show good alignment on the 1:1 line between the public and private labs in the case of filtered samples, even if very few data were available for the comparison (11 samples). On the contrary, the unfiltered data, available in a larger number (115 samples), indicate a much larger variability. In most cases, the unfiltered concentration from the private labs is lower than that of the public lab. This highlights the poor reliability of the unfiltered data.

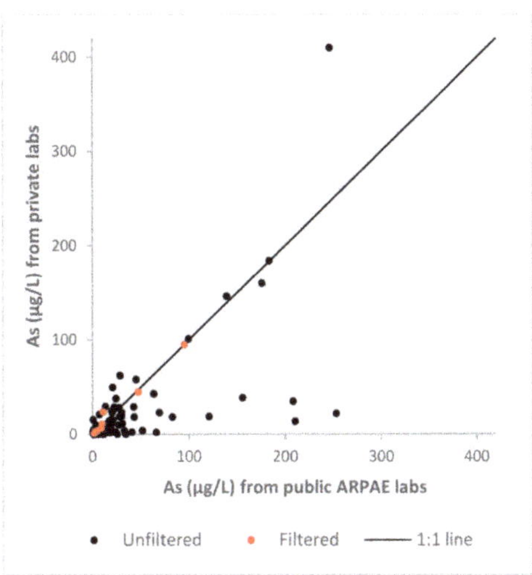

Figure 3. Scatter plot of counter-analyzed samples by private and public labs.

The Mann–Whitney U test [77] applied to the two groups of filtered and unfiltered data from double samples rejects the null hypothesis that the two groups pertain to the same population, with a p-value of 6.8×10^{-10}. This result confirms that filtered and unfiltered data should be considered separately for the calculation of NBLs.

Given the higher variability and uncertainty (i.e., lower data quality) that appears to be associated with the unfiltered samples, only the dataset of As concentrations from filtered samples was considered as the working dataset for the further steps of NBL calculation. Samples for which the information on filtration was missing were also excluded, since these may be affected by the same limitations as for unfiltered data. These choices cause a significant decrease in the number of available sites (from 45 to 25), monitoring points (from 980 to 392), and samples (from 4305 to 815; see Figure 1 and the SM for details on the availability of filtered samples). On the other hand, the reliability of the final NBLs is expected to increase since filtered concentrations represent higher quality data, as demonstrated by the above observations. Also, the exclusive use of filtered data allows defining background values that are consistent with national guidelines [84], which suggests field filtration of samples for the analysis of metals and metalloids.

3.2. Identification and Removal of Outliers Likely Representing Anthropogenic Influences

A total of 21 arsenic concentration values from 5 out of 25 sites (S02, S06, S11, S15, and S45) were identified as outliers from the IQR of the time series of each monitoring point (Table 2). Each outlier was critically evaluated, according to the conceptual model of the site, to assess whether the anomaly was most likely related to natural or anthropic causes. Six outliers were identified in the piezometers of site S02, a sugar factory in which some hydrocarbons spills/leaks were reported (see Section 2.1). The anomalous values ranged between 8.8 and 594.0 µg/L, pertaining to six different piezometers classified as "internal" and thus plausibly affected by the hydrocarbon contamination at the site. Moreover, the site is located in the western sector of the study area, which is made up of fluvial deposits that are expected to contain only rare peat deposits and vegetal material. For these reasons, the six outliers were considered to be the effect of anthropogenic influences on As concentrations and were deleted from the working database. The five outliers of site S06, another sugar factory in the western sector, were identified in five piezometers whose location with respect to the source of contamination is unknown. The outliers ranged from 2.4 to 66.0 µg/L. Two of the five outliers had As <5 µg/L and thus could not be related to relevant anthropogenic impacts (see Section 2.3.3). In the same site, four piezometers labeled as blanks showed As concentrations up to 40.2 µg/L, comparable to those of the outliers. For this reason, the outliers of this site were considered as natural anomalies and kept in the working database. The same criterion was applied to site S11, a small engine company in the western sector, showing three outliers ranging between 7.8 and 29.5 µg/L in three piezometers with unknown location, whereas blank piezometers at the same site had higher concentrations (up to 123.0 µg/L). Site S15 is a disused municipal landfill located in the central sector, made of deltaic deposits enriched in peats. The four outliers of this site range from 15.0 to 76.9 µg/L and pertain to four piezometers, three of which were downgradient to the source of contamination and one upgradient. The values were kept in the working database since the anomalous concentrations downgradient to the source (15.0 to 51.0 µg/L) were lower than those from the upgradient piezometer (76.9 µg/L) and other blank samples from the same site (up to 78.0 µg/L). The last site that showed anomalously high concentrations was S45, a large petrochemical complex located in the western sector. The three outliers (ranging from 11.6 to 27.8 µg/L) detected in three piezometers were cautiously excluded from the working database since, at this site, some hydrocarbons spills/leaks most likely occurred.

Table 2. Summary of the temporal As outliers.

Site ID	Geological Sector	No. Outliers	No. Piezometers	Location of Piezometers
S02	W	6	6	Internal
S06	W	5	5	Undefined
S11	W	3	3	Undefined
S15	E	4	4	3 downgradient, 1 upgradient
S45	W	3	3	Undefined

The Mann–Kendall test [80,81] applied to all the time series with sample size ≥ 3 (the minimum to let the test working) did not allow for detecting any significant temporal trend. This may be partly due to the small number of values available for each series, i.e., up to six, which is below the recommended minimum sample size of 10 [86–88], possibly spanned over at least 5–10 years [89,90], to obtain a statistically acceptable trend identification. Notwithstanding the above limitation, the absence of temporal trends agrees with the hypothesized mainly natural origin of arsenic in groundwater in the study area (Section 2.1).

For the analysis of spatial outliers, the median value of the time series, cleaned from the temporal outliers likely representing anthropogenic influences, was calculated for each piezometer. Spatial outliers were identified in 13 sites (S01, S02, S06, S11, S15, S18, S20, S23, S26, S28, S35, S40, and S45), in a variable number of piezometers, between 1 and 10 (Table 3). The outliers found in sites S06, S11, and S15 (ranging from 9.6 to 73.7 µg/L) were kept in the working database since the highest concentrations were detected in blank piezometers, and thus all values can be considered to reflect natural As contents. Sites S26, S28, S35, and S40 showed one to two spatial outliers. These four sites are located in the central (deltaic) sector. The values identified as outliers in these sites, ranging from 5.0 to 27.3 µg/L, were lower than the values observed in other sites in the same sector that did not display spatial outliers, such as 95.0 µg/L in site S17, or 95.0, 157.0, and 181.0 µg/L in site S21, or 90.6 µg/L in S34. Thus, the spatial outliers in sites S26, S28, S35, and S40 were kept in the working database because they were considered to be consistent with the natural background. Site S18 in the eastern sector is a former sugar factory showing six spatial outliers with the maximum value of 18.5 µg/L. This value is lower than the values calculated in other sites of the eastern sector for blank piezometers, e.g., 73.7 µg/L in site S15. For this reason, the outliers of site S18 were kept in the working database. One outlier was identified in each of the sites S01 and S20 in the central sector. In both cases, the anomaly is clearly related to one single sample of the temporal series, that the IQR criterium was unable to detect due to the small size (two samples) of the time series. Here, the choice was to delete these two anomalous samples, in order to increase data consistency within the two sites. The same reasoning was applied to the outlier in site S23 in the western sector, which was related to one anomalous value in the temporal series. The outliers of sites S02 and S45 in the western sector (3 and 10 values, respectively) were cautiously deleted from the database because some anthropogenic influences on As concentrations may have occurred in these two sites (see Section 2.1).

Table 3. Summary of the spatial As outliers.

Site ID	Geological Sector	No. Outliers/Piezometers	Location of Piezometers
S01	C	1	Undefined
S02	W	3	Internal
S06	W	2	1 undefined, 1 blank
S11	W	7	5 undefined, 2 blank
S15	E	4	1 downgradient, 2 upgradient, 1 blank
S18	E	6	Internal
S20	C	1	Undefined
S23	W	1	Undefined
S26	C	1	Undefined
S28	C	1	Internal
S35	C	2	Undefined
S40	C	2	Undefined
S45	W	10	Undefined

3.3. Derivation of Natural Background Levels

The population of median arsenic concentrations purged of temporal and spatial outliers was employed for the calculation of NBLs. Data were split into three sub-populations corresponding to three distinct (western, central, and eastern) geological sectors. These populations are represented in the probability plots of Figure 4. The range of median concentrations is larger in the central sub-population (0.005 to 181.0 µg/L) compared to the western and eastern sub-populations that show similar ranges between 0.005 and 108 µg/L and between 0.005 and 73.7 µg/L, respectively. The probability plots of Figure 4 show unimodal asymmetric distributions in the eastern and western sectors, whereas a bimodal distribution characterizes the central sector. All the three distributions have one or two extreme values which, following outlier analysis, can be considered as natural hot spots. The upper segment of the bimodal distribution of the central sector, identifiable between 40 and 100 µg/L, is missing in the plots of the other two sectors and likely represents, together with the two hot spots at 157 and 181 µg/L, the effect of stronger natural arsenic release triggered by the local occurrence of peaty layers and vegetal material in the shallow sub-surface. In the eastern and western sectors, there is also evidence of natural mobilization of arsenic, reaching values above the regulatory limit (10 µg/L) in several instances. However, the lower ranges and different shapes of the distributions compared to the central sector suggest the occurrence of conditions less favorable to arsenic mobilization in these two sectors, consistently with the conceptual model of Section 2.1.

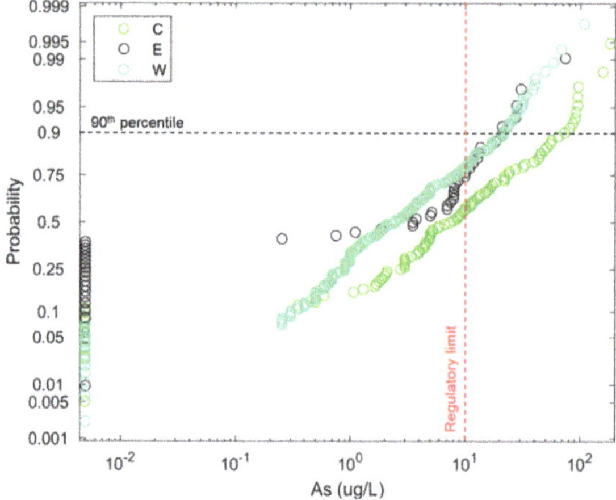

Figure 4. Probability plots of median As concentrations for each piezometer from all the sites (filtered As data), after the removal of outliers likely representing anthropogenic influences; the data are divided with respect to the three geological sectors (C: central, E: eastern, W: western).

A similar shape of data distribution (Figure 4) and results of a Mann–Whitney U test (p-value of 4.03×10^{-6}) suggest that the eastern and western datasets pertain to the same population, confirming the lithologic similarities (paucity of buried OM), whereas the data from the central sector represent a distinct population.

For these reasons, two NBLs were calculated for the study area, one pertaining to the central sector and corresponding to the 90th percentile of the distribution of the central dataset, whereas a second one was calculated for the eastern and western sectors, as the 90th percentile of the distribution of the two combined datasets. The two resulting NBLs correspond to 68 and 21 µg/L, respectively, confirming the serious issue of arsenic in shallow groundwater in the central sector.

3.4. Comparison with NBLs Derived from the Regional Monitoring Network

The 18 monitoring points of the regional network (Figure 1) were divided into two populations, corresponding to the eastern plus western sectors and the central sector, consistently with observations in Section 3.3. Thus, two NBLs were calculated following the methodology described in Section 2.3.5. The NBL for the eastern and western sectors was calculated on 15 out of the total 18 monitoring points of the network and was equal to 21 µg/L. The same value was obtained using the aggregation of site-specific datasets. The NBL for the central sector was calculated on 3 out of 18 monitoring points and was 6 µg/L, slightly below the regulatory limit of 10 µg/L and one order of magnitude below the NBL of 68 µg/L obtained from the aggregation of site-specific datasets.

The fact that distinct NBLs obtained from the regional network and from the assemblage of site-specific datasets had the same value for the eastern and western sectors (21 µg/L) suggests the following considerations: (1) the NBL value is robust, since it was obtained by two different and independent datasets; (2) the regional network is dense enough to catch the natural geochemical heterogeneity influencing As distribution in these two sectors; (3) the use of an aggregation of site-specific datasets derived from sites under remediation represents a valid alternative for calculating NBLs, after proper depuration from anthropogenic influences.

On the other hand, the significantly different NBLs obtained in the central zone (6 µg/L from the regional network and 68 µg/L from the assemblage of site-specific datasets) highlights two important aspects: (1) the inadequacy of the regional network to grab the geochemical heterogeneity of this sector, which led to an unreliably low NBL value. Such a value may lead to erroneous evaluations of potentially contaminated sites, attributing high As concentrations found in this sector to anthropogenic sources instead of natural sources; (2) the importance and the validity of the methodology proposed here, involving the use of site-specific datasets to fill the gap left by regional monitoring networks, which proved to be unable to catch the local-scale heterogeneity.

4. Conclusions

The present study involved the derivation of NBLs for groundwater As at the meso-scale (2600 km^2, corresponding to the administrative Province of Ferrara in the Po Plain, N Italy) for the shallow aquifer, using an aggregation of site-specific datasets collected from a public registry of sites under remediation. The use of this kind of data was motivated by the lack of an adequate number of monitoring points in the regional groundwater quality network able to grab the local-scale hydrochemical heterogeneity.

In the study area, the NBLs obtained for As were:

- 68 µg/L in the central sector, characterized by the abundance of buried OM, and thus, by a stronger potential for release of As to groundwater due to the reductive dissolution mechanism;
- 21 µg/L in the eastern and western sectors, characterized by a lower content of buried OM.

More general conclusions of this study are:

- Site-specific datasets can represent a cost-effective source of data useful for the derivation of NBLs, when regional monitoring networks fail to catch local-scale variability; however, the main disadvantage of using an assemblage of site-specific datasets is limited data quality, due to the likely application of different sampling and analytical methodologies;
- The lack of complete information on major ions and specific pollutants/contaminants for the sites, which prevents the application of conventional methodologies (e.g., pre-selection), can be overcome through a critical analysis of outliers that allows identification of possible anthropogenic influences; the analysis of outliers, however, must be supported by a robust conceptual model of each site, which must contain a description of the site (a) geology, (b) hydrogeology (type and depth of the aquifer involved, groundwater flow direction), and (c) contamination/pollution (chemical species and compounds involved, location of monitoring points with respect to the contamination/pollution sources);
- The design of regional monitoring networks of groundwater quality must consider local-scale geological heterogeneities that can generate local and high natural concentrations of particular chemical species, such as arsenic, in order to avoid the calculation

of unreliably low NBL values that might lead to erroneous evaluation of potentially contaminated sites.

The approach presented in this study is based on a local case study, but it can be reproduced in other areas worldwide where the abundance of different site-specific datasets can counterbalance the limitation of regional monitoring networks in detecting local-scale heterogeneity.

Supplementary Materials: The following are available online at https://www.mdpi.com/2073-4441/13/4/452/s1.

Author Contributions: Conceptualization, M.F. and M.R.; Data curation, V.G.S.; Formal analysis, M.F., C.Z., and M.R.; Methodology, M.F. and M.R.; Visualization, M.F. and M.R.; Writing—original draft, M.F. and M.R.; Writing—review & editing, C.Z., T.B., A.A. and E.D. All authors have read and agreed to the published version of the manuscript.

Funding: This research received no external funding.

Data Availability Statement: The data presented in this study is available in the supplementary materials.

Acknowledgments: We acknowledge Gabriella Dugoni, Gaia Boldrini, and Igor Villani (ARPAE), who allowed the investigation by providing their support on data collection and interpretation, and Alessandro Gargini for providing the basic idea for this work.

Conflicts of Interest: The authors declare no conflict of interest.

References

1. Fernandez-Luqueno, F.; López-Valdez, F.; Gamero-Melo, P.; Luna-Suárez, S.; Aguilera-González, E.N.; Martínez, A.I.; García-Guillermo, M.; Hernández-Martínez, G.; Herrera-Mendoza, R.; Álvarez-Garza, M.A.; et al. Heavy metal pollution in drinking water-a global risk for human health: A review. *Afr. J. Environ. Sci. Technol.* **2013**, *7*, 567–584.
2. Panagos, P.; Van Liedekerke, M.; Yigini, Y.; Montanarella, L. Contaminated sites in Europe: Review of the current situation based on data collected through a European network. *J. Environ. Public Health* **2013**, *2013*, 158764. [CrossRef]
3. Ravenscroft, P.; Brammer, H.; Richards, K. *Arsenic pollution: A global synthesis*; John Wiley & Sons: Chichester, UK, 2011; Volume 94, ISBN 978-1-405-18601-8.
4. Shankar, S.; Shanker, U. Arsenic Contamination of Groundwater: A Review of Sources, Prevalence, Health Risks, and Strategies for Mitigation. *Sci. World J.* **2014**, *2014*, 304524. [CrossRef] [PubMed]
5. De Caro, M.; Crosta, G.B.; Frattini, P. Hydrogeochemical characterization and Natural Background Levels in urbanized areas: Milan Metropolitan area (Northern Italy). *J. Hydrol.* **2017**, *547*, 455–473. [CrossRef]
6. Luzzadder-Beach, S. Evaluating the effects of spatial monitoring policy on groundwater quality portrayal. *Environ. Manag.* **1995**, *19*, 383–392. [CrossRef]
7. Gibbons, R.D.; Bhaumik, D.; Aryal, S. *Statistical Methods for Groundwater Monitoring*; Wiley Online Library: Hoboken, NJ, USA, 2009; Volume 2, ISBN 9780470164969.
8. Rotiroti, M.; Di Mauro, B.; Fumagalli, L.; Bonomi, T. COMPSEC, a new tool to derive natural background levels by the component separation approach: Application in two different hydrogeological contexts in northern Italy. *J. Geochem. Explor.* **2015**, *158*, 44–54. [CrossRef]
9. Marini, L.; Canepa, M.; Cipolli, F.; Ottonello, G.; Zuccolini, M.V. Use of stream sediment chemistry to predict trace element chemistry of groundwater. A case study from the Bisagno valley (Genoa, Italy). *J. Hydrol.* **2001**, *241*, 194–220. [CrossRef]
10. Edet, A.E.; Merkel, B.J.; Offiong, O.E. Trace element hydrochemical assessment of the Calabar Coastal Plain Aquifer, southeastern Nigeria using statistical methods. *Environ. Geol.* **2003**, *44*, 137–149. [CrossRef]
11. Preziosi, E.; Parrone, D.; Del Bon, A.; Ghergo, S. Natural background level assessment in groundwaters: Probability plot versus pre-selection method. *J. Geochem. Explor.* **2014**, *143*, 43–53. [CrossRef]
12. Peña Reyes, F.A.; Crosta, G.B.; Frattini, P.; Basiricò, S.; Della Pergola, R. Hydrogeochemical overview and natural arsenic occurrence in groundwater from alpine springs (upper Valtellina, Northern Italy). *J. Hydrol.* **2015**, *529*, 1530–1549. [CrossRef]
13. Barranquero, R.S.; Varni, M.; Vega, M.; Pardo, R.; De Galarreta, A.R. Arsenic, fluoride and other trace elements in the Argentina Pampean plain. *Geol. Acta* **2017**, *15*, 187–200.
14. Marandi, A.; Karro, E. Natural background levels and threshold values of monitored parameters in the Cambrian-Vendian groundwater body, Estonia. *Environ. Geol.* **2008**, *54*, 1217–1225. [CrossRef]
15. Wendland, F.; Berthold, G.; Blum, A.; Elsass, P.; Fritsche, J.G.; Kunkel, R.; Wolter, R. Derivation of natural background levels and threshold values for groundwater bodies in the Upper Rhine Valley (France, Switzerland and Germany). *Desalination* **2008**, *226*, 160–168. [CrossRef]

16. Mendizabal, I.; Baggelaar, P.K.; Stuyfzand, P.J. Hydrochemical trends for public supply well fields in The Netherlands (1898–2008), natural backgrounds and upscaling to groundwater bodies. *J. Hydrol.* **2012**, *450–451*, 279–292. [CrossRef]

17. Gunnarsdottir, M.J.; Gardarsson, S.M.; Jonsson, G.S.; Armannsson, H.; Bartram, J. Natural background levels for chemicals in Icelandic aquifers. *Hydrol. Res.* **2014**, *46*, 647–660. [CrossRef]

18. Gao, Y.; Qian, H.; Wang, H.; Chen, J.; Ren, W.; Yang, F. Assessment of background levels and pollution sources for arsenic and fluoride in the phreatic and confined groundwater of Xi'an city, Shaanxi, China. *Environ. Sci. Pollut. Res.* **2019**. [CrossRef]

19. Ducci, D.; Sellerino, M. Natural background levels for some ions in groundwater of the Campania region (southern Italy). *Environ. Earth Sci.* **2012**, *67*, 683–693. [CrossRef]

20. Molinari, A.; Guadagnini, L.; Marcaccio, M.; Guadagnini, A. Natural background levels and threshold values of chemical species in three large-scale groundwater bodies in Northern Italy. *Sci. Total Environ.* **2012**, *425*, 9–19. [CrossRef] [PubMed]

21. Molinari, A.; Guadagnini, L.; Marcaccio, M.; Guadagnini, A. Geostatistical multimodel approach for the assessment of the spatial distribution of natural background concentrations in large-scale groundwater bodies. *Water Res.* **2019**, *149*, 522–532. [CrossRef]

22. Parrone, D.; Ghergo, S.; Preziosi, E. A multi-method approach for the assessment of natural background levels in groundwater. *Sci. Total Environ.* **2019**, *659*, 884–894. [CrossRef]

23. Sellerino, M.; Forte, G.; Ducci, D. Identification of the natural background levels in the Phlaegrean fields groundwater body (Southern Italy). *J. Geochem. Explor.* **2019**, *200*, 181–192. [CrossRef]

24. Guadagnini, L.; Menafoglio, A.; Sanchez-Vila, X.; Guadagnini, A. Probabilistic assessment of spatial heterogeneity of natural background concentrations in large-scale groundwater bodies through Functional Geostatistics. *Sci. Total Environ.* **2020**, *740*, 140139. [CrossRef] [PubMed]

25. Preziosi, E.; Rossi, D.; Parrone, D.; Ghergo, S. Groundwater chemical status assessment considering geochemical background: An example from Northern Latium (Central Italy). *Rend. Lincei* **2016**, *27*, 59–66. [CrossRef]

26. Dalla Libera, N.; Fabbri, P.; Mason, L.; Piccinini, L.; Pola, M. Geostatistics as a tool to improve the natural background level definition: An application in groundwater. *Sci. Total Environ.* **2017**, *598*, 330–340. [CrossRef] [PubMed]

27. Müller, D.; Blum, A.; Hart, A.; Hookey, J.; Kunkel, R.; Scheidleder, A.; Tomlin, C.; Wendland, F. D18: Final proposal for a methodology to set up groundwater threshold values in Europe. *BRIDGE project, Background Criteria for the Identification of Groundwater Thresholds, 6th Framework Programme Contract.* 2006. Available online: http://www.hydrologie.org/BIB/Publ_UNESCO/SOG_BRIDGE/Deliverables/WP3/D18.pdf (accessed on 9 February 2021).

28. Hinsby, K.; Condesso de Melo, M.T.; Dahl, M. European case studies supporting the derivation of natural background levels and groundwater threshold values for the protection of dependent ecosystems and human health. *Sci. Total Environ.* **2008**, *401*, 1–20. [CrossRef] [PubMed]

29. Rotiroti, M.; Fumagalli, L.; Frigerio, M.C.; Stefania, G.A.; Simonetto, F.; Capodaglio, P.; Bonomi, T. Natural background levels and threshold values of selected species in the alluvial aquifers in the Aosta Valley Region (N Italy). *Rend Online Soc. Geol. It.* **2015**, *35*, 256–259. [CrossRef]

30. Coetsiers, M.; Blaser, P.; Martens, K.; Walraevens, K. Natural background levels and threshold values for groundwater in fluvial Pleistocene and Tertiary marine aquifers in Flanders, Belgium. *Environ. Geol.* **2009**, *57*, 1155–1168. [CrossRef]

31. Gemitzi, A. Evaluating the anthropogenic impacts on groundwaters; a methodology based on the determination of natural background levels and threshold values. *Environ. Earth Sci.* **2012**, *67*, 2223–2237. [CrossRef]

32. Preziosi, E.; Giuliano, G.; Vivona, R. Natural background levels and threshold values derivation for naturally As, V and F rich groundwater bodies: A methodological case study in Central Italy. *Environ. Earth Sci.* **2010**, *61*, 885–897. [CrossRef]

33. Rotiroti, M.; Fumagalli, L. Derivation of preliminary natural background levels for naturally Mn, Fe, As and NH4+ rich groundwater: The case study of Cremona area (Northern Italy). *Rendiconti Online Soc. Geol. Ital.* **2013**, *24*, 284–286.

34. Serianz, L.; Cerar, S.; Šraj, M. Hydrogeochemical characterization and determination of natural background levels (NBL) in groundwater within the main lithological units in Slovenia. *Environ. Earth Sci.* **2020**, *79*, 373. [CrossRef]

35. Wendland, F.; Hannappel, S.; Kunkel, R.; Schenk, R.; Voigt, H.J.; Wolter, R. A procedure to define natural groundwater conditions of groundwater bodies in Germany. *Water Sci. Technol.* **2005**, *51*, 249–257. [CrossRef]

36. Voigt, H.J.; Hannappel, S.; Kunkel, R.; Wendland, F. Assessment of natural groundwater concentrations of hydrogeological structures in Germany. *Geologija* **2005**, *50*, 35–47.

37. Chidichimo, F.; Biase, M.D.; Costabile, A.; Cuiuli, E.; Reillo, O.; Migliorino, C.; Treccosti, I.; Straface, S. GuEstNBL: The Software for the Guided Estimation of the Natural Background Levels of the Aquifers. *Water* **2020**, *12*, 2728. [CrossRef]

38. Ducci, D.; de Melo, M.T.C.; Preziosi, E.; Sellerino, M.; Parrone, D.; Ribeiro, L. Combining natural background levels (NBLs) assessment with indicator kriging analysis to improve groundwater quality data interpretation and management. *Sci. Total Environ.* **2016**, *569–570*, 569–584. [CrossRef]

39. Dalla Libera, N.; Fabbri, P.; Mason, L.; Piccinini, L.; Pola, M. A local natural background level concept to improve the natural background level: A case study on the drainage basin of the Venetian Lagoon in Northeastern Italy. *Environ. Earth Sci.* **2018**, *77*, 487. [CrossRef]

40. Avila-Sandoval, C.; Júnez-Ferreira, H.; González-Trinidad, J.; Bautista-Capetillo, C.; Pacheco-Guerrero, A.; Olmos-Trujillo, E. Spatio-Temporal Analysis of Natural and Anthropogenic Arsenic Sources in Groundwater Flow Systems. *Int. J. Environ. Res. Public Health* **2018**, *15*, 2374. [CrossRef] [PubMed]

41. Meyer, P.D.; Valocchi, A.J.; Eheart, J.W. Monitoring network design to provide initial detection of groundwater contamination. *Water Resour. Res.* **1994**, *30*, 2647–2659. [CrossRef]

42. Papapetridis, K.; Paleologos, E.K. Sampling Frequency of Groundwater Monitoring and Remediation Delay at Contaminated Sites. *Water Resour. Manag.* **2012**, *26*, 2673–2688. [CrossRef]

43. Preziosi, E.; Frollini, E.; Zoppini, A.; Ghergo, S.; Melita, M.; Parrone, D.; Rossi, D.; Amalfitano, S. Disentangling natural and anthropogenic impacts on groundwater by hydrogeochemical, isotopic and microbiological data: Hints from a municipal solid waste landfill. *Waste Manag.* **2019**, *84*, 245–255. [CrossRef]

44. Chidichimo, F.; De Biase, M.; Straface, S. Groundwater pollution assessment in landfill areas: Is it only about the leachate? *Waste Manag.* **2020**, *102*, 655–666. [CrossRef]

45. Vyas, V.M.; Roy, A.; Georgopoulos, P.G.; Strawderman, W.; Kosson, D.S. Development and Application of a Methodology for Determining Background Groundwater Quality at the Savannah River Site. *J. Air Waste Manag. Assoc.* **2006**, *56*, 159–168. [CrossRef] [PubMed]

46. Reimann, C.; Filzmoser, P. Normal and lognormal data distribution in geochemistry: Death of a myth. Consequences for the statistical treatment of geochemical and environmental data. *Environ. Geol.* **2000**, *39*, 1001–1014. [CrossRef]

47. Molinari, A.; Chidichimo, F.; Straface, S.; Guadagnini, A. Assessment of natural background levels in potentially contaminated coastal aquifers. *Sci. Total Environ.* **2014**, *476–477*, 38–48. [CrossRef] [PubMed]

48. Bulut, O.F.; Duru, B.; Çakmak, Ö.; Günhan, Ö.; Dilek, F.B.; Yetis, U. Determination of groundwater threshold values: A methodological approach. *J. Clean. Prod.* **2020**, *253*, 120001. [CrossRef]

49. Legislative Decree 152/2006. Decreto Legislativo n. 152 del 3 aprile 2006 sulle norme in materia ambientale (Legislative De-cree on Environmental Regulations). Available online: https://www.isprambiente.gov.it/it/garante_aia_ilva/normativa/normativa-ambientale/Dlgs_152_06_TestoUnicoAmbientale.pdf (accessed on 9 February 2021).

50. Carraro, A.; Fabbri, P.; Giaretta, A.; Peruzzo, L.; Tateo, F.; Tellini, F. Effects of redox conditions on the control of arsenic mobility in shallow alluvial aquifers on the Venetian Plain (Italy). *Sci. Total Environ.* **2015**, *532*, 581–594. [CrossRef] [PubMed]

51. Rotiroti, M.; Bonomi, T.; Sacchi, E.; McArthur, J.M.; Jakobsen, R.; Sciarra, A.; Etiope, G.; Zanotti, C.; Nava, V.; Fumagalli, L.; et al. Overlapping redox zones control arsenic pollution in Pleistocene multi-layer aquifers, the Po Plain (Italy). *Sci. Total Environ.* **2020**, *758*, 143646. [CrossRef]

52. Garzanti, E.; Vezzoli, G.; Andò, S. Paleogeographic and paleodrainage changes during Pleistocene glaciations (Po Plain, Northern Italy). *Earth-Sci. Rev.* **2011**, *105*, 25–48. [CrossRef]

53. Regione, E.R.; ENI-AGIP. *Riserve idriche sotterranee della Regione Emilia-Romagna (In Italian Transl.: Groundwater Resources of the Emilia-Romagna Region)*; Di Dio, G., Ed.; S.EL.CA. printer: Florence, Italy, 1998.

54. Molinari, F.C.; Boldrini, G.; Severi, P.; Dugoni, G.; Rapti Caputo, D.; Martinelli, G. *Risorse Idriche Sotterranee Della Provincia di Ferrara (In Italian; Transl. Groundwater Resources of the Ferrara Province)*; DB-MAP printer: Florence, Italy, 2007; pp. 1–62.

55. Amorosi, A.; Bruno, L.; Campo, B.; Morelli, A.; Rossi, V.; Scarponi, D.; Hong, W.; Bohacs, K.M.; Drexler, T.M. Global sea-level control on local parasequence architecture from the Holocene record of the Po Plain, Italy. *Mar. Pet. Geol.* **2017**, *87*, 99–111. [CrossRef]

56. Amorosi, A.; Bruno, L.; Cleveland, D.M.; Morelli, A.; Hong, W. Paleosols and associated channel-belt sand bodies from a continuously subsiding late Quaternary system (Po Basin, Italy): New insights into continental sequence stratigraphy. *GSA Bull.* **2017**, *129*, 449–463. [CrossRef]

57. Giacomelli, S.; Rossi, V.; Amorosi, A.; Bruno, L.; Campo, B.; Ciampalini, A.; Civa, A.; Hong, W.; Sgavetti, M.; de Souza Filho, C.R. A mid-late Holocene tidally-influenced drainage system revealed by integrated remote sensing, sedimentological and stratigraphic data. *Geomorphology* **2018**, *318*, 421–436. [CrossRef]

58. Campo, B.; Amorosi, A.; Vaiani, S.C. Sequence stratigraphy and late Quaternary paleoenvironmental evolution of the Northern Adriatic coastal plain (Italy). *Palaeogeogr. Palaeoclimatol. Palaeoecol.* **2017**, *466*, 265–278. [CrossRef]

59. Caschetto, M.; Colombani, N.; Mastrocicco, M.; Petitta, M.; Aravena, R. Estimating groundwater residence time and recharge patterns in a saline coastal aquifer. *Hydrol. Processes* **2016**, *30*, 4202–4213. [CrossRef]

60. Filippini, M.; Parker, B.L.; Dinelli, E.; Wanner, P.; Chapman, S.W.; Gargini, A. Assessing aquitard integrity in a complex aquifer—aquitard system contaminated by chlorinated hydrocarbons. *Water Res.* **2020**, *171*, 115388. [CrossRef]

61. Filippini, M.; Stumpp, C.; Nijenhuis, I.; Richnow, H.H.; Gargini, A. Evaluation of aquifer recharge and vulnerability in an alluvial lowland using environmental tracers. *J. Hydrol.* **2015**, *529*, 1657–1668. [CrossRef]

62. Calmistro, M.; Salemi, E.; Mastrocicco, M.; Colombani, N.; Brunelli, P.; Loberti, R.; Bellonzi, V.; Veronese, F.; Catozzo, L.; Carraro, G. Transnational Integrated Management of Water Resources in Agriculture for European Water Emergency Control (EU-WATER)—WP3 Regional Report: Inter-Regional Basin of the Po River, Italy. Available online: http://www.eu-water.eu/images/regionalreports/EU.WATER_abstract%20interbasin_EN.pdf (accessed on 15 September 2020).

63. Corbau, C.; Simeoni, U.; Zoccarato, C.; Mantovani, G.; Teatini, P. Coupling land use evolution and subsidence in the Po Delta, Italy: Revising the past occurrence and prospecting the future management challenges. *Sci. Total Environ.* **2019**, *654*, 1196–1208. [CrossRef] [PubMed]

64. Masetti, M.; Nghiem, S.V.; Sorichetta, A.; Stevenazzi, S.; Fabbri, P.; Pola, M.; Filippini, M.; Brakenridge, G.R. Urbanization affects air and water in Italy's Po plain. *Eos (United States)* **2015**, *96*, 13–16. [CrossRef]

65. Rotiroti, M.; Sacchi, E.; Fumagalli, L.; Bonomi, T. Origin of Arsenic in Groundwater from the Multilayer Aquifer in Cremona (Northern Italy). *Environ. Sci. Technol.* **2014**, *48*, 5395–5403. [CrossRef]
66. Molinari, A.; Ayora, C.; Marcaccio, M.; Guadagnini, L.; Sanchez-Vila, X.; Guadagnini, A. Geochemical modeling of arsenic release from a deep natural solid matrix under alternated redox conditions. *Environ. Sci. Pollut. Res.* **2014**, *21*, 1628–1637. [CrossRef]
67. Dalla Libera, N.; Fabbri, P.; Piccinini, L.; Pola, M.; Mason, L. Natural Arsenic in groundwater in the drainage basin to the Venice lagoon (Brenta Plain, NE Italy): The organic matter's role. *Rendiconti Online Soc. Geol. Ital.* **2016**, *41*, 30–33. [CrossRef]
68. Giambastiani, B.M.S.; Colombani, N.; Mastrocicco, M. Detecting Small-Scale Variability of Trace Elements in a Shallow Aquifer. *Water Air Soil Pollut.* **2015**, *226*, 7. [CrossRef]
69. Caschetto, M.; Colombani, N.; Mastrocicco, M.; Petitta, M.; Aravena, R. Nitrogen and sulphur cycling in the saline coastal aquifer of Ferrara, Italy. A multi-isotope approach. *Appl. Geochem.* **2017**, *76*, 88–98. [CrossRef]
70. Colombani, N.; Mastrocicco, M. Geochemical evolution and salinization of a coastal aquifer via seepage through peaty lenses. *Environ. Earth Sci.* **2016**, *75*, 798. [CrossRef]
71. Mastrocicco, M.; Giambastiani, B.M.S.; Severi, P.; Colombani, N. The Importance of Data Acquisition Techniques in Saltwater Intrusion Monitoring. *Water Resour. Manag.* **2012**, *26*, 2851–2866. [CrossRef]
72. Filippini, M.; Amorosi, A.; Campo, B.; Herrero-Martìn, S.; Nijenhuis, I.; Parker, B.L.; Gargini, A. Origin of VC-only plumes from naturally enhanced dechlorination in a peat-rich hydrogeologic setting. *J. Contam. Hydrol.* **2016**, *192*, 129–139. [CrossRef]
73. US EPA. *Data Quality Assessment: Statistical Methods for Practitioners, EPA QA/G-9S*; US Environmental Protection Agency: Washington, DC, USA, 2006.
74. Polya, D.A.; Watts, M.J. Sampling and analysis for monitoring arsenic in drinking water. In *Best Practice Guide on the Control of Arsenic in Drinking Water*; Bhattacharya, P., Polya, D., Jovanovic, D., Eds.; IWA Publishing: London, UK, 2017; ISBN 9781780404929.
75. Cidu, R.; Frau, F. Distribution of trace elements in filtered and non filtered aqueous fractions: Insights from rivers and streams of Sardinia (Italy). *Appl. Geochem.* **2009**, *24*, 611–623. [CrossRef]
76. Erickson, M.L.; Malenda, H.F.; Berquist, E.C. How or When Samples Are Collected Affects Measured Arsenic Concentration in New Drinking Water Wells. *Groundwater* **2018**, *56*, 921–933. [CrossRef]
77. Mann, H.B.; Whitney, D.R. On a Test of Whether one of Two Random Variables is Stochastically Larger than the Other. *Ann. Math. Stat.* **1947**, *18*, 50–60. [CrossRef]
78. Wang, C.; Caja, J.; Gómez, E. Comparison of methods for outlier identification in surface characterization. *Measurement* **2018**, *117*, 312–325. [CrossRef]
79. Tukey, J.W. Some graphic and semigraphic displays. In *Statistical Papers in Honor of George W. Snedecor*; Bancroft, T.A., Ed.; Iowa State University Press: Ames, IA, USA, 1972; Volume 5, pp. 293–316.
80. Mann, H.B. Nonparametric tests against trend. *Econometrica* **1945**, 245–259. [CrossRef]
81. Kendall, M.G. *Rank Correlation Methods*; Griffin: London, UK, 1955.
82. ISPRA. *Linee Guida Recanti la Procedura da Seguire per il Calcolo dei Valori di Fondo Naturale per i Corpi Idrici Sotterranei (DM 6 Luglio 2016)*; Manuali e Linee Guida; ISPRA: Rome, Italy, 2017; Volume 155/2017. Available online: https://www.isprambiente.gov.it/files2017/pubblicazioni/manuali-linee-guida/MLG_155_17.pdf (accessed on 20 September 2020).
83. Fumagalli, L.; Rotiroti, M.; Bonomi, T.; Zanotti, C.; Stefania, G.A.; Leoni, B. *Valutazione dei Valori di Fondo per le Acque Sotterranee "Assessment of Natural Background Levels in Groundwaters"*; Technical Report; Università degli Studi di Milano-Bicocca & Regione Lombardia: Milan, Italy, 2019.
84. ISPRA. Linee per la Determinazione dei Valori di Fondo per i Suoli e per le Acque Sotterranee (Guidelines for the Determination of Background Values in Soil and Groundwater). Manuali e Linee Guida. 174/2018. Available online: https://www.isprambiente.gov.it/files2018/pubblicazioni/manuali-linee-guida/MLG_174_18.pdf (accessed on 20 September 2020). (In Italian)
85. Bretzler, A.; Stolze, L.; Nikiema, J.; Lalanne, F.; Ghadiri, E.; Brennwald, M.S.; Rolle, M.; Schirmer, M. Hydrogeochemical and multi-tracer investigations of arsenic-affected aquifers in semi-arid West Africa. *Geosci. Front.* **2019**, *10*, 1685–1699. [CrossRef]
86. Gilbert, R.O. *Statistical Methods for Environmental Pollution Monitoring*; John Wiley & Sons: New York, NY, USA, 1987.
87. Şen, Z. Hydrological trend analysis with innovative and over-whitening procedures. *Hydrol. Sci. J.* **2017**, *62*, 294–305. [CrossRef]
88. Hu, Z.; Liu, S.; Zhong, G.; Lin, H.; Zhou, Z. Modified Mann-Kendall trend test for hydrological time series under the scaling hypothesis and its application. *Hydrol. Sci. J.* **2020**, *65*, 2419–2438. [CrossRef]
89. Gourcy, L.; Lopez, B. Technical Report on Groundwater Quality Trend and Trend Reversal Assessment—Procedures Applied by Member States for the First RBMP Cycle. Available online: https://circabc.europa.eu/ui/group/9ab5926d-bed4-4322-9aa7-9964bbe8312d/library/006b0646-6340-4233-8769-564fec15474a/details (accessed on 31 January 2021).
90. Frollini, E.; Preziosi, E.; Calace, N.; Guerra, M.; Guyennon, N.; Marcaccio, M.; Menichetti, S.; Romano, E.; Ghergo, S. Groundwater quality trend and trend reversal assessment in the European Water Framework Directive context: An example with nitrates in Italy. *Environ. Sci. Pollut. Res.* **2021**. [CrossRef] [PubMed]

water

MDPI

Article

Assessing Natural Background Levels in the Groundwater Bodies of the Apulia Region (Southern Italy)

Rita Masciale [1], Stefano Amalfitano [2], Eleonora Frollini [2], Stefano Ghergo [2], Marco Melita [2], Daniele Parrone [2,*], Elisabetta Preziosi [2], Michele Vurro [1], Annamaria Zoppini [2] and Giuseppe Passarella [1]

[1] Water Research Institute, National Research Council (IRSA-CNR), 70132 Bari, Italy; rita.masciale@cnr.it (R.M.); michele.vurro@cnr.it (M.V.); giuseppe.passarella@cnr.it (G.P.)

[2] Water Research Institute, National Research Council (IRSA-CNR), Monterotondo, 00015 Rome, Italy; amalfitano@irsa.cnr.it (S.A.); frollini@irsa.cnr.it (E.F.); ghergo@irsa.cnr.it (S.G.); melita@irsa.cnr.it (M.M.); preziosi@irsa.cnr.it (E.P.); zoppini@irsa.cnr.it (A.Z.)

* Correspondence: parrone@irsa.cnr.it

Abstract: Defining natural background levels (NBL) of geochemical parameters in groundwater is a key element for establishing threshold values and assessing the environmental state of groundwater bodies (GWBs). In the Apulia region (Italy), carbonate sequences and clastic sediments host the 29 regional GWBs. In this study, we applied the Italian guidelines for the assessment of the NBLs, implementing the EU Water Framework Directive, in a south-European region characterized by the typical Mediterranean climatic and hydrologic features. Inorganic compounds were analyzed at GWB scale using groundwater quality data measured half-yearly from 1995 to 2018 in the regional groundwater monitoring network (341 wells and 20 springs). Nitrates, chloride, sulfate, boron, iron, manganese and sporadically fluorides, boron, selenium, arsenic, exceed the national standards, likely due to salt contamination along the coast, agricultural practices or natural reasons. Monitoring sites impacted by evident anthropic activities were excluded from the dataset prior to NBL calculation using a web-based software tool implemented to automate the procedure. The NBLs resulted larger than the law limits for iron, manganese, chlorides, and sulfates. This methodology is suitable to be applied in Mediterranean coastal areas with high anthropic impact and overexploitation of groundwater for agricultural needs. The NBL definition can be considered one of the pillars for sustainable and long-term groundwater management by tracing a clear boundary between natural and anthropic impacts.

Keywords: groundwater; natural background levels; Italian guidelines; pre-selection method

Citation: Masciale, R.; Amalfitano, S.; Frollini, E.; Ghergo, S.; Melita, M.; Parrone, D.; Preziosi, E.; Vurro, M.; Zoppini, A.; Passarella, G. Assessing Natural Background Levels in the Groundwater Bodies of the Apulia Region (Southern Italy). *Water* **2021**, *13*, 958. https://doi.org/10.3390/w13070958

Academic Editor: Frédéric Huneau

Received: 5 March 2021
Accepted: 29 March 2021
Published: 31 March 2021

Publisher's Note: MDPI stays neutral with regard to jurisdictional claims in published maps and institutional affiliations.

1. Introduction

The EU Water Directives [1–3], implemented in Italy by Acts [4–6], require Member States that good status of groundwater bodies (GWBs) is achieved or maintained to protect all dependent ecosystems. Chemical status is determined based on quality standards set in Europe for nitrates and pesticides, and threshold values (TVs) that have to be set by Member States for all contaminants or groups of contaminants characterizing the GWBs as being at risk of failing the environmental objectives. The NBL can be defined as the concentration of a substance, or the value of an indicator, in a given GWB or group of GWBs, with no or extremely limited anthropogenic modifications. In this perspective, the TV of a considered natural substance may be redefined by the local authorities based on the NBL. The EU Directives do not provide any methodology for the TVs and NBLs establishment in the groundwater and the Member States are left free to apply their own approach considering the conceptual models of the groundwater bodies.

Chemical composition of groundwater depends on a wide range of natural factors such as water-rock interaction, residence time, rainfall input, chemical and biological processes that occur in the unsaturated and saturated zone, interactions with other water bodies [7].

Distinguishing between natural and anthropogenic sources is not a simple task since it requires high-level knowledge of all processes involved and the availability of a proper amount of monitoring data. The simplest methodologies for the NBL estimation within a GWB are based on the temporal and spatial approaches or on a combination of them [8,9]. The temporal approach uses historical groundwater quality data, which are assumed to be representative of the existing conditions before supposed human impacts. However, historical chemical data are usually scarce and have been determined using different methods and protocols compared to the modern sampling and analytical standards. The spatial approach uses monitoring data collected from areas in which the anthropogenic activity is assumed low and with no impact on the water quality. In this case, some care should be paid, when extending the NBLs values calculated at unspoiled areas to those impacted since they could have different physical, chemical, and biological characteristics. The difficulty of finding historical datasets and identifying pristine portions of the GWBs in populated areas has often led to developing other approaches.

The guidance issued within the EU 6th FP BRIDGE project [10,11] suggests methods based on either a statistical or a pre-selection approach. The partition of populations method [12–16] is a statistical approach based upon the idea that the concentration distribution of a chemical species in a groundwater system can be considered as the combination of two or more components of natural and anthropogenic origin.

On the other hand, a pre-selection based approach eliminates samples clearly affected by human activities on the base of specific markers such as nitrate, ammonium, chloride, synthetic organic compounds, etc. The samples exceeding a fixed value of the chosen markers are eliminated and the NBL is chosen as the 90th, 95th, or 97.7th percentile of the remaining water samples dataset. The main disadvantage of this approach is that the samples are eliminated according to qualitative criteria [17], e. g. the amount of available data and the shape of the related statistical distribution, and that the removal of samples might result in a too scarce data set.

Urresti-Estala et al. [18] assessed the NBLs using the objective criteria of two statistical techniques: the iterative 2σ technique, and the distribution function. Differently from the commonly employed statistical techniques, these Authors proposed two different approaches that do not require the elimination of samples, nor specific prior distributions, or the presence of large data sets. Both techniques are capable to discriminate between high values due to natural background populations from those induced by human influence.

Different attempts to apply a multi-method approach for the NBLs assessment have been made generally combining the pre-selection method and statistical techniques [19–23]. These studies pointed out that their integration reinforces the validity of the assessment, particularly when the hydrogeological setup and geochemical characteristic of groundwater are properly considered, confirming the greater importance of the conceptual model of the system than the choice of a statistical method. Molinari et al. [24] proposed other methodologies integrating partition population and numerical modeling of flow and transport processes to estimate NBLs of manganese and sulfate in potentially contaminated coastal aquifers. This study highlighted the need to validate the NBLs assessment procedures with accurate hydro-geochemical modeling results to identify external forcing components influencing the NBLs, such as the effects of tides and seawater intrusion, and avoid the possibility of interpreting naturally induced concentrations as anthropogenic effects.

Recently, IRSA-CNR collaborated with the Italian Institute for Environmental Protection and Research (ISPRA) in defining scientifically based guidelines for the NBLs assessment and clarifying some methodological aspects [25]. These guidelines rely on the pre-selection method but leave to the local water managers the task of identifying the specific pressure indicators while the only mentioned parameters are nitrate (for oxidizing conditions) and ammonium (for the reducing environments). Further, the guidelines do not mention how to deal with saline groundwater; hence, a procedure for the definition of a boundary between contaminated and uncontaminated groundwater in coastal aquifers has still to be defined. Assessing the origin of groundwater salinization in coastal aquifers is an issue mainly because it can depend on many concurrent natural or anthropic causes. Besides the possible contribution of seawater intrusion driven by overexploitation and sea-level rise, the scientific literature reports a detailed list of potential sources [26,27].

This work proposes a tentative application of the national guidelines for the assessment of the NBL values of some inorganic compounds in the GWBs of the Apulia region (southeast of Italy). In the case study, nitrate and ammonium concentrations have been considered as marker of anthropic contamination of the water samples. This work has been carried out within the framework of the research project on the assessment of background levels in the groundwater bodies of the Apulia region (VIOLA) jointly developed with the Department of Water Resources Management of the Apulia Region.

2. Materials and Methods

The Italian guidelines for the NBLs assessment describe an integrated approach based on the pre-selection and statistical methods [25]. The procedure consists of ten steps, some of which (steps 1–4) apply on the monitoring sites (MSs) and others (steps 5–10) on the single parameters:

1. Conceptual model definition.
2. Data exploration (validation of analytical data and handling of non-detect data).
3. Hydrochemical facies identification, based e.g., on dissolved oxygen concentration and/or redox potential. The facies analysis, when performed, leads to the separation of the dataset into separate subsets (e.g., reducing/oxidizing datasets) with different geochemical characteristics to be processed separately.
4. Pre-selection of data using the chosen anthropic markers (e.g., NO_3^- concentrations in oxidizing facies and NH_4^+ concentrations in reducing facies); the resulting data are compiled in the so-called "pre-selected dataset".
5. Analysis of the time series per monitoring site (MS): identification and handling of the outliers by graphical (box and whisker plot) and statistical method (Huber's test), trend analysis, selection of the median (or single value when only one measure is available) to be used as "representative value" in the following step.
6. Analysis of the data at the GWB scale (one data per monitoring station): identification and handling of the outliers of the preselected datasets, assessment of the existence of multiple populations and consequent identification of sub-sets.
7. Analysis of the statistical distribution (using the Shapiro-Wilk test) of the target parameter(s).
8. Assessment of pre-selected dataset size (significant spatial size: monitoring stations ≥ 15, significant temporal size: at least 8 half yearly observations);
9. Identification of specific path types based on pre-selected dataset size;
10. Assessment of the NBLs by statistical methods according to the specific path.

The pre-selected dataset size is a key element because it determines the procedure for NBL assessment and the robustness of the results. The guidelines identify four possible scenarios of dataset size (see point 9 of the previous list) considering both spatial and temporal dimension (Table 1).

The obtained NBLs are characterized by different confidence level (high, medium, low, very low) with regard to: the number of total observations or the number of total monitoring sites, the extension of GWB, and the aquifer type (confined or unconfined) (Table 2). When the confidence level is low or very low, the NBL is considered provisional. The confidence level is used to select a NBL at GWB scale from those calculated for different facies.

Table 1. Type of dataset following the size in space (number of monitoring stations) and time (number of observations in time) and relative procedure for NBL calculation (steps 8 to 10). MS = monitoring site.

Type	Spatial size	Temporal Size	Distribution of the Parameter Dataset in Space	NBL from the Spatial Analsyis
A	$\geq$15 MS	$\geq$8 half yearly obs.	Normal	Maximum value
			Not normal	95th percentile
B	$\geq$15 MS	<8 half yearly obs.	Normal	Maximum value
			Not normal	95th percentile
C	<15 MS	$\geq$8 half yearly obs.	-	Maximum value
D	<15 MS	<8 half yearly obs.	Total obs. (space, time) $\geq$ 10	(PROVISIONAL) 90th percentile
			Total obs. (space, time) < 10	(PROVISIONAL) Estimated by analogy with a similar GWB

Table 2. Confidence level of the obtained NBL. U. = unconfined aquifer; C. = confined aquifer; H = high; M = medium; L = low; LL very low.

Path Type	N. Total obs.	N. Total MS	Dimension of GWB or Portion of GWB Represented by Dataset (km^2)							
			<10		10–70		70–700		>700	
			Aquifer type							
			U.	C.	U.	C.	U.	C.	U.	C.
A		15–25	H	H	H	H	M	H	M	H
		>25	H	H	H	H	H	H	H	H
B		15–25	M	H	M	H	L	M	L	M
		>25	H	H	H	H	M	H	M	M
C	$\leq$15		M	M	L	M	LL	L	LL	LL
	16–30		M	H	M	M	L	M	LL	L
	>30		H	H	M	H	M	H	M	M
D	<10		L	L	LL	LL	LL	LL	LL	LL
	$\geq$10		L	L	L	L	LL	L	LL	LL

The logical phases and the requirements of the guidelines have been implemented at the IRSA-CNR into an automated tool capable of standardizing NBLs assessment operations. This tool, eNaBLe [28], thanks to its modularity, has been successfully integrated into the VIOLA project web site (http://www.irsa.cnr.it/Viola/, accessed on 30 March 2021) which hosts different statistical, graphical elaboration, and visualization procedures. eNaBLe and the other VIOLA web site procedure are written in PHP [29] and use MySQL [30] as relational databases manager.

The database is structured in 5 main tables (Monitoring Points, Wells, Springs, Samples, Chemical Analyses), 2 ancillary tables (GWBs, Parameters) and one table for user authentication and privileges. The user interface of the eNaBLe tool is organized in three logical phases corresponding to three different associated web pages: (i) Selection of the GWB data to be processed, analytical parameters to take into account and selection of the calculation options to be used; (ii) NBLs assessment and relative confidence levels; (iii) output of results in graphical, tabular and report form.

By performing a query on the selected dataset, eNaBLe produces a preliminary table in which all the analytical parameters and their basic statistical description are listed, proposing for the NBLs evaluation those analytical parameters in which at least one exceedance of the relative TV has been found. Next, it is possible to select the options for the NBL assessment, such as the appropriate time interval for the data to be processed, the parameters to be used for the redox facies separation and the limits to apply for the preselection process to nitrates or ammonium in relation to the oxidizing or reducing facies, among others.

A screenshot of the GWB selection and parameters configuration page is shown in Figure 1.

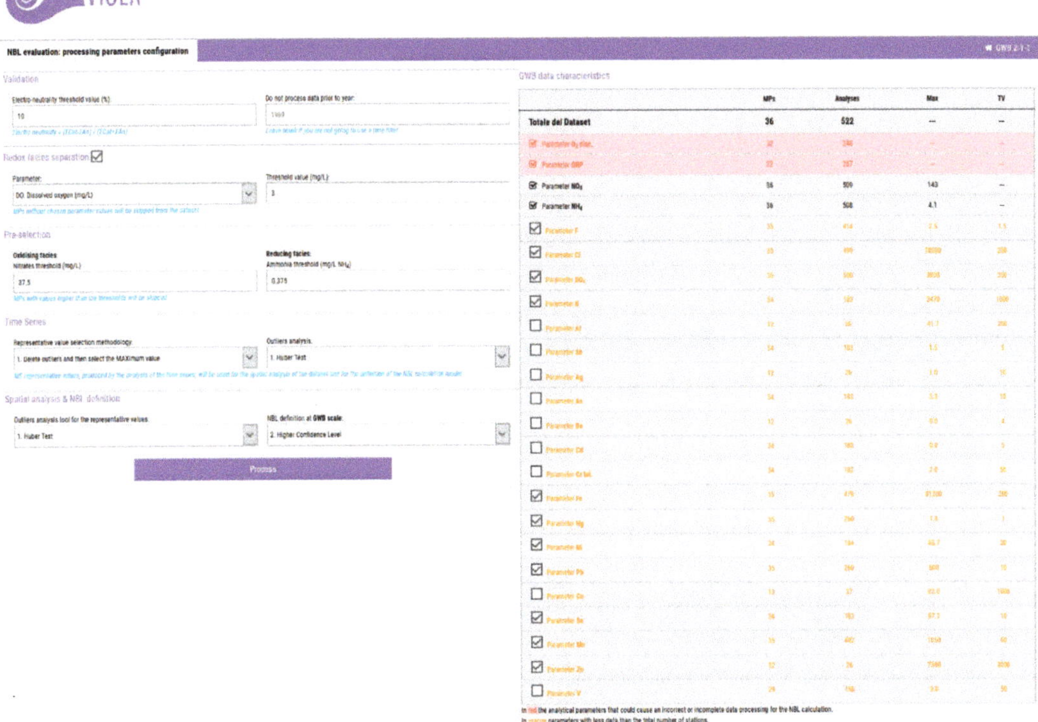

Figure 1. The page of the VIOLA Project web site where the parameters for the NBLs assessment can be configured. On the right side the procedure provides general information about the GWB analytical dataset while, on the left, it allows the settings for the NBL evaluation.

The results of the calculation as well as the settings used are listed in the second page of the procedure (Figure 2). In addition, the result page provides the links to the intermediate products of the calculation such as the validated dataset (built after time filtering, non-detected values handling and electrical balance verification), pre-selected and facies separated datasets, final datasets used for the calculation and the path type choice. The calculated NBLs are reported by redox facies, if this option has been chosen, and at GWB scale. The calculation path type used (A, B, C, D, Table 1) and the confidence level of the resulting NBL (Table 2) are also specified.

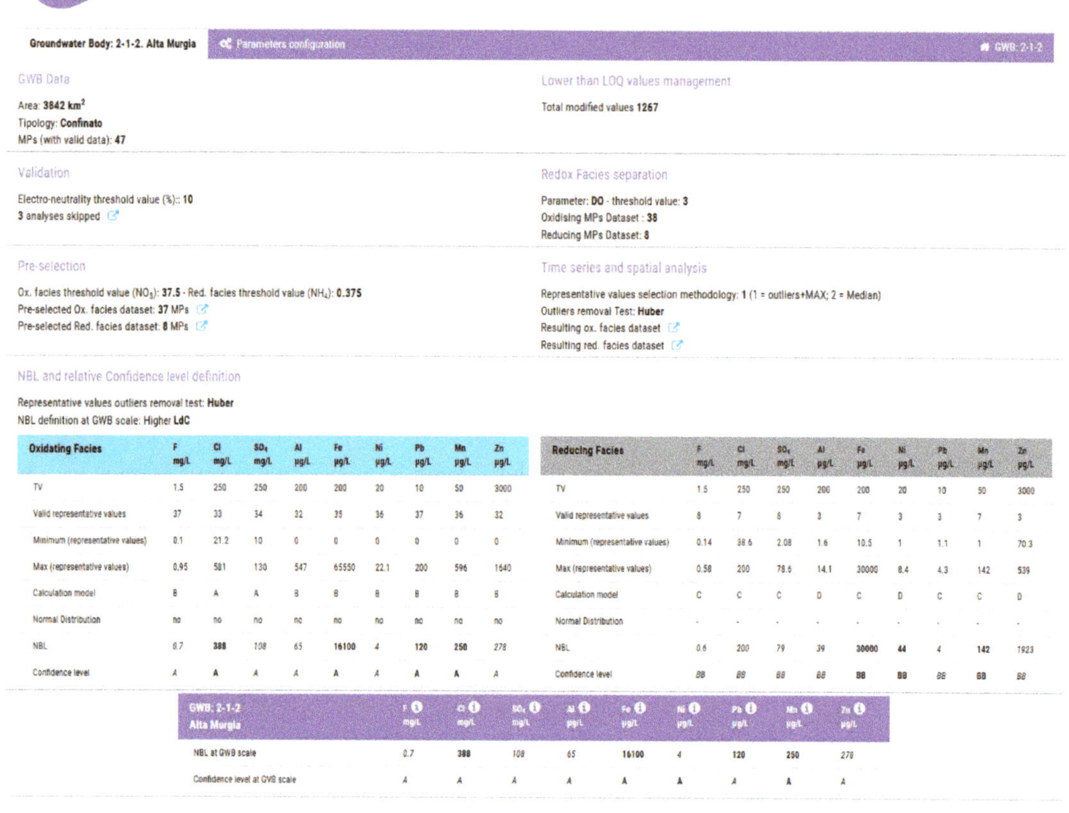

Oxidating Facies	F mg/L	Cl mg/L	SO₄ mg/L	Al µg/L	Fe µg/L	Ni µg/L	Pb µg/L	Mn µg/L	Zn µg/L
TV	1.5	250	250	200	200	20	10	50	3000
Valid representative values	37	33	34	32	35	36	37	36	32
Minimum (representative values)	0.1	21.2	10	0	0	0	0	0	0
Max (representative values)	0.95	581	130	547	65550	22.1	200	596	1640
Calculation model	B	A	A	B	B	B	B	B	B
Normal Distribution	no	no	no	no	no	no	no	no	no
NBL	0.7	**388**	108	65	**16100**	4	**120**	**250**	278
Confidence level	A	A	A	A	A	A	A	A	A

Reducing Facies	F mg/L	Cl mg/L	SO₄ mg/L	Al µg/L	Fe µg/L	Ni µg/L	Pb µg/L	Mn µg/L	Zn µg/L
TV	1.5	250	250	200	200	20	10	50	3000
Valid representative values	8	7	8	3	7	3	3	7	3
Minimum (representative values)	0.14	38.6	2.08	1.6	10.5	1	1.1	1	70.3
Max (representative values)	0.58	200	78.6	14.1	30000	8.4	4.3	142	539
Calculation model	C	C	C	D	C	D	C	C	D
Normal Distribution	-	-	-	-	-	-	-	-	-
NBL	0.6	200	79	39	**30000**	**44**	4	**142**	1923
Confidence level	BB	BB	BB	BB	BB	BB	BB	BB	BB

GWB: 2-1-2 Alta Murgia	F mg/L	Cl mg/L	SO₄ mg/L	Al µg/L	Fe µg/L	Ni µg/L	Pb µg/L	Mn µg/L	Zn µg/L
NBL at GWB scale	0.7	**388**	108	65	**16100**	4	**120**	**250**	278
Confidence level at GWB scale	A	A	A	A	A	A	A	A	A

Figure 2. Results of the NBLs assessment. The GWD characteristics, the computational settings and the link to the intermediate and processed datasets are listed in the upper part, while the values of the NBLs for the selected parameters and the relative confidence levels appear in the lower part.

3. Study Area

Apulia is a southeastern region of Italy (Figure 3) that covers an area of about 19.500 km² hosting more than four millions inhabitants [31]. This region faces the Adriatic and Ionian seas with a coastline extension of about 900 km.

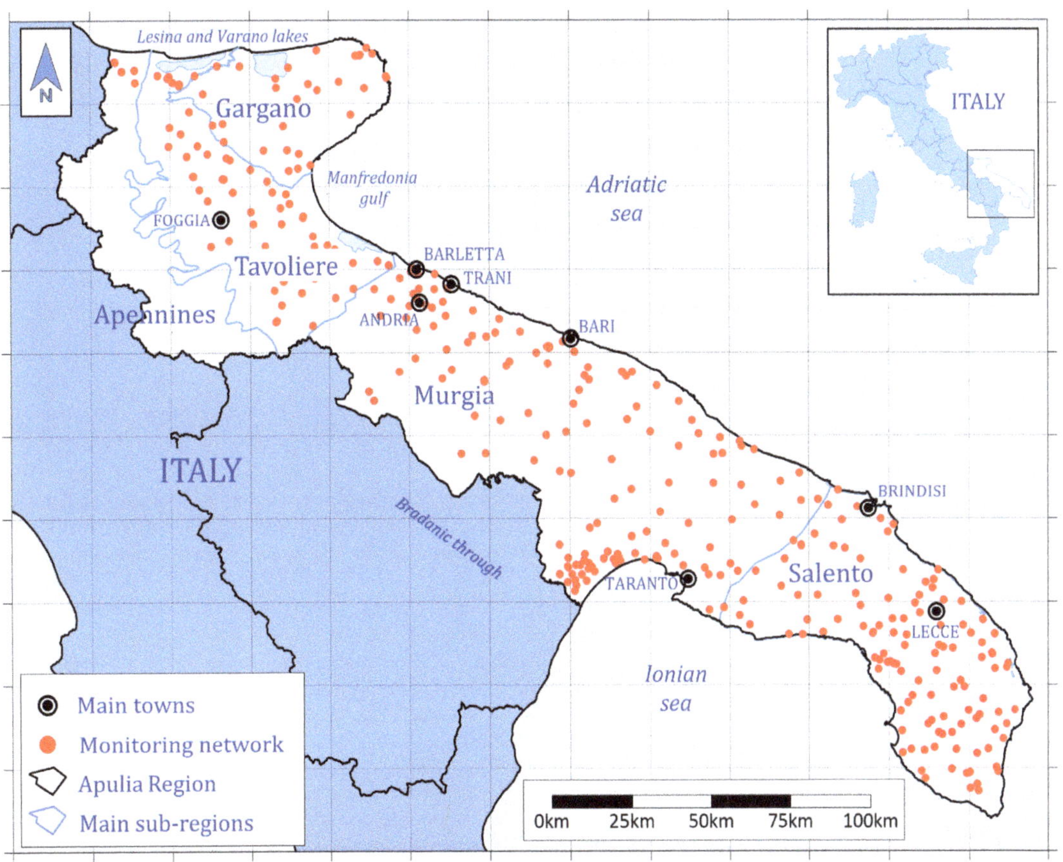

Figure 3. Map of the Apulia region, showing the main geographical sub-regional areas (Gargano, Tavoliere, Murgia and Salento) and the regional groundwater monitoring network.

Apulia region is characterized by dry sub-humid (Mediterranean) climate with hot and dry summers and mild, wet winters [32]. Precipitation are on annual average around 600 mm, and mainly concentrated between November and February.

The region economy is mainly driven by the agricultural sector followed by service and industry. Agricultural areas cover more than 70% of the region and the main crops are wheat, olive, grapes, citrus, and vegetables.

Apulia region mostly corresponds to the exposed area of the Apulian carbonate platform (Apulia foreland), separated from the Southern Apennines by the Bradanic Through (foredeep). The main geological complexes are the Mesozoic carbonate platform and the Paleocene to Pleistocene sedimentary covers [33] (Figure 4).

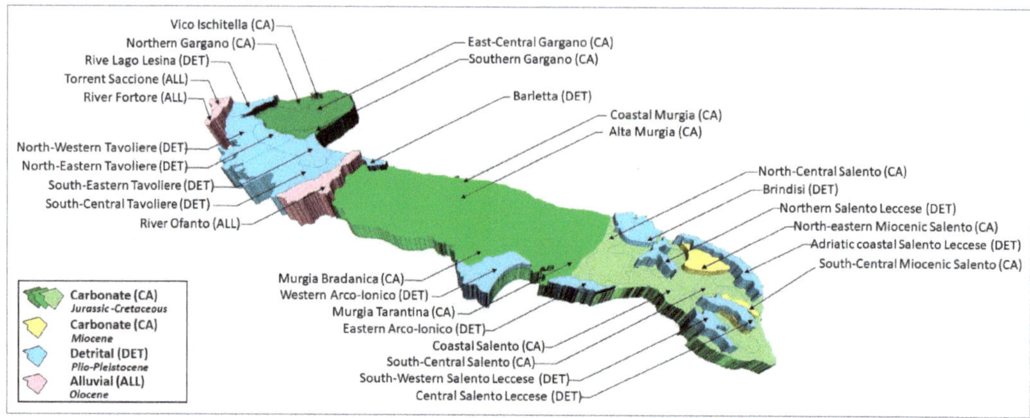

Figure 4. Sketch map showing the geographic position of Apulia GWBs.

The carbonate platform consists of a several thousand-meter-thick succession of limestone, dolomite and limestone dolomite (Middle Jurassic–Cretaceous) diffusely outcropping in the Gargano area in the North, in the Murgia in the center, and in the Salento in the South of the region.

The sedimentary covers consist of heterogeneous marine and alluvial deposits of different ages [34].

All the calcareous and calcareous-dolomite rocks forming the Mesozoic succession are influenced by karst phenomena of different evolution degree. This causes an almost total absence of surface water and a widespread karst landscape. Karstic processes combined to the presence of fractures affecting the Mesozoic succession, give rise to a secondary permeability, which is always markedly discontinuous in space and may vary from very low to high [35,36].

For all of the above, the groundwater resources hosted in the three Mesozoic hydrogeological structures of Gargano, Murgia and Salento, representing the largest coastal karst Italian aquifers, are the main regional water supply of Apulia [37]. They are exposed to contact with the sea and the groundwater systems are largely governed by the freshwater-saltwater equilibria.

The hydrogeological structure of the Gargano is generally characterized by confined groundwater flow, except along a narrow coastal belt. Along the south-western boundary, groundwater interacts with warm waters rising from the deepest part of the carbonate platform [38]. These deep groundwater fluxes are drained by springs in the SW portion of the promontory (Manfredonia Gulf) and in the NW portion toward Lesina and Varano lakes. Particularly, in these areas they assume typical fossil water characteristics [39].

In the Murgia hydrostructure, the presence at different depths of thick dolomitic layers, practically impermeable, causes the carbonate aquifer to be under pressure and with irregular geometry. Only near the coast, and mainly in the southern part of the Murgia, the groundwater flow becomes unconfined. The watershed for groundwater flow approximately overlaps the highest topographic altitudes along the NW-SE direction and allows to identify the Bradanic and the Adriatic sectors.

The Salento is the most permeable carbonate hydrostructure of the Apulia. Groundwater flows mainly under phreatic conditions and the water table heights reach only 3–4 m above sea level. Due to the Salento geographical shape, the groundwater assumes a lenticular shape, floating on seawater. The aquifer mainly is recharged by rainfall, but in the north-west sector, it is also laterally fed by groundwater coming from the Murgia aquifer [36].

The fourth main hydrogeological structure of the Apulia Region is the Tavoliere, a shallow aquifer consisting of a complex alternation of different grain size alluvial sediments. This aquifer is unconfined in the western part of the plain, where the coarse-grained sediments prevail, and confined in the central-eastern part where fine-grained material tend to prevail in the upper part of the sequence [40].

Following the criteria established by Italian law for defining the GBWs [5], three main type of hydrogeological complexes (HCs) have been recognized in the Apulia: carbonate, detrital, and alluvial [41,42]. Moreover, on the basis of hydrogeological features of the formation hosting groundwater and its physical-chemical characteristics different GWBs were distinguished within some aquifer for a total of 29 GWBs (Figure 4).

Since the first regional groundwater monitoring network was established in 1995, different redesign have been done until 2015 to fit the WFD provisions and optimize the network configuration [43]. It follows that the surveys have been affected by some discontinuities in time. Currently, the qualitative groundwater monitoring network consists of 341 wells and 20 springs distributed almost uniformly over the regional territory (Figure 3).

The dataset used in this study consists of more than 3600 measured values of the main hydrogeochemical parameters, discontinuously measured from 1995 to 2018. In general, two sampling campaigns were carried out per year, one at the end of the recharging period and the other at the end of the dry season, for a total of 22.

Since the measurements refer to a very long time interval, the sampling and analytical methods could have likely changed over time. Inhomogeneities are therefore possible, but the NBL assessment procedures have been designed to minimize their effect on the results (use of median values, outliers removal, statistical distribution analysis). Detailed information is available only for data collected after 2000. From this date on, official monitoring protocols and field forms indicate the physical parameters (pH, Eh, EC, temperature and DO) have been measured onsite by a multiparametric probe. Water samples have been filtered through 0.45 μm pore-size filters immediately upon collection and the aliquot for cations and trace elements has been acidified with supra pure HNO_3 1%. All the chemical parameters have been measured and validated at certified public and private laboratories by ion chromatography for anions and by Inductively Coupled Plasma-Mass Spectrometry for major cations and trace elements.

Severe impacts on regional GWBs are due to overexploitation and qualitative degradation [41,42]. Quantitative pressures produce impacts in terms of piezometric lowering, reduction of springs outflow and decrease of groundwater contribution to surface aquatic ecosystems. The impact from quantitative significant pressures can result in an increase of saline intrusion with a consequent reduction of freshwater availability and soil salinization in agriculture.

4. Results

4.1. GW Geochemistry and Descriptive Statistics

A summary of the qualitative status of the Apulia GWBs is provided, based on selected monitoring data.

In Table 3, the main statistics of the physical-chemical parameters relating to the Apulian monitoring network are represented. Groundwater shows a wide range of pH values, with conditions ranging from acidic to frankly alkaline, and slightly alkaline mean values. Oxidizing conditions predominate in the region, although reducing environments are also found within some GWBs, highlighted by DO deficiency and negative ORP values. The variability from 0.1 to tens mS/cm, of the electrical conductivity (EC) is typical of wide aquifers that spread from inland to the coastline.

Table 3. Main statistical indicators of the physical-chemical parameters measured in the Apulia monitoring network.

	T (°C)	pH	EC (µS/cm)	DO (mg/L)	ORP (mV)
N Observations	3463	1936	3251	1804	1802
Min	7.2	4.71	129	0.01	−473
Max	34.5	11.90	55,760	15.2	504
Mean	18.8	7.30	3982	5.6	73
Median	18.5	7.25	1500	6.1	112
Std Dev	2.8	0.5	8401.9	5.4	128.5

Table 4 reports the exceedances, with respect to the values prescribed by the Italian water-related Legislative Decrees [4,5], of those parameters resulting as main indicators of significant anthropic pressures and related impacts arisen in the regional GWBs characterization [44,45]. The most evident result is the high percentage of exceedances for chloride (49.0%) and sulphates (16.8%) that can be related to saline contamination phenomena, especially in coastal areas. The percentages relating to nitrate (20.3%) and ammonium (6.4%), classical indicators of anthropogenic contamination, can be linked to the extensive agricultural land use and, partially, to livestock activities. More than a quarter of the dataset shows exceedances of the TVs for iron (32.2%) and manganese (26.2%). Both these parameters are sensitive to the redox reducing conditions of the aquifers which are quite frequent, as mentioned previously. Several values exceeding the legal limits were found for selenium (3.7%) and fluoride (3.5%), while sporadic exceedances were found for boron (0.8%) and arsenic (0.3%).

Table 4. Statistics and exceedances of the main inorganic parameters compared to the corresponding TVs.

	NO$_3$ (mg/L)	NH$_4$ (mg/L)	NO$_2$ (mg/L)	F (mg/L)	Cl (mg/L)	SO$_4$ (mg/L)	B (µg/L)	Mn (µg/L)	Fe (µg/L)	As (µg/L)	Se (µg/L)
N Obs.	3609	3510	3663	3663	3381	3382	2539	2539	2974	2539	2539
Max value	716	53.9	21.9	5.87	24,106	6600	4446	12,800	178,000	79.0	282.7
TVs	50	0.5	0.5	1.50	250	250	1000	50	200	10	10
N Exceed.	732	225	68	130	1655	568	21	664	957	7	95
% Exceed.	20.3	6.4	1.9	3.5	49.0	16.8	0.8	26.2	32.2	0.3	3.7

In more detail, Figure 5 shows the Piper's classification diagram distinguished by HC. This type of representation provides information on the geochemical facies and therefore on the processes that led to the observed main chemistry. Each point represents the median value of Piper's diagram characteristic parameters in each considered GWB, distinguished by color.

Groundwater of the Gargano and Murgia-Salento carbonate HCs (Figure 5a) shows a clear transition from frankly calcium-bicarbonate terms to chloride-alkaline terms, with increasing salinity. The points of the detrital HCs (Figure 5b) show more variable compositions, generally ranging between the chloride-alkaline and the sulphate-chloride-earth alkaline facies. Finally, groundwater from the Miocene carbonate HCs has a clear calcium-bicarbonate composition, while the alluvial HCs are characterized by mixed terms, without dominant ions (Figure 5c). The presence of clear bicarbonate-earth alkaline terms is mainly attributable to water-rock interaction processes in the carbonate aquifers, while the enrichment in chlorides, sulfates, and alkalis (sodium in particular), reflecting on a substantial increase in groundwater salinity, are found in coastal GWBs.

The NBLs at the GWB scale were assessed for all the parameters in Table 4, except for nitrogen species, which were used as anthropic indicators. To ensure a certain significance in the statistical processing, the calculation was carried out for the GWBs with at least 7 monitoring points, for a total of 15 investigated GWBs and 294 MSs.

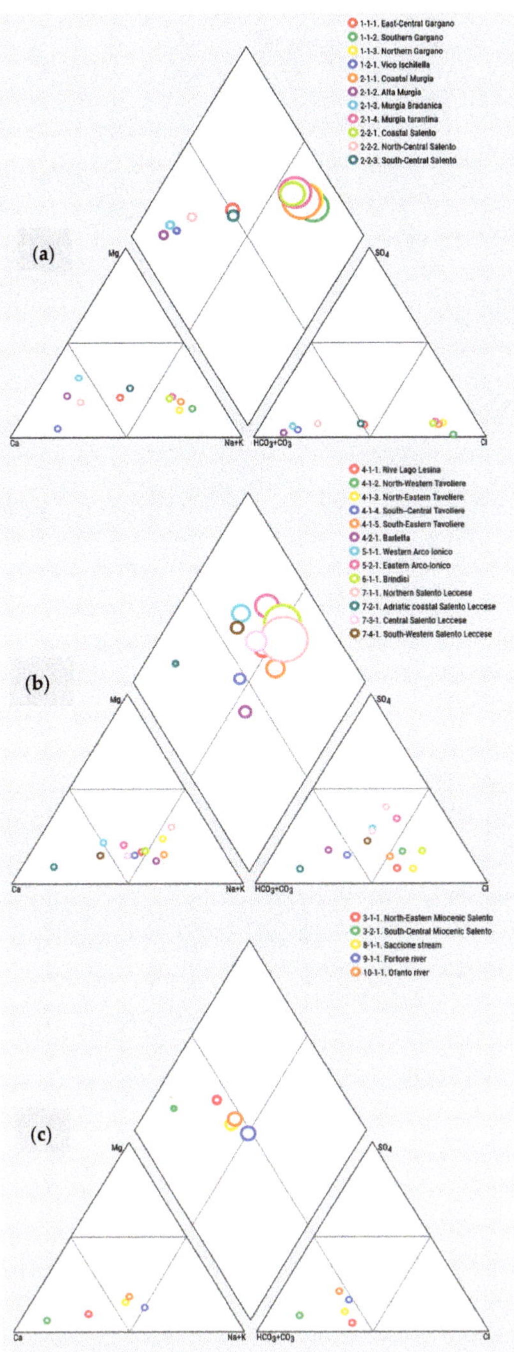

Figure 5. Piper diagrams showing major groundwater components (**a**) Gargano and Murgia-Salento carbonate HCs; (**b**) detrital HCs; (**c**) Alluvial and Miocene carbonate HCs.

4.2. Preliminary Analysis on the Monitoring Stations

The proposed methodology requires some preliminary processing in order to rearrange the dataset for the following steps. At first, all the analytical values below their own limit of quantification (LOQ) must be replaced with a value equal to half the limit itself. Nearly 6500 values have been replaced with regard to the 8 investigated parameters and the 15 considered GWBs.

Successively, the dataset is treated by removing the analyses whose electrical balance error (EBE) exceeds the given threshold. This validation step resulted in the removal of 55 analyses from the overall dataset, whose EBE was exceeding the threshold of 10%.

A third important data preprocessing step concerns the redox facies separation. In the case at hand, once set an ORP threshold of 50 mV (selected following a comparison with the values of Fe and Mn, tendentially showing higher concentrations below this limit), 154 MSs out of 294 were identified in the oxidizing facies and 63 MSs in the reducing one, with distribution and numerical consistency different by GWB. Finally, due to the lack of ORP and DO measurements, the redox facies assignment has not been practicable in 77 MSs. The facies separation was applied only for iron and manganese, which are particularly affected by reducing conditions that increase the solubility of their oxy-hydroxides. Therefore, only for these two metals, the NBLs were first calculated separately for the two redox facies, then a unique value (the highest of the two) was chosen from them. For all the other parameters the NBLs were calculated from the total dataset of 294 MSs.

The last step of the preliminary analysis is the pre-selection of the MSs. In this phase, some MSs and the related datasets are removed based on an anthropogenic contamination test. The test operates using two different markers of anthropogenic contamination, according to the redox conditions: a threshold of 37.5 mg/L of nitrate for the oxidizing facies and 0.375 mg/L of ammonium for the reducing one. Only those MSs in oxidizing/reducing facies, where nitrates/ammonium do not exceed the respective concentration threshold, have been retained for next processing; all the others have been considered somehow contaminated by anthropic activities and discarded. The selected concentration limits correspond to 75% of the expected quality standards/TVs. When the methodology does not involve the facies separation, both markers were used for the MSs pre-selection. On average, 66% of MSs complies with the adopted pre-selection criteria, but with very diversified distributions within the different GWBs.

In particular, the pre-selection criterion allowed retaining a number of monitoring stations (pre-selected monitoring sites = PS MSs) ranging from 40% to 90% of the initial number in 10 of the considered GWBs. The percentage of pre-selected stations falls down to values smaller than 38% in the remaining 5 GWBs. Unfortunately, these latter five GWBs, mostly belonging to the Tavoliere aquifer, were initially affected by a poor number of MSs, so the related computed NBLs are expected to be characterized by a very low confidence level. In conclusion, 194 monitoring sites were selected for the NBLs assessment in 15 GWBs of which 104 in the oxidizing and 54 in the reducing facies (Table 5).

Table 5. Data validation and pre-selection of the monitoring stations. EBE = electrical balance error; MS = monitoring site; PS MSs = pre-selected monitoring sites.

GWB	Data Validation		Total Dataset			Oxidizing Facies			Reducing Facies		
	<LOQ	EBE > 10%	MSs	PS MSs	%PS MSs	MSs	PS MSs	%PS MSs	MSs	PS MSs	%PS MSs
East-Central Gargano	234	1	16	14	87.5	8	7	87.5	2	2	100.0
Southern Gargano	106	1	7	3	42.9	1	1	100.0	3	1	33.3
Northern Gargano	209	2	8	7	87.5	4	4	100.0	2	2	100.0
Coastal Murgia	1102	5	35	26	74.3	15	9	60.0	14	14	100.0
Alta Murgia	1230	3	46	41	89.1	31	30	96.8	7	6	85.7
Murgia Bradanica	383	6	16	11	68.8	12	8	66.7	2	2	100.0
Murgia Tarantina	355	9	16	13	81.3	3	3	100.0	7	5	71.4

Table 5. *Cont.*

GWB	Data Validation		Total Dataset			Oxidizing Facies			Reducing Facies		
	<LOQ	EBE > 10%	MSs	PS MSs	%PS MSs	MSs	PS MSs	%PS MSs	MSs	PS MSs	%PS MSs
Coastal Salento	900	9	47	32	68.1	15	12	80.0	15	12	80.0
North-Central Salento	190	2	10	5	50.0	3	2	66.7	3	3	100.0
South-Central Salento	845	9	43	28	65.1	30	22	73.3	4	3	75.0
North-Eastern Tavoliere	131	2	10	2	20.0	6	1	16.7	1	1	100.0
North-Western Tavoliere	118	0	6	1	16.7	4	1	25.0	1	1	100.0
South–Central Tavoliere	143	1	6	2	33.3	5	2	40.0	0	-	-
South-Eastern Tavoliere	221	2	13	5	38.5	8	1	12.5	2	2	100.0
Western Arco Ionico	302	3	15	4	26.7	9	1	11.1	0	-	-
All GWBs	6469	55	294	194	66.0	154	104	67.5	63	54	85.7

4.3. NBLs Calculation for the Non-Redox Sensitive Elements

For each PS MSs, a representative value was defined as the median of all available measurements for each investigated parameter (fluoride, chloride, sulphate, boron, arsenic, selenium). The outliers were then analyzed using the Huber's non-parametric test and the identified anomalous data were eliminated from the dataset, before performing the calculation of the NBLs. Consequently, the number of the representative values is not equal for all parameters, resulting in part from the elimination of the outliers and in part from any missing analysis for that specific parameter.

According to the guidelines, the specific path types (A, B, C, D) were identified based on the pre-selected dataset size. For type A and B the Shapiro–Wilk test was then applied to verify the normality of the dataset. Data relating to the performed elaborations (including the specific path types and the number of representative values for each parameter) are shown in Table S1 (Supplementary Material).

The most frequent type of dataset were C and D, then A and finally B were the least represented (see Table S1 in the Supplementary Materials).

The defined NBLs for all the parameters at the GWBs scale and the associated confidence levels defined considering the aquifer type (confined or unconfined), its extension and the number of data used for the NBL assessment, are shown in Table 6 and in Figure 6. Chlorides show diffusely high background values, often above the corresponding TV (80% of the selected GWBs). Values sometimes higher than the legal limits have also been calculated for sulphate (6 GWBs) and selenium (4 GWBs), while the NBLs defined for fluoride, boron and arsenic are always lower than the respective TVs. Unfortunately, only for 4 GWBs it was possible to associate a high confidence level to the calculated NBLs, while it resulted almost always low or very low. The low/very low confidence levels are attributable to a strong reduction of MS number due to the high percentage of MS exceeding nitrate/ammonia and/or the lack of the specific analytical value.

4.4. NBLs Calculation for the Redox Sensitive Elements (Iron and Manganese)

As already mentioned, for iron and manganese the NBLs were calculated separately for the oxidizing and reducing facies, and then the higher of the two values was chosen as NBL of the entire GWB. When the number of data did not allow the NBL calculation for both facies, the only available value was assigned to the GWB. The calculated NBLs and the associated confidence levels are shown in Table 7 and Figure 7, while the other data relating to the performed elaborations are shown in Table S2 (Supplementary Materials).

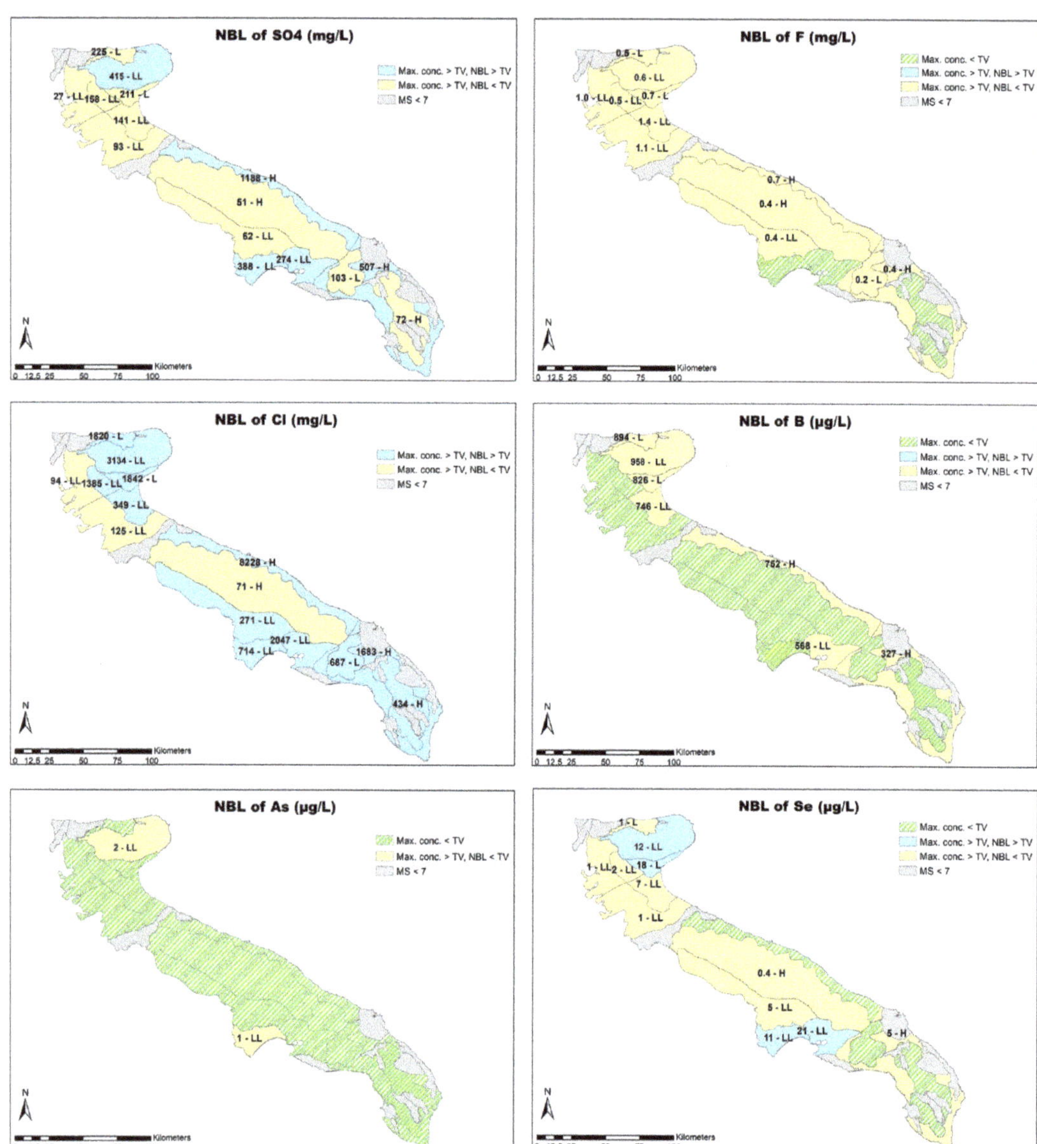

Figure 6. NBLs defined for sulfates, fluorides, chlorides, boron, arsenic and selenium at the GWBs scale and the associated confidence levels (H = high, L = low, LL = very low). The NBLs obtained for GWBs with maximum concentration observed during monitoring below TV were not shown in the figure. In grey are shown the GWBs where the NBL assessment was not performed (MS < 7).

Table 6. Defined NBLs and associated confidence levels for the non-redox sensitive elements. NBLs above the threshold values are shown in bold. H = high; L = low; LL = very low. As indicated in the guidelines, the NBL for each parameter was expressed with the same measurement unit and rounded up with the same number of decimals as the relative TV set by Legislative Decree 152/2006.

GWB	F (mg/L)	Cl (mg/L)	SO4 (mg/L)	B (µg/L)	As (µg/L)	Se (µg/L)	Confidence Level
East-Central Gargano	0.6	**3134**	**415**	958	2	**12**	LL
Southern Gargano	0.7	**1842**	**211**	826	1	**18**	L
Northern Gargano	0.5	**1820**	225	894	1	1	L
Coastal Murgia	0.7	**8228**	**1188**	752	2	9	H
Alta Murgia	0.4	71	51	86	1	2	H
Murgia Bradanica	0.4	**271**	62	156	1	5	LL
Murgia Tarantina	0.6	**2047**	**274**	568	1	**21**	LL
Coastal Salento	0.4	**1683**	**507**	327	2	5	H
North-Central Salento	0.2	**687**	103	114	1	2	L
South-Central Salento	0.3	**434**	72	99	1	4	H
North-Eastern Tavoliere	1.0	**1385**	158	94	1	2	LL
North-Western Tavoliere	0.5	94	27	150	1	1	LL
South-Central Tavoliere	1.1	125	93	236	1	1	LL
South-Eastern Tavoliere	1.4	**349**	141	746	2	7	LL
Western Arco Ionico	0.3	**714**	**388**	410	1	**11**	LL
TV	1.5	250	250	1000	10	10	-

Table 7. Defined NBLs and associated confidence levels for iron and manganese. NBLs above the threshold values are shown in bold. L = low; LL = very low.

GWB	Fe (µg/L)				Mn (µg/L)			
	Ox	Red	GWB	Conf. Lev.	Ox	Red	GWB	Conf. Lev.
East-Central n Gargano	**2096**	**834**	**2096**	LL	**86**	48	**86**	LL
Southern Gargano	-	**18,060**	**18,060**	L	-	**86**	**86**	L
Northern Gargano	3	**10,335**	**10,335**	L	1	**207**	**207**	L
Coastal Murgia	194	**9817**	**9871**	LL	16	**289**	**289**	LL
Alta Murgia	**234**	**1907**	**1907**	LL	43	47	47	LL
Murgia Bradanica	**1172**	8	**1172**	LL	21	2	21	LL
Murgia Tarantina	6	**12,626**	**12,626**	LL	2	**278**	**278**	LL
Coastal Salento	132	**7759**	**7759**	LL	**86**	**207**	**207**	LL
North-Central Salento	50	96	96	L	36	**262**	**262**	L
South-Central Salento	**241**	**1010**	**1010**	LL	24	**78**	**78**	LL
North-Eastern Tavoliere	72	43	72	LL	10	7	10	LL
North-Western Tavoliere	50	48	50	LL	9	3	9	LL
South-Central Tavoliere	**430**	-	**430**	LL	**88**	-	**88**	LL
South-Eastern Tavoliere	-	**2466**	**2466**	LL	-	**76**	**76**	LL
Western Arco Ionico	9	-	9	LL	3	-	3	LL
TV	200	200	200	-	50	50	50	-

For both elements, the exceedances of the TVs concern almost all GWBs and are mainly due, as expected, to values defined for the reducing facies. Particularly, iron shows severe exceedances, with NBLs even two orders of magnitude higher than the corresponding TV (200 µg/L).

The confidence levels associated with the defined NBLs are always low or very low. This is due to the reduced number of initial MSs (217), as well as to the redox facies separation, which produces two different datasets with reduced numerical consistency.

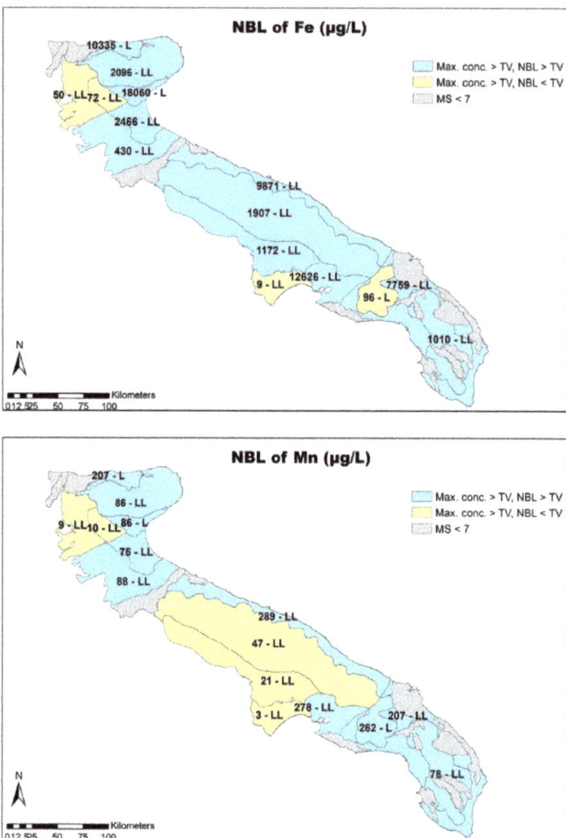

Figure 7. NBLs defined for iron and manganese at the GWBs scale and the associated confidence levels (L = low, LL = very low). In grey are shown the GWBs where the NBL assessment was not performed (MS < 7).

5. Discussion and Conclusions

The Apulian GWBs characterization, carried out according to the rules set by the national regulation [5], pointed out the exceedance of the threshold values for some inorganic parameters, identifying a group of groundwater bodies at risk of failing to meet the Water Framework Directive environmental objectives. Consequently, the assessment of the natural background levels becomes an urgent need, in order to establish the natural or anthropogenic origin of the high values encountered for such parameters.

This work reports a tentative application of the recently published national guidelines for the NBL assessment to the inorganic compounds monitored in the GWBs of the Apulia region. The NBLs values calculated resulted often higher than the related threshold values.

Concerning the redox-sensitive elements iron and manganese, the estimated NBLs exceed, even of 11 and 10 orders of magnitude respectively, the TVs at 15 GWBs. The lack of redox-related parameter measurements (i.e., ORP and DO) forced us to eliminate many MSs from the available database, making the pre-selected dataset rather small and the estimated NBLs confidence levels low or very low. Furthermore, the resulting NBLs almost always correspond to the values estimated in the reducing facies confirming the extreme sensitivity of these two elements in a negative redox environment. This outcome is not surprising considering that Apulian GWBs, and particularly the carbonate ones, are often very deep and confined with the roof at even more than 100 m under the sea level.

The abundance of iron and manganese in the carbonate Apulian GWBs can be associated both to clay-rich lenses interbedded in the carbonate sequence and to the widespread red soils covering the limestone and filling karst cavities and fractures. Their presence is also justified in some detrital and alluvial aquifers that derive from the weathering and erosion of carbonate rocks [44,45].

The estimated NBLs of chloride and sulphate exceed the corresponding TVs in 12 and in 6 of the 15 investigated GWBs respectively, with a wide range of variability. This result was easily predictable for the coastal GWBs but unexpected in the Murgia Bradanica, being this latter very far from the coastline. As known, chloride concentration in coastal GWBs can be particularly affected by the current seawater intrusion, nevertheless different concurrent saline sources can overlap to seawater contributing to groundwater salinization independently from the GWB distance from the coastline [26,27].

In some cases, the mobilization of paleo-seawaters intruded into the aquifers during previous marine transgressions can occur. These saline waters, isolated from active flow, reside for a period long enough to allow water to assume typical connate water characteristics. This is the case of the deep groundwater flux exchanges that occur along the main tectonic elements as an effect of the Apennine convergence towards the Apulia foreland giving rise to upward flow of warm waters drained by springs [38,39]. Previous studies revealed that the chemical and isotopic features of salt groundwater, sampled in different point of the Apulian karstic aquifers or drained from coastal springs, are very different from modern seawater [46–49]. Nevertheless, currently it is not clear to what extent this phenomenon can be considered natural or forced by anthropogenic activity (e.g., excessive pumping in coastal area).

In 4 of the 15 considered GWBs, the calculated NBL for selenium exceed the threshold value set for this element. Selenium can be present in trace in the carbonate rocks and the argillaceous sediments that fill up the karst cavities [50,51]. In many cases, high values of selenium concentration have been measured in springs draining deep water whose composition results from a high degree of water-rock interaction. Nevertheless, given the evident spatial and temporal fragmentation of high selenium concentration measurements, some Authors proposed the hypothesis that it may originate from its use in agriculture as a nutrient, which, in particular environmental circumstances (e.g., soil temperature, humidity and acidity), leaches from the ground into the underlying groundwater [52].

The application of the Italian Guideline to the estimation of the NBLs of the Apulian GWBs, as proposed in this work, highlighted some problems related to the procedure itself. The Guidelines require an initial preselection of the monitoring sites and the related sets of analysis to retain for the NBLs estimation, based on several validation tests. Unfortunately, this preliminary step often reduces consistently the available information making the estimations unreliable or impossible. This is particularly true in those groundwater bodies characterized by a bounded extension and usually monitored with a few sampling sites. It is not a case that the NBLs estimation has not been possible in almost half of the regional groundwater bodies.

The data spatial consistency and the temporal continuity of monitoring are critical, but the existing groundwater monitoring networks have been designed for different purposes, such as the evaluation of the environmental state of the groundwater body or keeping under control local-specific issues.

The used dataset covers a time span of 24 years during which the monitoring has been suspended even for years and then restarted, sampling and analytical methods have changed with the scientific and technical advance. As a consequence, spatial consistency and temporal continuity of the available data can result deficient concerning some parameter or the whole groundwater body.

All the above requires a deep technical and economical evaluation of the opportunity of redesigning or modifying the existing monitoring network in order to reduce the weakness highlighted with regard to the natural background levels estimation in the regional groundwater bodies.

This topic is the subject of specific further investigations in an experimental coastal area, whose final goal is to solve the key question of source of salinization, and define specific protocols for NBLs assessment in coastal aquifers [53–55].

In conclusion, the aim of this work was to apply and verify the national guidelines for the assessment of the NBLs, which implement the EU Water Framework Directive, in a south-European, regional contest characterized by the typical Mediterranean climatic and hydrologic features. A frequent issue of such areas is the high anthropization of the coastal areas and the overexploitation of groundwater for agricultural needs in almost total lacking surface resources. Even though some local specific issues arose during the application of the methodology, the results are suitable to be applied for the next steps of the EU-WFD implementation. In general, the NBL definition can be considered, at any latitude, one of the pillars for sustainable and long-term groundwater management, by tracing a clear boundary between what is supposed to be natural and what is clearly anthropic.

Supplementary Materials: The following are available online at https://www.mdpi.com/article/10.3390/w13070958/s1, Table S1: Results of the procedure for the NBLs calculation at the GWB scale, applied to the non-redox sensitive elements, Table S2: Results of the procedure for the NBLs calculation at the GWB scale, applied to the redox sensitive elements (Fe and Mn).

Author Contributions: Conceptualization, R.M., S.A., E.F., S.G., M.M., D.P., E.P., M.V., A.Z. and G.P.; methodology, E.P., D.P. and S.G.; software, S.G.; formal analysis, D.P., R.M.; data curation, G.P., S.G., R.M.; writing—original draft, R.M., G.P., S.G. and D.P.; writing—review and editing, R.M., G.P., S.G., D.P., E.F., E.P., S.A., A.Z.; supervision, G.P.; project administration, G.P. and E.P.; funding acquisition, G.P. and R.M. All authors have read and agreed to the published version of the manuscript.

Funding: The VIOLA project has been funded by Apulia Region, Regional Department of Water Resources Management, within the frame of the Regional Operational Program (ROP) E.R.D.F.-E.S.F. 2014/2020.

Institutional Review Board Statement: Not applicable.

Informed Consent Statement: Not applicable.

Data Availability Statement: Paper experimental data are not currently publicly available but can be supplied on request from the corresponding author.

Conflicts of Interest: The authors declare no conflict of interest. The funders had no role in the design of the study; in the collection, analyses, or interpretation of data; in the writing of the manuscript, or in the decision to publish the results.

References

1. Directive 2000/60/EC. Water Framework Directive (WFD). Directive of the European Parliament and of the Council of 23 October 2000 establishing a framework for Community action in the field of water policy. *OJL* **2000**, *327*, 1–73.
2. Directive 2006/118/EC. GroundWater Daughter Directive (GWDD). Directive of the European Parliament and of the Council of 12 December 2006 on the protection of groundwater against pollution and deterioration. *OJL* **2006**, *372*, 19–31.
3. Commission Directive 2014/80/EU of 20 June 2014 amending Annex II to Directive 2006/118/EC of the European Parliament and of the Council on the protection of groundwater against pollution and deterioration. *OJL* **2014**, *182*, 52–55.
4. D.Lgs. 152/06—Decreto Legislativo n. 152 del 3 aprile. *Norme in Materia Ambientale*; Pubblicato nella Gazzetta Ufficiale n. 88 del 14 Aprile 2006. (Supplemento Ordinario n. 96); Istituto Poligrafico e Zecca dello Stato s.p.a.: Rome, Italy, 2006.
5. D.Lgs. 30/09—Decreto Legislativo n. 30 del 16 marzo. *Attuazione della Direttiva 2006/118/CE, Relativa alla Protezione delle Acque Sotterranee Dall'inquinamento e dal Deterioramento*; Pubblicato nella Gazzetta Ufficiale n. 79 del 4 Aprile 2009; Istituto Poligrafico e Zecca dello Stato s.p.a.: Rome, Italy, 2009.
6. *Ministerial Decree of 6 July 2016–Recepimento della Direttiva 2014/80/UE Della Commissione del 20 Giugno 2014 che Modifica L'allegato II Della Direttiva 2006/118/CE del PARLAMENTO Europeo e del Consiglio Sulla Protezione Delle Acque Sotterranee Dall'inquinamento e dal Deterioramento*; Pubblicato nella Gazzetta Ufficiale n.165 del 16 Luglio, 2016; Istituto Poligrafico e Zecca dello Stato s.p.a.: Rome, Italy, 2016.
7. Appelo, C.A.J.; Postma, D. *Geochemistry, Groundwater and Pollution*, 2nd ed.; CRC Press: Boca Raton, FL, USA, 2005; p. 668.
8. Banks, D.; Reimann, C.; Royset, O.; Skarphagen, H.; Saether, O.M. Natural concentrations of major and trace elements in some Norwegian bedrock groundwaters. *Appl. Geochem.* **1995**, *10*, 1–16. [CrossRef]

9. Kelly, W.R.; Panno, S.V. Some considerations in applying background concentrations to ground water studies. *Ground Water* **2008**, *46*, 790–792. [CrossRef] [PubMed]
10. Muller, D.; Blum, A.; Hart, A.; Hookey, J.; Kunkel, R.; Scheidleder, A.; Tomlin, C.; Wendland, F. Final proposal for a methodology to set up groundwater threshold values in Europe. In *Report to the EU project "BRIDGE"*; Deliverable D18; 2006; Available online: http://nfp-at.eionet.europa.eu/Public/irc/eionet-circle/bridge/library?l=/deliverables/bridge_groundw-205pdf/_EN_1.0_&a=d (accessed on 11 January 2021).
11. European Commission. *Guidance on Groundwater Status and Trend Assessment, Guidance Document no 18*; Technical Report; European Communities: Luxembourg, 2009; ISBN 978-92-79-11374-1.
12. Molinari, A.; Guadagnini, L.; Marcaccio, M.; Guadagnini, A. Natural background levels and threshold values of chemical species in three large-scale groundwater bodies in Northern Italy. *Sci. Total Environ.* **2012**, *425*, 9–19. [CrossRef]
13. Walter, T. Determining natural background values with probability plots. In Proceedings of the EU Groundwater Policy Developments Conference UNESCO, Paris, France, 13–15 November 2008.
14. Panno, S.V.; Kelly, W.R.; Martinsek, A.T.; Hackley, K.C. Estimating background and threshold nitrate concentrations using probability graphs. *Ground Water* **2006**, *44*, 697–709. [CrossRef] [PubMed]
15. Wendland, F.; Hannappel, S.; Kunkel, R.; Schenk, R.; Voigt, H.J.; Wolter, R. A procedure to define natural groundwater conditions of groundwater bodies in Germany. *Water Sci. Technol.* **2005**, *51*, 249–257. [CrossRef]
16. Edmunds, W.M.; Shand, P.; Hart, P.; Ward, R.S. The natural (baseline) quality of groundwater: A UK pilot study. *Sci. Total Environ.* **2003**, *310*, 25–35. [CrossRef]
17. Nakić, Z.; Kovač, Z.; Parlov, J.; Perković, D. Ambient Background Values of Selected Chemical Substances in Four Groundwater Bodies in the Pannonian Region of Croatia. *Water* **2020**, *12*, 2671. [CrossRef]
18. Urresti-Estala, B.; Carrasco-Cantos, F.; Vadillo-Pérez, I.; Jiménez-Gavilán, P. Determination of background levels on water quality of groundwater bodies: A methodological proposal applied to a Mediterranean River basin (Guadalhorce River, Málaga, southern Spain). *J. Environ. Manag.* **2013**, *117*, 121–130. [CrossRef] [PubMed]
19. Parrone, D.; Ghergo, S.; Preziosi, E. A multi-method approach for the assessment of natural background levels in groundwater. *Sci. Total Environ.* **2019**, *659*, 884–894. [CrossRef] [PubMed]
20. Biddau, R.; Cidu, R.; Lorrai, M.; Mulas, M.G. Assessing background values of chloride, sulfate and fluoride in groundwater: A geochemical-statistical approach at a regional scale. *J. Geochem. Explor.* **2017**, *181*, 243–255. [CrossRef]
21. Dalla Libera, N.; Fabbri, P.; Mason, L.; Piccinini, L.; Pola, M. Geostatistics as a tool to improve the natural background level definition: An application in groundwater. *Sci. Total Environ.* **2017**, *598*, 330–340. [CrossRef] [PubMed]
22. Ducci, D.; Condesso de Melo, M.T.; Preziosi, E.; Sellerino, M.; Parrone, D.; Ribeiro, L. Combining natural background levels (NBLs) assessment with indicator kriging analysis to improve groundwater quality data interpretation and management. *Sci. Total Environ.* **2016**, *569–570*, 569–584. [CrossRef] [PubMed]
23. Preziosi, E.; Parrone, D.; Del Bon, A.; Ghergo, S. Natural background level assessment in groundwaters: Probability plot versus pre-selection method. *J. Geochem. Explor.* **2014**, *143*, 43–53. [CrossRef]
24. Molinari, A.; Chidichimo, F.; Straface, S.; Guadagnini, A. Assessment of natural background levels in potentially contaminated coastal aquifers. *Sci. Total Environ.* **2014**, *476–477*, 38–48. [CrossRef]
25. ISPRA—Istituto Superiore per la Protezione e la Ricerca Ambientale. *Linee Guida Recanti la Procedura da Seguire per il Calcolo dei Valori di Fondo Naturale per i Corpi Idrici Sotterranei (DM 6 Luglio 2016)*; Manuali e Linee Guida 155/2017; ISPRA: Rome, Italy, 2017; ISBN 978-88-448-0830-3. Available online: https://www.isprambiente.gov.it/it/pubblicazioni/manuali-e-linee-guida/linee-guidarecanti-la-procedura-da-seguire-per-il-calcolo-dei-valori-di-fondo-per-i-corpi-idrici-sotterranei-dm-6-luglio-2016 (accessed on 11 January 2021).
26. Mirzavand, M.; Ghasemieh, H.; Sadatinejad, S.J.; Bagheri, R. An overview on source, mechanism and investigation approaches in groundwater salinization studies. *Int. J. Environ. Sci. Technol.* **2020**, *17*, 2463–2476. [CrossRef]
27. Emblanch, C.; Kapelj, S.; Lambrakis, N.; Morell, I.; Petalas, C. *Sources of Aquifer Salinisation. Chapter 3-WG3 Environmental Tracing. Part II. Final Report of Action COST 621 "Groundwater Management of Karst Coastal Aquifers"*; European Commission Publication Office: Luxemburg, 2005; ISBN 92-894-0015-1.
28. Parrone, D.; Frollini, E.; Preziosi, E.; Ghergo, S. eNaBLe, an On-Line Tool to Evaluate Natural Background Levels in Groundwater Bodies. *Water* **2021**, *13*, 74. [CrossRef]
29. HP: Hypertext Preprocessor. Available online: https://www.php.net (accessed on 11 January 2021).
30. MySQL. Available online: https://www.mysql.com (accessed on 11 January 2021).
31. ISTAT (Istituto Nazionale di Statistica). Bilancio Demografico Nazionale. 2019. Available online: http://www.istat.it/it/archivio/245466 (accessed on 11 January 2021).
32. Passarella, G.; Bruno, D.; Lay-Ekuakille, A.; Maggi, S.; Masciale, R.; Zaccaria, D. Spatial and temporal classification of coastal regions using bioclimatic indices in a Mediterranean environment. *Sci. Total Environ.* **2020**, *700*, 134415. [CrossRef] [PubMed]
33. Ciaranfi, N.; Pieri, P.; Ricchetti, G. Note alla carta geologica delle Murge e del Salento (Puglia centromeridionale). *Mem. Soc. Geol. Ital.* **1988**, *41*, 449–460.
34. Ricchetti, G.; Ciaranfi, N.; Sinni, E.L.; Mongelli, F.; Pieri, P. Geodinamica ed evoluzione sedimentaria e tettonica dell'Avampaese apulo. *Mem. Soc. Geol. Ital.* **1988**, *41*, 57–82.

35. Maggiore, M.; Pagliarulo, P. Groundwater vulnerability and pollution sources in the Apulian region (Southern Italy). In Proceedings of the 2nd Symposium Protection of Groundwater from Pollution and Seawater Intrusion, Bari, Italy, 27 September–1 October 1999; pp. 9–20.
36. Cotecchia, V.; Grassi, D.; Polemio, M. Carbonate aquifers in Apulia and seawater intrusion. *G. Geol. Appl.* **2005**, *1*, 219–231.
37. Maggiore, M.; Pagliarulo, P. Circolazione idrica ed equilibri idrogeologici negli acquiferi della Puglia. In Proceedings of the Conference Uso e Tutela dei Corpi Idrici Sotterranei Pugliesi, Bari, Italy, 21 June 2002. *Geol. Territ.* **2004**, *1*, 13–35.
38. Mongelli, F.; Ricchetti, G. Heat flow along the Candelaro fault, Gargano headland (Italy). *Geothermics* **1970**, *2*, 450–458. [CrossRef]
39. Maggiore, M.; Mongelli, F. Hydrogeothermal model of groundwater supply to San Nazario Spring (Gargano, Southern Italy). In Proceedings of the International Conference on Environmental Changes in Karst Areas, ICECKA, Bari, Italy, 15–27 September 1991; Quaderni del Dip.di Geografia—Università di Padova: Padova, Italy, 1991; Volume 13, pp. 307–324.
40. Masciale, R.; Barca, E.; Passarella, G. A methodology for rapid assessment of the environmental status of the shallow aquifer of "Tavoliere di Puglia" (Southern Italy). *Environ. Monit. Assess.* **2011**, *177*, 245–261. [CrossRef]
41. Passarella, G.; Vurro, M.; Barca, E.; Sollitto, D.; Pedalino, M.; Iannarelli, M.A.; Campana, C.; Casale, V. *Identificazione e Caratterizzazione dei Corpi Idrici della Puglia ai Sensi del D.Lgs.30/2009*; IRSA Research Report: Regione Puglia, Bari, Italy, 2013; p. 94.
42. Passarella, G.; Vurro, M.; Barca, E.; Bruno, D.E. *Attività Complementari ed Integrative della Caratterizzazione dei Corpi Idrici Sotterranei*; IRSA Research Report: Regione Puglia, Bari, Italy, 2015; p. 117.
43. Barca, E.; Passarella, G.; Vurro, M.; Morea, A. MSANOS: Data-driven, multi-approach software for optimal redesign of environmental monitoring networks. *Water Res. Manag.* **2015**, *29*, 619–644. [CrossRef]
44. Boni, M.; Rollinson, G.; Mondillo, N.; Balassone, G.; Santoro, L. Quantitative mineralogical characterization of karst bauxite deposits in the Southern Apennines, Italy. *Econ. Geol.* **2013**, *108*, 813–833. [CrossRef]
45. Moresi, M.; Mongelli, G. The relation between the terra rossa and the carbonate-free residue of the underlying limestones and dolostones in Apulia, Italy. *Clay Miner.* **1988**, *23*, 439–446. [CrossRef]
46. Mr, B.; Mz, B.; Fidelibus, M.D.; Morotti, M.; Sappa, G.; Tulipano, L. First results of isotopic ratio 87Sr/86Sr in the characterization of seawater intrusion in the coastal karstic aqui-fer of Murgia (Southern Italy). In Proceedings of the 15th SWIM, Ghent, Belgium, 1998. *Natuurwet. Tijdschr.* **1999**, *79*, 132–139.
47. Fidelibus, M.; Calò, G.; Tinelli, R.; Tulipano, L. Salt ground waters in the Salento karstic coastal aquifer (Apulia, Southern Italy). In *Advances in the Research of Aquatic Environment*; Lambrakis, N., Stournaras, G., Katsanou, K., Eds.; Springer: Berlin/Heidelberg, Germany, 2011. [CrossRef]
48. Fidelibus, M.D.; Tulipano, L. Mixing phenomena owing to seawater intrusion for the inter- pretation of chemical and isotopic data of discharge waters in the Apulian coastal carbonate aquifer (Southern Italy). In *Hydrogeology of Salt Water Intrusion*; De Breuck, W., Ed.; A Selection of SWIM Papers; Heise: Hannover, Germany, 1991.
49. Fidelibus, M.D.; Tulipano, L. Regional flow of intruding sea water in the carbonate aquifers of Apulia (Southern Italy). In *Rapporter och Meddelanden-Proc. 14th SWIM, Malmo, Sweden*; Geological Survey of Sweden, Gotab, Stockholm: Uppsala, Sweden, 1996; Volume 86, pp. 230–240. ISBN 91-7158-572-9.
50. Bassil, J.; Naveau, A.; Fontaine, C.; Grasset, L.; Bodin, J.; Porel, G.; Razac, M.; Kazpard, V.; Popescu, S.M. Investigation of the nature and origin of the geological matrices rich in selenium within the Hydrogeological Experimental Site of Poitiers, France. *Comptes Rendus Geosci.* **2016**, *348*, 598–608. [CrossRef]
51. Renarda, F.; Montes-Hernandeza, G.; Ruiz-Agudod, E.; Putnis, C.V. Selenium incorporation into calcite and its effect on crystal growth: An atomic force microscopy study. *Chem. Geol.* **2013**, *340*, 151–161. [CrossRef]
52. Al Kuisi, M.; Abdel-Fattah, A. Groundwater vulnerability to selenium in semi-arid environments: Amman Zarqa Basin, Jordan. *Environ. Geochem. Health* **2010**, *32*, 107–128. [CrossRef] [PubMed]
53. Masciale, R.; Amalfitano, S.; Frollini, E.; Ghergo, S.; Melita, M.; Parrone, D.; Preziosi, E.; Vurro, M.; Zoppini, A.; Passarella, G. The VIOLA Project: Natural background levels for the groundwater bodies of Apulia Region (Southern Italy). In Proceedings of the EGU General Assembly 2020, Online, 4–8 May 2020. [CrossRef]
54. Parrone, D.; Amalfitano, S.; Frollini, E.; Ghergo, S.; Masciale, R.; Melita, M.; Passarella, G.; Vurro, M.; Zoppini, A.; Preziosi, E. The VIOLA project: Geochemical characterization and natural background levels in a coastal GWB of the Apulia Region (southern Italy). In Proceedings of the EGU General Assembly 2020, Online, 4–8 May 2020. [CrossRef]
55. Melita, M.; Amalfitano, S.; Frollini, E.; Ghergo, S.; Masciale, R.; Daniele, P.; Passarella, G.; Preziosi, E.; Vurro, M.; Zoppini, A. The VIOLA Project: Functional responses of groundwater microbial community across the salinity gradient in a coastal karst aquifer. In Proceedings of the EGU General Assembly 2020, Online, 4–8 May 2020. [CrossRef]

Article

Natural Background Levels of Potentially Toxic Elements in Groundwater from a Former Asbestos Mine in Serpentinite (Balangero, North Italy)

Elisa Sacchi [1,*], Massimo Bergamini [2], Elisa Lazzari [2], Arianna Musacchio [1], Jordi-René Mor [1] and Elisa Pugliaro [2]

1 Department of Earth and Environmental Sciences, University of Pavia, Via Ferrata 9, I-27100 Pavia, Italy; arianna.musacchio01@universitadipavia.it (A.M.); jrene.mor@irsa.cnr.it (J.-R.M.)
2 R.S.A. Srl, Viale Copperi 15, I-10070 Balangero, Italy; bergamini@rsa-srl.it (M.B.); lazzari@rsa-srl.it (E.L.); pugliaro@rsa-srl.it (E.P.)
* Correspondence: elisa.sacchi@unipv.it; Tel.: +39-0382-985880

Abstract: The definition of natural background levels (NBLs) for potentially toxic elements (PTEs) in groundwater from mining environments is a real challenge, as anthropogenic activities boost water–rock interactions, further increasing the naturally high concentrations. This study illustrates the procedure followed to derive PTE concentration values that can be adopted as NBLs for the former Balangero asbestos mine, a "Contaminated Site of National Interest". A full hydrogeochemical characterisation allowed for defining the dominant $Mg-HCO_3$ facies, tending towards the $Mg-SO_4$ facies with increasing mineralisation. PTE concentrations are high, and often exceed the groundwater quality thresholds for Cr VI, Ni, Mn and Fe (5, 20, 50 and 200 μg/L, respectively). The Italian guidelines for NBL assessment recommend using the median as a representative concentration for each monitoring station. However, this involves discarding half of the measurements and in particular the higher concentrations, thus resulting in too conservative estimates. Using instead all the available measurements and the recommended statistical evaluation, the derived NBLs were: Cr = 39.3, Cr VI = 38.1, Ni = 84, Mn = 71.36, Fe = 58.4, Zn = 232.2 μg/L. These values are compared to literature data from similar hydrogeochemical settings, to support the conclusion on their natural origin. Results highlight the need for a partial rethink of the guidelines for the assessment of NBLs in naturally enriched environmental settings.

Keywords: trace metals; Lanzo Massif; ultramafic rocks; ophiolites; chromium; hexavalent chromium; nickel; neutral mine drainage

Citation: Sacchi, E.; Bergamini, M.; Lazzari, E.; Musacchio, A.; Mor, J.-R.; Pugliaro, E. Natural Background Levels of Potentially Toxic Elements in Groundwater from a Former Asbestos Mine in Serpentinite (Balangero, North Italy). *Water* **2021**, *13*, 735. https://doi.org/10.3390/w13050735

Academic Editor: Elisabetta Preziosi

Received: 30 January 2021
Accepted: 2 March 2021
Published: 8 March 2021

Publisher's Note: MDPI stays neutral with regard to jurisdictional claims in published maps and institutional affiliations.

1. Introduction

Potentially toxic elements (PTEs), also known as trace or heavy metals [1], are always present in soils, being derived from the weathering of parent rocks that may contain them, and, depending on the hydrogeochemical context, may be found in groundwater too [2]. In the Anthropocene, high levels of PTEs have been introduced into the environment, affecting even ecosystems in remote areas [3]. In addition, acid rain, resulting from a change in atmospheric chemistry, has boosted soil and rock weathering and enhanced the solubility of metals in surface and groundwater [4], ultimately leading to increasing contents in water resources. As the human population heavily relies on groundwater for drinking water supply, threshold values (TVs) need to be established, corresponding to concentrations which should not be exceeded in order to protect human health and the environment [5,6].

Naturally high concentrations of PTEs in groundwater, exceeding the established TVs, represent a serious problem in many areas worldwide, affecting both hard rock [7,8] and sedimentary aquifers [9,10]. The anthropogenic contribution is less frequently observed, as generally PTEs form cations or complexes, and are retained by the soil or the aquifer matrix. However, some PTEs with high ionic potential (e.g., As, Cr, V) can form oxyanions, especially in their high oxidation state. This is the case of hexavalent chromium (Cr VI), which has many industrial applications and originates well-identified contamination plumes in urban and industrial areas [11]. Great concern is therefore raised when Cr VI groundwater concentrations exceed the TV by reason of its high toxicity [12].

To protect the population from the negative impact of PTEs, the European Commission has promulgated the Groundwater Directive [13], as part of the more general Water Framework Directive [14]. This directive requires member states to set up monitoring networks and to define the groundwater quality status, based on defined TVs for potentially toxic substances, taking into account the natural background levels (NBLs). The TVs are valid nationwide and if, during monitoring activities, some groundwater samples exceed one or more TVs, the presence of anthropogenic contamination should be investigated [15].

The assessment of NBLs at the local scale can be performed using different procedures, that can be broadly classified into two approaches [16]. The pre-selection method requires the identification of groundwater monitoring stations (MSs) unaffected by anthropogenic pollution, based on a list of common contaminants (nitrates, ammonia and hydrocarbons). Once an adequate number of MSs are identified with "pristine" hydrochemical features, the NBLs of the dissolved substances are determined using statistical parameters such as the median concentration values or other selected percentiles. The second approach requires a statistical data treatment that recognises the presence of one or more data populations and, through component separation methods, extracts the "unaffected" MSs that are used to define the NBLs. An overview of the different procedures used to calculate NBLs [16], and of the common issues encountered in defining TVs and NBLs, is reported elsewhere [16,17].

The determination of NBLs is a difficult task in areas such as plain aquifers, generally affected by anthropogenic contamination, where uncontaminated MSs, but still representative of the natural groundwater hydrogeochemistry, are almost impossible to find [18]. Even more complicated is the determination of NBLs in hard rock aquifers, where the low and discontinuous permeability may give rise to groundwater characterised by different degrees of equilibration with the matrix, and consequently different mineralisation and PTE contents [19]. However, deriving NBLs for PTEs in mining environments represents a real challenge. Here, naturally high background concentrations may be present, which may be further increased by anthropogenic activities [20]. These, other than introducing exogenous substances or contaminants in groundwater, fuel up the natural water–rock interaction process. Indeed, rock excavation and crushing increase the specific surface area available for interactions with percolating waters, reduce the groundwater transit times and perturb the natural equilibrium by altering the pH-Eh conditions [21].

The main aim of this work is to illustrate the procedure followed to derive some PTE concentration values that can be reasonably adopted as NBLs in a former mining area. For legal reasons, the guidelines in force in Italy were followed [22], as far as possible. However, to account for the hydrogeological complexity of the area and to better exploit all the available monitoring data, some modifications of the procedure were necessary. In the paper, we discuss the derived NBLs based on the literature from similar hydrogeological settings, to support the conclusion on the natural origin of these high PTE concentrations.

2. Study Area

Located about 30 km NW of Turin, in the Piedmont region of North Italy (Figure 1), the former Balangero and Corio asbestos mine operated the extraction of chrysotile from the 1920s to the 1990s, becoming the largest asbestos mine in Western Europe. The mine is hosted in the serpentinised outer rim of the Lanzo Massif, one of the largest world outcrops of mantle peridotite, at the contact with the Sesia–Lanzo zone. The Lanzo Massif is a portion of subcontinental mantle that was exhumed to the seafloor during the opening of the Mesozoic Ligurian–Piedmontese oceanic basin [23,24]. The Sesia–Lanzo zone is a portion of continental crust from the African margin that, during the Alpine orogenesis, was intensively folded and thrusted, originating the present-day tectonic slices, hectometric to kilometric in size and E–W directed [25].

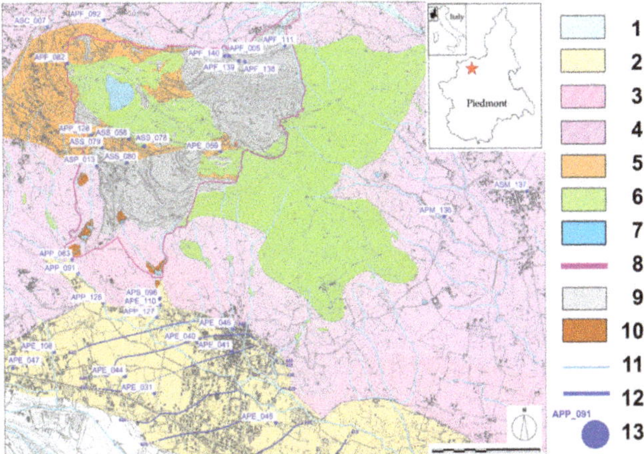

Figure 1. Location of the investigated area. 1 = recent alluvial deposits; 2 = alluvial deposits of the Balangero Plain (Middle–Upper Pleistocene); 3 = fluvioglacial deposits (Middle Pleistocene); 4 = fluvial deposits (Lower Pleistocene); 5 = Sesia-Lanzo zone; 6 = Lanzo Massif serpentinite; 7 = mine pit lake; 8 = remediation site; 9 = tailing piles; 10 = waste sludge; 11 = rivers; 12 = piezometric map of the phreatic aquifer from water table levels measured on Nov. 10–11, 2008 (m a.s.l.) [26]; 13 = monitoring station.

In the study area, the Lanzo Massif serpentinite is constituted by antigorite, diopside, chlorite, olivine and magnetite, with accessory Ni-Fe-Co alloys and sulphides [23,27,28]. The Sesia–Lanzo rocks are metabasites with glaucophane, epidote and garnet, associated with variable amounts of albite, chlorite, phengite, rutile, titanite, apatite, magnetite and sulphides (pyrite, chalcopyrite and covellite) [25]. Along the contact between the two units, a rodingite rim up to 6 m thick is present, originated by the metamorphism of original gabbro or basaltic intrusions. Rodingite is constituted by fine- to medum-grained diopside, irregularly alternated to garnet ± chlorite, and hosts ore minerals (Fe-Cu-Zn-Pb-Ni-Co sulphides, Ni-Co-Fe arsenides and minor Fe-oxides) [27,28]. Finally, serpentinite is characterised by the presence of veins of different origin and composition, and variably associated metallic ores [29], among which are the stockwork chrysotile-rich veins (up to 15% in volume) that were exploited in the past [30]. In 1998, after the abandoning of the mining activities and following the increased evidence of human health effects among mine workers and the local population [31], the area was declared a "Contaminated Site of National Interest", and its remediation is under the responsibility of the Italian Ministry of the Environment, Land and Sea.

The remediation site is located in a mountain area, extends for about 400 ha and includes the open mine pit, now occupied by a lake, the former industrial plant, now dismantled, and the mountain sides where mine tailings and waste sludge were dumped (Figure 1). Previous remediation activities included the stabilisation of the mine tailing piles, the removal of dumped waste sludge, and specific hydrogeological interventions to control runoff. These activities required the handling of large quantities of asbestos-rich materials, and were conducted following strict protocols to limit the potential health risk for workers and the surrounding inhabitants [32]. Since 2012, the remediation site was subject to an extensive characterisation of the presence of asbestos and of PTEs in soils, rivers and groundwater, to define the confinement and remediation strategies. During these monitoring activities, some PTE concentrations exceeding the established TVs were detected in groundwater [15], in particular for Ni and Cr VI, triggering further investigation on their origin [22].

Hydrogeology

In the study area, four formations are potentially suitable to host groundwater. These are:

1. the crystalline rocks of the Sesia–Lanzo zone (henceforth indicated as SL);
2. the serpentinite of the Lanzo Massif (henceforth indicated as SERP);
3. the fluvial deposits, the glacial basal and ablation deposits of Middle Pleistocene, the fluvial deposits of Lower Pleistocene (grouped together as fluvioglacial deposits) and their colluvial–eluvial covers (henceforth indicated as FGD);
4. the alluvial deposits constituting the Balangero Plain (henceforth indicated as BPA).

The first two formations are constituted by massive rock bodies affected by fractures that can host groundwater mostly during the rainy seasons, in spring and in autumn. The fracture permeability, discontinuous by nature, does not allow considering these formations as "aquifers" according to the definition of ISPRA [22] (i.e., "one or more rock layers or other geological layers with a sufficiently high porosity and permeability to permit a significant groundwater flow or the extraction of significant amounts of groundwater" [15]). As such, the guidelines for the definition of NBLs would not be strictly applicable [22]. However, some wells, boreholes and springs located in these formations were also included. The water scarcity posed some difficulties in finding a sufficient number of MSs, located far from potential contamination sources related to the past mining activity. In addition, as the water-bearing fractures develop at the contact between the SL and the SERP formations, the attribution of each MS to one or to the other was based on borehole stratigraphy and on the geological expertise.

The fluvioglacial deposits and their colluvial–eluvial covers are located at the foothills of the mountain range. Although their origin is different, they share the same hydrogeological characteristics. Generally, these deposits have a limited thickness and low permeability, and therefore cannot be considered as "aquifers" [22]. No MSs are indicated on the geological map in this formation (Figure 1); however, the boreholes and wells located in the Balangero Plain, at the mountain foothills, are mostly developed in this formation, as indicated by their stratigraphy.

Finally, the Balangero alluvial deposits, located between the FGD and the Stura River, constitute the Lanzo valley floor. They originated from the dismantling of the SL and SERP rocks, mixed with the alluvial material transported by the river. They constitute a porous aquifer that can host phreatic groundwater. The piezometric map of this area shows a groundwater flow directed SSE from the mountain foothills towards the Stura River (Figure 1). Although located outside of the remediation site, this formation was included since it represents the natural outflow of the water recharged in the former mining site, it is the closest and the only aquifer of the area and it hosts villages and human settlements where monitoring stations can easily be found. Furthermore, the inhabitants of the area could be the target of the potential groundwater contamination leaking out of the remediation site.

3. Materials and Methods

3.1. The Italian Legislative Framework

The Italian TVs for the PTEs of concern are Cr = 50, Cr VI = 5, Co = 50, Ni = 20, Mn = 50, Fe = 200 and Zn = 3000 µg/L [15]. The guidelines for the evaluation of NBLs [22] indicate a complex multistep procedure that combines both the pre-selection and the component separation approaches. In the first phase of the procedure, based on the hydrogeological conceptual model, an adequate number of MSs should be identified. MSs should belong to the same hydrogeochemical context or facies, as defined by physico-chemical parameters and major ions' concentrations. The MSs potentially affected by anthropogenic contamination, as indicated by the occurrence of nitrates, ammonia or organic contaminants, should be disregarded. In the second phase, the investigation concentrates on the identification of temporal trends at those MSs with measurements performed at least 2 times per year over at least four years. This allows selecting, for each MS, the representative concentration values, generally corresponding to the median. These values are then examined to identify outliers, the presence of multiple populations and the statistical distribution of the MS dataset. In the third phase, the dataset size is assessed for its significance at the spatial (based on the number of MSs) and temporal scale (at least 8 half-yearly measurements for at least 80% of the MSs). Four cases could be present (A = significant size in space and time; B = significant size in space but not in time; C = significant size only in time; D = non-significant size), affecting the percentiles that can be used as NBLs and the confidence level attributed to this evaluation. For NBLs with low confidence levels, additional monitoring is required. More detailed information on the procedure is reported elsewhere [22,33]. In particular, the paper is organised in different sections, clearly referring to the different steps of the workflow, as reported by [33].

3.2. Sampling and Analytical Methods

Starting from 2012, several sampling campaigns were conducted on boreholes and wells from all the investigated formations, to assess the potential effects of the former mining activities on groundwater quality. However, the monitored parameters only included the main PTEs of concern (Cr, Cr VI, Co, Ni, Mn, Fe, Zn). These elements were determined at the Nuovi Servizi Ambientali S.r.l. laboratory (now Lifeanalytics Torino S.r.l.) by inductively coupled plasma optical emission spectroscopy (ICP-OES), with the exception of Cr VI, analysed by ultraviolet–visible spectroscopy (UV–VIS).

The complete hydrogeochemical characterisation requested by the guidelines for NBL assessment [22] was conducted in 2019 and included 30 MSs. Given the hydrogeological features of the study area, sampling was performed in spring (25 June to 4 July 2019), following a rainy period. Wells were purged prior to the collection of the water sample by pumping 3–4 times the well volume with a low-flow pump, until the stabilisation of the physico-chemical parameters. In some cases, where purging was not possible, the sample collection was performed with a bailer. Temperature, redox potential, electrical conductivity (E.C.), pH and dissolved oxygen were measured in the field with the HACH HQ40D multiparameter probe and a flow chamber to avoid any contact with the atmosphere. The field redox potential was corrected to report the values relative to the standard hydrogen electrode. Alkalinity was also measured in the field with the HACH Digital Titrator and standardized titration cartridges. Water samples for anion analyses were collected in high-density polyethylene (HDPE) pre-washed containers. The sample aliquot for cation and trace element analyses was filtered in the field to 0.45 µm and acidified with ultrapure chloridric acid. Samples for hydrocarbon analyses were collected in dark glass containers.

The parameters determined in the laboratory included the main cations and anions (Ca^{2+}, Mg^{2+}, Na^+, K^+, NH_4^+, SO_4^{2-}, Cl^-, HCO_3^-, NO_3^-); the elemental concentrations of Si, Al and PTEs (Cr, Cr VI, Co, Ni, Mn, Fe, Zn); and the total hydrocarbon content (light and heavy fractions). All the analyses were performed by the Nuovi Servizi Ambientali S.r.l. (now Lifeanalytics Torino S.r.l.) laboratory, following approved standard analytical methods [34]. In particular, PTEs were determined by the same techniques adopted during previous investigations. The results are reported in Table S1 (field parameters, major ions and ionic balance) and Table S2 (minor and trace elements), both in the Supplementary Materials.

3.3. Data Validation and Selection

The analytical quality check was performed by calculating the charge balance error using major ions (Ca^{2+}, Mg^{2+}, Na^+, K^+, SO_4^{2-}, Cl^-, HCO_3^-, NO_3^-), and considering as acceptable a value (Err. %) lower than 10% (Table S1) [6]. All the samples have an acceptable ionic balance error except APF140. This sample has a very low mineralisation (calculated Total Dissolved Solids = 71 mg/L) and a low Eh (191 mV), allowing high Fe contents in solution (Fe = 1583 µg/L). If this is accounted for (as Fe^{2+}), the ionic balance fits within the limits.

Among major cations, Na^+ and K^+ sometimes showed contents below the limit of detection (LOD) of 0.5 mg/L. As these were necessary to define the hydrochemical facies, the values < LOD were substituted with a concentration value equal to half the LOD. Concerning NH_4^+, only five samples had detectable concentrations of this parameter (LOD = 0.026 mg/L). Si is ubiquitous (LOD = 1 mg/L), as only six samples are below the LOD; on the other hand, Al (LOD = 0.001 mg/L) could be analysed in eight samples only. By contrast, the PTEs of concern in the NBL evaluation (Cr, Cr VI, Co, Ni, Mn, Fe, Zn) commonly show concentrations < LOD (LOD = 1 µg/L, for Cr VI LOD = 3 µg/L), as observed in a consistent and variable number of samples (Tables S2 and S3 in the Supplementary Materials). A concentration of half the LOD was substituted for all concentrations < LOD, as this choice has little influence on the upper tail of the distribution curve (i.e., percentiles $\geq$ 90p) [22]. The section dedicated to NBL calculations discusses in detail this choice and its consequences.

4. Results

4.1. Hydrochemical Facies

The graphical representation of hydrochemical data was performed to describe the water quality, compare the samples to identify similarities and differences and suggest the specific processes affecting the composition during its evolution (i.e., water–rock interaction processes). As this study deals with PTEs, water samples were first classified based on Eh and pH (Figure 2) [35].

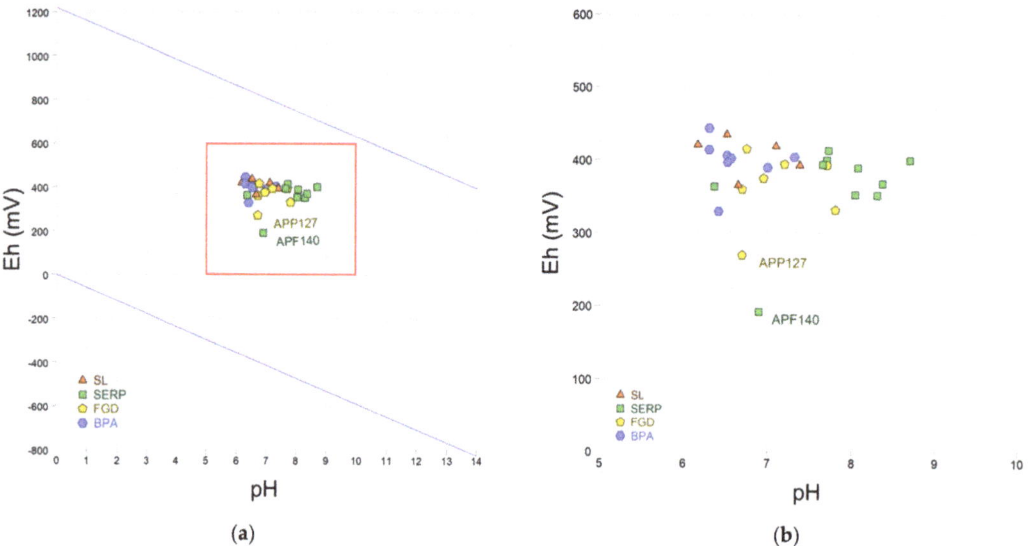

Figure 2. Eh–pH diagrams: (a) full plot, where the blue lines indicate the stability field of liquid water; (b) an enlargement of the portion indicated in red.

Samples are characterised by a circumneutral pH (6.17 to 8.70) in line with the natural range of most natural groundwater [36] unaffected by anthropogenic contamination or acid mine drainage [21]. Eh values (mV) range between 191 and 444, corresponding to dissolved oxygen contents from 1.90 to 10.0 mg/L. Most of the samples can be classified as oxic, except for APP127 (FGD) and APF140 (SERP) that are closer to the limit of reducing waters (Figure 2).

The hydrochemical facies was defined using a Piper diagram [37] (Figure 3), where the major anions and cations are reported in meq/L (%).

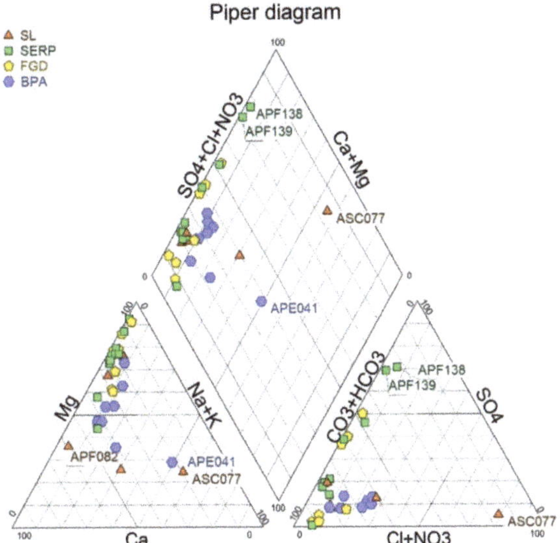

Figure 3. Piper diagram.

Concerning cations, most of the samples have a Mg facies, this element being very abundant in rocks from the SERP and SL formations. Only sample APF082 (SL), collected close to a rodingite vein, which is rich in Ca-bearing silicates [28], has a Ca facies. All the other samples fall in the field of non-dominant cations. Sample APE041 (BPA), close to the field of dominant Na and K, shows high concentrations of K^+, even greater than Na^+, which is uncommon in natural waters. This high K^+ concentration could be due to anthropogenic contamination as the sample belongs to a well located in the Balangero village. Finally, sample ASC077 (SL), a spring collected following some intense precipitation events, has a Na-K facies. As for the anions, most of the samples are of HCO_3 facies, with a tendency towards the SO_4 facies with increasing mineralisation, as indicated by the significant correlation between E.C. and SO_4^{2-} (N = 30; R^2 = 0.67; p < 0.01). Samples APF138 and APF139 (SERP) are, instead, of SO_4 facies. These were collected at the foothills of the Fandaglia mine waste dump, where rock debris were accumulated. The presence of sulphides in the SERP formation is well known [38], and sulphates could derive from the oxidation of these minerals, as is commonly observed in mining areas [21]. Finally, ASC077 (SL) is the only sample of Cl facies.

In conclusion, the dominant hydrochemical facies is Mg-HCO_3, and no systematic changes in hydrochemistry are observed between samples from different formations. Waters tend to evolve towards a sulphate-rich facies, turning to Mg-HCO_3 (SO_4) facies. Two samples from the SERP formation are of Mg-SO_4 facies and one from the SL formation is of Na-Cl facies.

These observations are confirmed by the Stiff diagrams [39], where samples are represented in the order of increasing mineralisation and with different colours according to their formation (Figure S1 in the Supplementary Materials). The shape of the polygons is very similar, regardless of the formation, and the dominant facies is Mg-HCO_3. The only exceptions are those already evidenced by the Piper diagram, i.e., the two low-mineralised samples ASC077 (Na-Cl) and APF082 (Ca-HCO_3), both of the SL formation, the medium-mineralised sample APE041 (PBA) for its K^+ content and the two highly mineralised samples APF138 and APF139 (SERP) for their SO_4^{2-} content.

Based on these results, the collected samples appear to belong to the same hydrogeochemical context characterised by silicate Mg-rich rocks and sediments, and their weathering products. The rocks are locally affected by ore deposits (e.g., rodingite and metal sulphides both in veins or disseminated in the matrix [28]), whose weathering may slightly change the hydrochemical facies. Consequently, the four formations host groundwater with very similar characteristics. The only exception is sample ASC077 with a low mineralisation and a Na-Cl facies. This sample was collected close to a road, and the Na:Cl molar ratio of 1:1 suggests that these ions may derive from de-icing salt. Nevertheless, the PTE concentrations of this sample do not differ from the others and its inclusion in the dataset does not modify its characteristics.

4.2. Pre-Selection

This step aims at eliminating from the dataset the MSs with evidence of anthropogenic contamination based on the presence of some indicators (i.e., nitrates, ammonia, hydrocarbons).

For nitrates and ammonia, only samples below the concentration of 37.5 mg/L and 0.375 mg/L, respectively, corresponding to 75% of the TV allowed in drinking water, were retained in the dataset [22]. Indeed, since the BPA hosts villages and housing, a lower threshold (e.g., 10 mg/L for NO_3^-) would have required the exclusion of numerous MSs from this formation. Only one MS, APE040 (BPA), exceeded the threshold level for nitrates and was therefore eliminated from the dataset. No sample exceeds the limit for ammonia. Concerning hydrocarbons, the regulations only indicate a TV of 350 µg/L for total hydrocarbons expressed as n-hexane. All the results are below this limit. Low concentrations (1–2 µg/L, expressed as n-hexane) of light hydrocarbons (C5–C10) are present in a few samples from all the formations with the exception of SL. However, based on the literature, it is supposed that they could be of natural abiotic origin [40].

Having demonstrated that the MSs belong to the same hydrogeochemical context, the PTE dataset to be used for the evaluation of the NBLs was integrated with the results of previous campaigns, in order to have a sufficient amount of data for statistical treatment. All the available PTE measurements are reported in Table S3 (Supplementary Materials), where the values marked in bold exceed the TVs for groundwater quality [15].

5. Data Treatment
5.1. Trend Analysis

The available dataset is unsuitable to evaluate the presence of temporal trends, as no MS has a series of 8 measurements with a regular frequency extending for at least 2 years (Table S3), as requested by the guidelines [22]. Nevertheless, the trend analysis was performed for the two MSs, APP126 (FGD) and APE031 (BPA), with a sufficient number of measurements (11 and 10, respectively, although with some missing values). The procedure first involves a check for the presence of outliers, followed by the application of the Mann–Kendall test, that does not assume a normal distribution and permits missing data [41,42]. However, the test requires a regular sampling frequency. Since in our case the number of samplings varies between zero and three per year, the year was selected as a time unit and, for the years where multiple sampling were present, a random selection was performed [43].

In the case of APP126 (FGD), the sampling dates randomly selected were 24 July 2015, 3 February 2016, 20 April 2017, 12 January 2018 and 25 June 2019, and no significant trends for Cr, Cr VI, Ni and Mn were detected (Table S4 in Supplementary Materials).

Concerning APE031 (BPA), this MS was sampled first in 2012, then was not monitored for two years, and then was sampled regularly starting from 2015 (Table S3). If the trend analysis is conducted for the period 2015–2019 and using the randomly selected samplings of 24 July 2015, 14 December 2016, 20 April 2017, 12 January 2018 and 25 June 2019, the trend hypothesis is rejected for Cr, Mn and Ni. However, if 2012 is also included in the analysis, and the missing measurements of 2013 and 2014 are estimated by linear interpolation, a significant trend for Ni and for Mn is detected (Table S4). Using Sen's slope estimator [44], a decreasing trend for Ni of −7.87 µg/L per year and an increasing trend for Mn of +0.27 µg/L per year are identified. In conclusion, for both stations, the trend analysis is not sufficiently sound to detect changes in time, and given the overall agreement of PTE concentrations with the other stations of the dataset, they will not be eliminated for subsequent treatment.

5.2. Spatial Analysis

The synthesis table with the PTEs' representative values for each MS is presented as Table 1. For the MSs only sampled once, the measured PTE concentrations are assumed as representative. For the MSs sampled more than once, the median concentration value of each element is considered as representative. In addition, Table 1 reports the total number of MSs, the number of values < LOD and the percentage of non-detected values (%ND). In this regard, Co and Cr VI, with 97% and 79% of ND data, respectively, are the more problematic.

Table 1. Synthesis table reporting the potentially toxic element (PTE) representative values for each monitoring station (MS). Unit: SL = Sesia-Lanzo; SERP = Serpentinite; FGD = Fluvioglacial Deposits; BPA = Balangero Plain Aquifer. Values in italic correspond to those < limit of detection (LOD). Values in bold exceed the threshold values (TVs) for water quality [15]. % ND = percentage of non-detected values.

Monitoring Station	Unit	Cr µg/L	Cr VI µg/L	Co µg/L	Ni µg/L	Mn µg/L	Fe µg/L	Zn µg/L
APE031	BPA	1.0	*1.5*	*0.5*	**71.3**	1.3	7.2	38.4
APS096	FGD	4.8	*1.5*	*0.5*	**23.9**	5.8	1.6	*0.5*
APP126	FGD	17.8	**14.1**	*0.5*	1.7	*0.5*	*0.5*	*0.5*
APF082	SL	11.0	3.3	*0.5*	**20.4**	4.4	**329.9**	15.5
APP128	SL	4.7	3.6	*0.5*	14.4	8.1	*0.5*	*0.5*
APP127	FGD	*0.5*	*1.5*	2.9	8.9	1571.5	966.0	*0.5*
APE059	SERP	4.5	*1.5*	*0.5*	**26.6**	17.7	3.4	*0.5*
ASS078	SERP	6.7	*1.5*	*0.5*	8.7	2.3	4.3	*0.5*
ASS058	SERP	1.4	*1.5*	*0.5*	2.5	*0.5*	8.6	*0.5*
ASS080	SERP	2.7	*1.5*	*0.5*	8.5	1.0	3.5	*0.5*
APE047	BPA	1.6	*1.5*	*0.5*	**56.6**	*0.5*	*0.5*	*0.5*
APE108	BPA	2.4	*1.5*	*0.5*	**32.3**	2.5	*0.5*	46.2
APE044	BPA	*0.5*	*1.5*	*0.5*	**37.8**	22.0	7.9	213.5
APE045	BPA	*0.5*	*1.5*	*0.5*	**37.5**	2.7	*0.5*	33.7
APE041	BPA	3.9	*1.5*	*0.5*	6.1	*0.5*	*0.5*	57.0
APE046	BPA	5.4	*1.5*	*0.5*	2.5	*0.5*	*0.5*	27.0
APP091	FGD	1.7	*1.5*	*0.5*	5.6	*0.5*	1.1	21.4
APE110	FGD	3.6	*1.5*	*0.5*	17.7	*0.5*	*0.5*	11.4
APF092	SL	*0.5*	*1.5*	*0.5*	1.4	6.8	30.0	0.9
ASP013	SERP	8.5	3.8	*0.5*	**34.1**	*0.5*	*0.5*	*0.5*
APP083	FGD	6.8	**6.7**	*0.5*	0.9	*0.5*	3.4	*0.5*
ASS79	SL	4.8	*1.5*	*0.5*	16.0	4.0	*0.5*	*0.5*
APF111	SERP	3.9	*1.5*	*0.5*	12.1	*0.5*	*0.5*	*0.5*
APF139	SERP	*0.5*	*1.5*	*0.5*	8.9	2.1	19.5	*0.5*
APF005	SERP	41.0	**38.1**	*0.5*	*0.5*	*0.5*	1.5	*0.5*
APF138	SERP	1.9	*1.5*	*0.5*	1.6	6.8	38.9	*0.5*
APF140	SERP	*0.5*	*1.5*	*0.5*	*0.5*	125.0	1583.0	*0.5*
ASM137	FGD	*0.5*	*1.5*	*0.5*	15.8	*0.5*	*0.5*	*0.5*
ASC077	SL	*0.5*	*1.5*	*0.5*	*0.5*	5.3	*0.5*	3.6
MS		29	29	29	29	29	29	29
< LOD		8	23	28	3	12	12	18
% ND		28	79	97	10	41	41	62

Graphical methods such as boxplots and normal quantile–quantile (Q–Q) plots were used to evaluate potential outliers among the MSs. In the boxplots, the MSs with values exceeding 1.5 times the difference between 25p and 75p were considered as potential outliers. In the normal Q–Q plots, a line represents the normal distribution, and the potential outliers are those plotting far from it. The results are discussed for each element, with the exception of Co that is detected in only one station, and the relative plots are reported in the Supplementary Materials (Figure S2).

5.2.1. Total and Hexavalent Chromium

The two parameters are jointly discussed, since numerous MSs have ND values of Cr VI, and since they are significantly correlated (N = 29; R^2 = 0.92; $p < 0.01$). Two MSs, APF005 (SERP) and APP126 (FGD), were identified as potential outliers in the boxplots. These two MSs have medium–low mineralisation and belong to the dominant Mg-HCO_3 facies. In the Cr VI boxplot, due to the large number of ND values, all the detected values are identified as outliers. Normal Q–Q plots indicate that the two parameters are not normally distributed, but do align in the plot, suggesting that the distribution can be normalised. For these reasons, the two potential outlier MSs do not seem to show any particular reason to be excluded from the dataset for calculation of NBLs.

5.2.2. Nickel

The Ni boxplot evidences only one MS as a potential outlier, APE031, that shows the highest representative value. The normal Q–Q plot suggests that this element could have a normal distribution but also indicates an additional MS as potential outlier, APE047. Both MSs belong to the BPA formation, have a Mg-HCO_3 facies and intermediate mineralisation and do not show any peculiarity supporting their elimination from the dataset for calculating the Ni NBL.

5.2.3. Manganese

Four MSs showing representative concentrations greater than 10 µg/L are identified in the boxplot as potential outliers. These are, in decreasing concentration order:

- APP127 (FGD), with a Mg-$HCO_3(SO_4)$ facies. This MS is the only one where Co was detected, and has the second highest Fe concentration. These elements could be present in solution because of the relatively low redox potential. This station has been repeatedly monitored (Table S3), and the well stratigraphy has identified Mn oxide deposits;
- APF140 (SERP), an MS with low mineralisation and Mg-HCO_3 facies. This MS shows the highest Fe concentration and is the only one with a reducing Eh (Figure 2);
- APE044 (BPA), with medium mineralisation and Mg-HCO_3 facies. This MS has elevated Zn concentrations, but average Ni concentrations;
- APE059 (SERP), with medium–low mineralisation and Mg-HCO_3 facies.

The normal Q–Q plot indicates that the data strongly deviate from a normal distribution. The two MSs plotted far from the normality line are the two first stations.

For the reasons discussed above, none of these stations can be reasonably eliminated from the dataset for the calculation of NBLs. Even those stations with low redox potential, which display the higher Mn concentrations, are likely reducing for natural reasons since no other organic indicator of contamination, such as hydrocarbons, was detected in these samples.

5.2.4. Iron

The Fe plots are very similar to those of Mn. Five MSs are indicated as potential outliers, in decreasing concentration order:

- APF140 (SERP) and APP127 (FGD), showing high Fe concentrations because of their low redox potential;
- APF082 (SL), with a low mineralisation and a Ca-HCO_3 facies. The MS is located close to a rodingite vein, constituted by Ca-rich silicates hosting metallic ores [28];

- APF138 (SERP), with high mineralisation and Mg-SO$_4$ facies. As previously discussed, this MS is located at the foothills of the Fandaglia mine waste dump and its Fe content could derive from oxidation of sulphides disseminated in serpentinite;
- APF092 (SL), with very low mineralisation and Mg(Ca)-HCO$_3$ facies. This MS is very similar in hydrochemistry to APF082, and could contain Fe for the same reason.

Therefore, also in the case of Fe, there are no reasons to eliminate any MS from the dataset for the determination of NBLs, as all the possible outliers could be derived from natural water–rock interaction processes.

5.2.5. Zinc

The Zn boxplot evidences two stations as potential outliers:

- APE044 (BPA), an MS already shown to have a high Mn concentration;
- APE041 (BPA), an MS with K$^+$ concentrations higher than Na$^+$, an uncommon feature in natural waters. Being located in the Balangero village, this well could reflect some anthropogenic contamination.

While the latter station could reasonably be eliminated from the dataset, this would have few consequences since the Zn concentration is not very high and the other PTEs are not anomalous as well. The normal Q–Q plot confirms the choice not to delete this MS, as the data do not show a normal distribution but are well aligned. Based on this plot, only APE044 would be a potential outlier.

5.3. Statistical Distribution

The statistical distribution was evaluated in the dataset of the representative PTE values for each MS (Table 1). The descriptive statistics are reported in Table 2.

Table 2. Descriptive statistics of the representative PTE values (in µg/L) for each MS.

Including ND Data		Cr	Cr VI	Co	Ni	Mn	Fe	Zn
N	Valid	29	29	29	29	29	29	29
	Missing	0	0	0	0	0	0	0
Mean		4.965	3.588	0.583	16.376	61.898	104.007	16.466
Median		2.700	1.500	0.500	8.900	2.100	1.500	0.500
Mode		0.5	1.5	0.5	0.5	0.5	0.5	0.5
Std. Deviation		7.9155	7.0997	0.4457	17.7297	291.25	340.02	41.0952
Skewness		3.730	4.511	5.385	1.554	5.333	3.768	4.245
Kurtosis		16.028	21.586	29.000	2.382	28.596	14.339	20.118
Minimum		0.5	1.5	0.5	0.5	0.5	0.5	0.5
Maximum		41.0	38.1	2.9	71.3	1571.5	1583.0	213.5
Percentiles	25	0.500	1.500	0.500	2.100	0.500	0.500	0.500
	50	2.700	1.500	0.500	8.900	2.100	1.500	0.500
	75	5.100	1.500	0.500	25.250	6.300	8.225	18.425
	90	11.000	6.650	0.500	37.800	21.950	329.90	46.200
	95	29.400	26.100	1.700	63.925	848.25	1274.0	135.25

All the parameters have a non-normal distribution, as indicated by the non-corresponding mean and median values, and by the positive skewness that suggests a tail extending towards high values. The mode for all parameters corresponds to the value adopted for the data < LOD (0.5 µg/L for all PTEs except Cr VI = 1.5 µg/L). Frequency histograms (Figure S3 in the Supplementary Materials) show the same pattern, with a mode corresponding to the lowest values. This suggests the absence of a normal distribution but the presence of a single population with a distribution that can be normalised.

Although the presence of numerous MSs with ND data (Table 1, Figure S3) likely affects the distribution, if these are eliminated, the distribution still shows a strong positive skewness (Table 3); however, some elements (Cr, Cr VI, Fe, Zn) display multiple modes, while Ni and Mn remain unimodal.

Table 3. Descriptive statistics of the representative PTE values (in µg/L) for each MS, excluding ND.

Excluding ND		Cr	Cr VI	Co	Ni	Mn	Fe	Zn
N	Valid	21	6	1	26	17	16	11
	Missing	8	23	28	3	12	13	18
Mean		6666	11.592	2.900	18.208	105.238	188.106	42.591
Median		4500	5.225	2.900	13.225	5.300	7.525	27.000
Mode		3.9 [a]	3.3 [a]	2.9	2.5	6.8	1.1 [a]	0.8 [a]
Std. Deviation		8.7581	13.6142		17.8417	378.98	446.08	59.3218
Skewness		3.373	2.017		1.480	4.083	2.672	2.810
Kurtosis		12.675	4.079		2.138	16.755	6.774	8.570
Minimum		1.0	3.3	2.9	0.8	1.0	1.1	0.8
Maximum		41.0	38.1	2.9	71.3	1571.5	1583.0	213.5
Percentiles	25	2.150	3.525	2.900	4.788	2.350	3.362	11.400
	50	4.500	5.225	2.900	13.225	5.300	7.525	27.000
	75	6.725	20.100	2.900	28.013	12.875	36.675	46.200
	90	16.440	38.100	2.900	43.425	414.30	1151.1	182.20
	95	38.680	38.100	2.900	66.137	1571.5	1583.0	213.50

a. Multiple modes exist. The smallest value is shown.

To test if the MS representative values have normal or lognormal distribution, the Shapiro–Wilk test was applied to the data and to their logarithm [45]. The test rejects the normal distribution hypothesis in all cases, except for Ni that shows an almost significant lognormal distribution. However, if we exclude the MSs with ND values, the Shapiro–Wilk test on log data accepts the hypothesis for Cr, Cr VI, Ni and Zn, indicating that their distribution is lognormal and therefore can be normalised.

The non-normal distribution of Fe and Mn, even eliminating the MSs with ND values, is likely related to the presence of two MSs with very high contents (APP127 and APF140), because of their low redox potential. If these are eliminated from the dataset (Table 1), Cr and Ni show a lognormal distribution; if, in addition, the ND values are omitted, all the PTEs display a lognormal distribution. These results are summarised in Table S5 (Supplementary Materials).

In conclusion, results indicate that for Cr, Cr VI, Ni and Zn, the MSs belong to one single population that can be treated to obtain the NBLs. For Fe and Mn, the exclusion of the MSs APP127 and APF140 because of their low redox potential should be further evaluated, based on all the available measurements.

5.4. Assessment of the Dataset

Since the different MSs constitute a statistically representative homogenous population, belonging to the same hydrogeochemical context, all the measurements available from previous campaigns can be used to evaluate the NBLs of total Cr, Cr VI, Co, Ni, Mn, Fe and Zn (Table S3 in the Supplementary Materials), so as to deal with sufficiently large amounts of data. Unfortunately, not all the MSs have been analysed more than once and for all the PTEs, leading to variable numbers of measurements, from 74 for Mn to 39 for Fe and Zn (Table 4). However, both the number of MSs sampled at least once (29) and the total number of PTE measurements (39–74) exceed the minimum number required for a significant spatial analysis. On the other hand, as already detailed, the measurements do not show an adequate temporal dimension (i.e., there are no MSs with at least 8 measurements with a regular frequency over at least two years, see Section 5.1). Therefore, the temporal dimension of the dataset is not significant. Accordingly, this dataset is classified as type B [22].

Table 4. Descriptive statistics of the total number of PTE measurements (in µg/L) (Table S3), including ND.

Including ND		Cr	Cr VI	Co	Ni	Mn	Fe	Zn
N	Valid	71	67	71	71	74	39	39
	Missing	3	7	3	3	0	35	35
Mean		4.965	6.979	5.473	0.654	21.332	183.1	87.249
Median		2.700	3.1	1.5	0.5	9.1	1.1	1.7
Mode		0.5	0.5	1.5	0.5	0.5	0.5	0.5
Std. Deviation		7.9155	9.51	7.909	0.915	25.790	844.760	305.937
Skewness		3.730	2.139	2.591	5.959	1.519	5.401	4.042
Kurtosis		16.028	4.584	7.392	35.02	1.764	30.038	16.787
Minimum		0.5	0.5	1.5	0.5	0.5	0.5	0.5
Maximum		41.0	41.4	38.1	6.6	109	5554	1583
Percentiles	25	0.5	1.5	0.5	1.8	0.5	0.5	0.5
	50	3.1	1.5	0.5	9.1	1.1	1.7	0.5
	75	7.4	6.5	0.5	34	6.05	8	11.4
	90	19.6	15.0	0.5	64.06	46.74	58.40	57.00
	95	31.1	21.4	0.5	78.78	1568.75	966	120

The descriptive statistics of the dataset are reported in Table 4.

Concerning Co, the statistics are not relevant as this element shows only two measurements > LOD, both from the same MS. All the parameters have a non-normal distribution, as indicated by the non-corresponding mean and median values. The positive skewness suggests a tail extending towards high values. As for the case of the MS dataset, the mode for all parameters corresponds to the value adopted for the results < LOD (0.5 µg/L for all PTEs except Cr VI = 1.5 µg/L).

The descriptive statistics of the dataset excluding the ND still show a positive skewness but also the presence of multiple modes (Table 5).

Table 5. Descriptive statistics of the total number of PTE measurements (in µg/L) (Table S3), excluding ND.

Excluding ND		Cr	Cr VI	Co	Ni	Mn	Fe	Zn
N	Valid	53	21	2	63	38	21	13
	Missing	21	53	72	11	36	53	61
Mean		6666	9.179	14.176	5.95	23.978	356.089	161.605
Median		4500	5.4	12.8	5.95	12.1	5.55	6.4
Mode		3.9 [a]	1.6	12.8 [a]	5.3 [a]	1.2[a]	1.8	1.6 [a]
Std. Deviation		8.7581	10.1134	9.5163	0.9192	26.2264	1159.6447	406.5181
Skewness		3.373	1.824	1.526		1.39	3.759	2.829
Kurtosis		12.675	3.085	2.253		1.355	14.046	7.774
Minimum		1.0	1	3.8	5.3	1.1	1	1.6
Maximum		41.0	41.4	38.1	6.6	109	5554	1583
Percentiles	25	1.85	6.65	5.3	3.4	2.475	2.65	7.5
	50	5.4	12.8	5.95	12.1	5.55	6.4	33.7
	75	14.75	18.1	6.6	34.3	21.1	29.85	66.65
	90	21.68	34.88	6.6	68.86	1567.1	900.6	232.2
	95	39.2	38.1	6.6	82.26	4392.15	1521.3	307

[a]. Multiple modes exist. The smallest value is shown.

Frequency histograms (Figure S4 in the Supplementary Materials) show the same pattern, with a mode corresponding to the lowest values and a marked positive skewness. The presence of a second mode is particularly evident for total Cr. The measurements >10 µg/L belong to the MSs APE031 (BPA), APP126 (FGD), APF082 (SL) and APF005 (SERP), all attributed to the same hydrogeochemical context.

Despite the presence of multiple modes, the D'Agostino (for a number of available measurements ≥50) or Shapiro–Wilk (for a lower number of measurements) statistical tests were applied to the data or to their logarithm to check for the presence of a normal distribution [45,46]. The results are shown in Table S6 (Supplementary Materials). The tests on all the PTE measurements indicate that they do not have a normal distribution, but Cr and Ni have a lognormal distribution. If ND values are excluded, all PTEs have a lognormal distribution except Fe and Mn. The non-normal distribution of Fe and Mn, even eliminating the ND values, is likely related to high concentrations recorded in two MSs (APP127 and APF140), because of their low redox potential. If the measurements from these two MSs are eliminated from the dataset, Cr and Ni show a lognormal distribution; if, in addition, the ND values are omitted, all the PTEs display a lognormal distribution, with the exception of Fe. This is due to the presence of one measurement (Fe = 639 μg/L, dated 17/04/2012, Table S3) from the MS APF082 (SL). This MS has a Ca-HCO$_3$ facies and is located close to a rodingite vein, enriched in metal ores. The MS was indicated as a potential outlier (see Section 5.2.4), but was not excluded because, using the median value to represent the MS, Fe displayed a lognormal distribution (Table S5 in the Supplementary Materials). Therefore, as a first approximation, we will keep this measurement in the dataset for further elaboration.

5.5. Evaluation of NBLs

As already indicated, the dataset has a significant spatial dimension, but its temporal dimension is not adequate, allowing us to classify it as B type [22]. This does not change the procedure to calculate the NBLs with respect to a dataset of the A type (significant for both spatial and temporal dimension), but only the confidence level attributed to the evaluation.

According to the guidelines [22], the evaluation should be based on the representative PTE values for each MS, corresponding to the median value for those MSs sampled more than once, and to the single measurement for the MSs measured only once (Table 1). Indeed, it was previously demonstrated that, for each PTE except for Fe and Mn, the data distribution can be normalised if ND values are excluded. However, ND values are numerous, especially for Cr VI, Co and Zn. The percentiles (90p and 95p) that could be used as NBLs are reported in Table 6, compared to the maximum concentrations recorded. For some PTEs with a low number of MSs showing detectable concentrations (e.g., Cr VI or Co), these percentiles correspond to the maximum measured concentration. For this reason, the guidelines suggest to use a lower percentile for MSs with a low number of detected values (N). In addition, if the two MSs with a low redox potential, APP127 (FGD) and APF140 (SERP), are eliminated from the dataset, Fe and Mn can be normalised too, but this also reduces N. The comparison is shown in Table 6. No differences are observed with respect to the previously calculated percentiles except for those of Fe and Mn that are noticeably lower. However, since APP127 (FGD) is the only station with detectable Co concentrations, the percentiles for this element cannot be determined.

Table 6. PTE percentiles calculated using the representative PTE values (in μg/L) for each MS (Table 1). In bold, the values exceeding the TVs for water quality [15].

	Data from Table 1 Without ND				Excluding the MSs APP127 and APF140			
	N	Max	95p	90p	N	Max	95p	90p
Cr	21	41	38.7	16.44	21	41	38.7	16.44
Cr VI	6	**38.1**	**38.1**	**38.1**	6	**38.1**	**38.1**	**38.1**
Co	1	2.9	2.9	2.9	0	NA	NA	NA
Ni	26	**71.3**	**66.1**	**43.42**	25	**71.3**	**66.9**	**45.3**
Mn	17	**1571.5**	**1571.5**	**414.3**	15	22.0	21.95	19.4
Fe	16	**1583**	**1583**	**1151.1**	14	**329.9**	**329.9**	184.4
Zn	11	213.5	213.5	182.2	11	213.5	213.5	182.2

Moreover, it must be noted that, when using the median value as representative for the MSs, half of the values are eliminated from the calculation of the percentiles, and particularly the higher ones. While this is necessary and justified when the aim is to eliminate from the dataset those MSs potentially affected by anthropogenic contamination, it is less acceptable when the aim is to define the highest concentrations that can be reached in a natural context, unaffected by anthropogenic contamination, such as that of the study area. For comparison, Table 7 reports the percentiles calculated based on all the available measurements (Table S3).

Table 7. PTE percentiles calculated using the total number of PTE measurements (in µg/L) (Table S3). In bold, the values exceeding the TVs for water quality [15].

	Data from Table S3 Without ND				Excluding Data from APP127 and APF140			
	N	Max	95p	90p	N	Max	95p	90p
Cr	52	41.4	39.2	21.68	52	41.4	39.3	21.86
Cr VI	21	**38.1**	**38.1**	**34.88**	21	**38.1**	**38.1**	**34.88**
Co	2	6.6	6.6	6.6	0	NA	NA	NA
Ni	63	**109**	**82.26**	**68.86**	59	**109**	**84.0**	**71.3**
Mn	38	**5554**	**4392**	**1567**	32	**92.1**	**71.36**	27.22
Fe	21	**1583**	**1521**	**900.6**	19	**639**	**630**	58.4
Zn	13	307	307	232.2	13	307	307	232.2

Comparing the results that exclude the measurements from the two MSs with low redox potential, the percentiles calculated for Cr and Cr VI are little affected, whereas changes are observed in the percentiles for Ni, Zn and especially for Fe and Mn. In addition, in this case, the 95p values for Cr VI, Ni, Mn and Fe largely exceed the TVs for water quality [15]. The rationale to select one or another value as an NBL will therefore be evaluated based on the literature from similar hydrogeological settings.

6. Discussion

The study area is characterised by the presence of ferromagnesian rocks (ultramafic serpentinite, stratified rodingite or in veins, metabasites of the SL complex), and by sediments derived from their erosion. In addition, the presence of metal ores has been assessed (oxides, sulphides and Fe-Ni-Co alloys together with other metals). In this geochemical context, the considered PTEs are mostly present in their reduced form (oxidation state +2 for Fe, Mn, Co, Ni and Zn, +3 for Cr). The weathering of primary minerals can mobilise and solubilise these elements, at different concentrations and ratios, depending on the environmental geochemical conditions. At circumneutral pH, Fe and Mn are only soluble in reducing environments, whereas in oxidising environments, they precipitate, forming oxy-hydroxides. The high specific surface of these minerals may favour the co-precipitation of other elements, such as Co, Cr III and Zn, whose concentration in oxic waters is generally low. By contrast, Ni and Cr VI are more mobile. In particular, the latter can form oxyanions that are poorly retained by the aquifer matrix [47].

In the following discussion, the 95p values evaluated for the study area are compared with the concentrations reported in other regional, national and international studies, focusing on groundwater circulating in ultramafic rocks, generally as fractured reservoirs, or in sedimentary aquifers derived from their dismantling.

6.1. Comparison with Similar Hydrogeochemical Settings

In the Piedmont region, ARPA has defined the PTE NBLs in groundwater from the shallow and the deep porous aquifers, mostly located in the Po plain area at the mouth of the Lanzo valleys, with a procedure similar to that adopted in this study [48]. Both aquifers host groundwater of Mg-HCO$_3$ facies, due to the presence of ferromagnesian minerals in the matrix, potentially leading to high concentrations of PTEs, especially Ni and Cr. Indeed, in the shallow aquifer (GWB-S3a, sub-area GWB-S3a-A), Ni generally exceeds the TV for groundwater quality (20 µg/L). According to ARPA, the estimation of the Ni NBL in this area is not possible, as it is not sufficiently homogenous. However, the available data treatment suggests that it could reach up to 100 µg/L [48], a value that is higher than the 95p evaluated at our site based on all the available measurements (84 µg/L), and similar to the maximum concentration recorded (109 µg/L). The presence of Cr and Cr VI has been assessed too, but the estimation of their NBLs is complicated by the presence of a clear anthropogenic input. In the deep aquifer (GWB-P2), the two MSs located closer to the study area record Cr concentrations reaching up to 50 µg/L, and traces of Cr VI (above the LOD of 5 µg/L). In addition, ARPA evaluates the presence of Mn in groundwater. At the regional scale and in shallow aquifers, this metal shows high and variable mean contents that can reach up to 3000 µg/L. As its behaviour in solution is determined by its oxidation state and by the mobilisation processes, it is not possible to calculate an NBL due to the high spatial and temporal variability of the monitoring data. However, at the mouth of the Lanzo valleys, one station reaches concentrations higher than 200 µg/L in historical data (2005–2009) [48]. Even in this case, the maximum concentration from all the available measurements (92.1 µg/L) could therefore be justified.

In Italy, ophiolite outcrops are common in the Alpine–Apennine chain, as these are the remnants of the Ligurian–Piedmontese basin of Jurassic age [24]. In these areas, studies on the abundance of Cr and Ni in groundwater suggest that serpentinite is a source of PTE contamination of non-anthropogenic origin [49]. A recent study [50] reports and re-interprets groundwater hydrochemistry and PTE concentrations of 598 samples from five areas where variably serpentinised ultramafic rocks outcrop. As this database is available, we extracted the results of the groundwater analyses with nitrate concentrations < 37.5 mg/L and performed a descriptive statistical treatment (Table 8).

Table 8. Descriptive statistics of the PTE concentrations in Italian groundwater (in µg/L) [50] unaffected by anthropogenic contamination (i.e., with nitrate concentrations < 37.5 mg/L).

		Cr	Cr VI	Fe	Mn	Ni
N	Valid	541	96	494	474	445
	Missing	18	463	65	85	114
Mean		7.58	15.26	25.78	2.93	13.73
Median		4.00	12.00	8.00	0.00	5.40
Mode		2.00	3.00	0.00	0.00	0.00
Std. Deviation		9.71	14.64	72.36	27.80	23.01
Skewness		2921.23	1718.94	8321.03	19147.22	4701.43
Kurtosis		11297.90	4002.48	91757.03	392675.03	35534.60
Minimum		0.02	0.05	0.00	0.00	0.00
Maximum		73.00	73.00	948.00	579.00	259.60
	25	2.00	3.00	0.000	0.00	1.15
	50	4.00	12.00	8.00	0.00	5.40
Percentiles	75	9.00	21.00	19.00	0.77	18.00
	95	28.90	42.75	127.00	5.98	52.77
	99	46.17	73.00	304.05	51.10	121.16

The 95p values calculated for our study site are between the 95p and the 99p values of the Italian dataset, for Cr and Ni, and are lower than the 95p for Cr VI. However, our Fe and Mn concentrations are much higher than the 99p values of the above dataset, but lower than the maximum values reported there [50]. In the paper, the authors discuss the potential role played by Fe^{3+} included in serpentine as an oxidising agent for Cr^{3+}, following the reaction:

$$Cr^{3+} + 3\,Fe^{3+} + 4\,H_2O \rightarrow CrO_4{}^{2-} + 3\,Fe^{2+} + 8\,H^+ \tag{1}$$

Based on geochemical modelling, this reaction would more accurately describe the formation of the chromate ion than the generally acknowledged reaction using Mn oxide [9,51]:

$$Cr^{3+} + 1.5\,MnO_2(s) + H_2O \rightarrow HCrO_4{}^- + 1.5\,Mn^{2+} + H^+ \tag{2}$$

The discussion about the processes involved in the formation of Cr VI is out of the scope of this paper. However, it should be noted that, during the characterisation of the

Balangero remediation site, soils were found to contain abundant Mn oxides, especially close to the station APP127 (FGD).

Studies on groundwater contamination by Cr VI of geogenic origin are very numerous worldwide [50], ranging from simple characterisation, to statistical data treatment, analysis of correlations with other parameters, modelling of the solubilisation process and evaluation of the potential human health impact. Although the complete review of these studies is not within the scope of this paper, some of them show strong similarities with our study case and are therefore summarised below.

A thorough statistical study on the presence of Cr and Cr VI in drinking waters from California is reported by [52]. This is attributed to the presence of ultramafic and serpentinite outcrops, and to soils and sediments derived from their erosion. The more relevant results of their study are:

1. the excellent linear correlation between Cr and Cr VI, with a slope close to one. Results of over 918 datapoints indicate that $90 \pm 1\%$ of Cr is in the Cr VI form, reaching concentrations very similar to those of our study site (Figure 4);
2. the more frequent presence and the higher Cr VI concentrations found in alkaline waters (pH > 8) with respect to neutral or acidic waters. This is in agreement with the experimental results showing that Cr VI can be desorbed from mineral surfaces at a pH higher than 7.5;
3. the more frequent presence and the higher Cr VI concentrations found in oxic waters ($O_2 > 0.5$ mg/L) with respect to reducing waters;
4. at comparable oxic conditions, the more frequent presence and the higher Cr VI concentrations found in deeper and older waters;
5. the negative correlation of Cr VI with other PTEs, such as Fe^{2+}, Mn^{2+} and Ni^{2+}, which are soluble at low redox potential but less soluble in oxic conditions. The negative correlation between Cr VI and Ni is also found at our study site (N = 67; r = -0.33; $p < 0.01$).

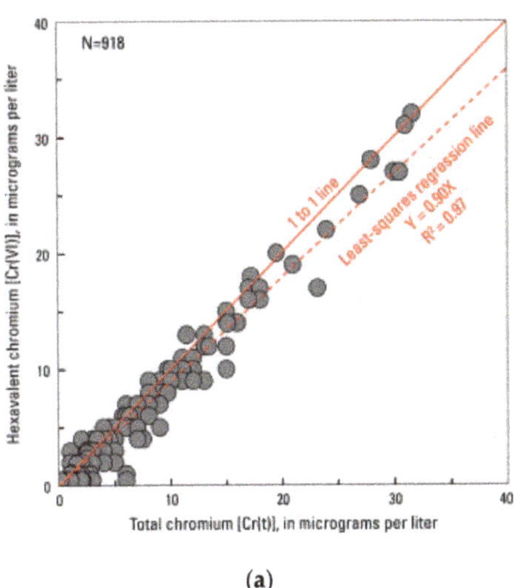

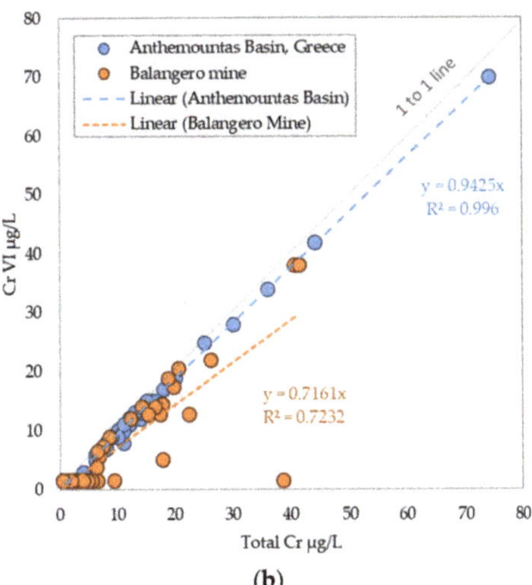

Figure 4. Relationship between Cr and Cr VI: (**a**) in drinking waters from California [52]; (**b**) in waters from the Anthemountas Basin, Greece [9] and at our study site (regression calculated using the total number of PTE measurements, including ND).

The significant correlation between Cr and Cr VI is also reported for groundwater from the Anthemountas Basin in Greece [9], reaching even higher concentrations with respect to those observed at our study site (Figure 4).

In Turkey, PTEs were analysed in an area characterised by the presence of ophiolites and of an ophiolitic mélange, sampled during the wet and the dry season [19]. These waters display E.C., dissolved oxygen and Mg-HCO_3 hydrochemical facies comparable to our study site, despite an even higher pH. The maximum concentrations recorded of Cr (11.3 µg/L), Ni (36.1 µg/L) and Zn (98.6 µg/L) are lower, whereas those of Fe (1018 µg/L) and Mn (78.6 µg/L), which are higher in the dry season, are comparable to the 95p values calculated at our study site.

High PTE concentrations are common in mining environments, especially when solutions are acidic. The processes governing the chemistry of these solutions are well known and identified under the broad name of acid mine drainage [21]. More recently, studies have addressed solutions enriched in sulphates and PTEs but at circumneutral pH (neutral mine drainage) [53,54]. Generally, the acidity related to the dissolution of sulphides in oxic environments is buffered by the host rock minerals, particularly carbonates but also silicates. However, the chemistry of the solution reflects complex processes that include the precipitation of Fe oxy-hydroxides, the co-precipitation or adsorption of other trace metals, the precipitation of Fe sulphate or gypsum and CO_2 degassing [55]. An example of such studies is reported for the Hitura mine (Finland) [20], exploiting an Fe, Ni and Cu sulphide deposit (mainly of pentlandite, chalcopyrite and pyrrhotite) hosted in serpentinite. The paper compares some parameters measured in mine waters, groundwater unaffected by mining activities and water circulating in the mine waste dumps. Uncontaminated groundwater is of Ca-HCO_3 facies, turning to Mg-SO_4 facies with increasing contamination. Low sulphates and circumneutral pH characterise unaffected waters that nevertheless show elevated Fe (mean 970 µg/L) and Mn (mean 80 µg/L) concentrations, attributed to the natural background of the area. These Fe and Mn values are comparable with the 95p values calculated for our study area.

6.2. Proposed NBLs

In Section 5.5, the percentiles that can be used as NBLs have been evaluated alternately by closely following the indications of guidelines [22] (i.e., using the MS representative values, Table 6), or by using all the available measurements (Table 7), having checked that the distribution for each element (except for Fe) can be normalised. We have also evidenced two MSs that are characterised by very high Fe and Mn concentrations due to their low redox potential. At the end of this discussion, we consider that these two stations should not be included in the evaluation of the NBLs for the following reasons:

- APP127 (FGD) was monitored from 2015 for its elevated Mn contents (Table S3). It is located in an area where, for natural reasons, abundant Mn oxides are present. FGD is characterised by a low permeability that likely prevents the diffusion of atmospheric oxygen from the surface and permits the establishment of low redox conditions, favouring the solubilisation of Fe and Mn from the matrix;
- APF140 (SERP) was only sampled once and, according to the field notes, could not be adequately purged. The elevated Fe and Mn concentrations could therefore be related to the presence of stagnant water.

Regarding the percentiles to be used as NBLs, the choice also depends on the number of MSs or measurements. If ND values are considered, the 95p can be used to estimate the NBLs, whereas if ND values are eliminated, the amount of available data should be taken into account. The guidelines suggest to select a probability level lower that 1-1/N (i.e., adopt the 90p when more than 10 datapoints are available; adopt the 95p when more than 20 datapoints are available; adopt the 99p when more than 100 datapoints are available). Therefore, if considering the MS representative values, only Cr and Ni would have enough data to allow for the use of the 95p, whereas for the other PTEs, the 90p (or even a lower percentile for Cr VI) should be used. By contrast, if using all the available measurements, the 95p could be used for all the PTEs, excluding Fe and Mn where the 90p should be used.

Considering Cr, if the two stations with low redox potential are excluded, the 95p values do not change much between the two calculation methods (from 38.7 to 39.3 µg/L), and do not exceed the TV for groundwater quality (50 µg/L) [15]. The 95p for Cr VI is independent from the dataset used (38.1 µg/L) and largely exceeds the regulatory limits (5 µg/L). Total Cr and Cr VI contents are significantly correlated both if considering the MSs (N = 29; R^2 = 0.92; $p < 0.01$) or the full dataset (N = 67; R^2 = 0.72; $p < 0.01$), with the slope of the regression line close to one, indicating that all the Cr in solution is hexavalent Cr (Figure 4). This is in agreement with the literature data and supports the natural origin of this element.

Co, if the MS APP127 (FGD) is eliminated from the dataset, is always < LOD and no NBL can be calculated for this element. However, based on the available information on this MS, a natural origin of this element can be inferred. Indeed, Mn oxides strongly adsorb Co and are abundant in soils near this station.

Considering Ni, the concentrations measured in the study area are comparable with those reported in the literature. Its natural origin is suggested by the negative correlation with Cr VI, and by the assessed presence of Ni in groundwater from the Po plain aquifers, at the mouth of the Lanzo valleys. The calculated 95p values vary from 66.9 µg/L (using the MS representative values) to 84 µg/L (using all the measurements), and always exceed the TV of 20 µg/L. However, both 95p values are lower than the NBL suggested for the sector of the Po plain aquifer at the mouth of the Lanzo valleys (> 100 µg/L) [48].

The 95p values substantially differ (Tables 6 and 7) for Mn (from 22 to 71.3 µg/L), Fe (from about 330 to 630 µg/L) and Zn (from 213.5 to 307 µg/L) according to the dataset used for the evaluation. In addition, while Zn 95p values are always lower than the TV (3000 µg/L), Fe values are always higher (200 µg/L) and Mn values change from lower to higher (50 µg/L). Literature data on ultramafic rocks characterised by the presence of metallic ores, especially sulphides, indicate that groundwater concentrations in this range are commonly observed, deriving from natural water–rock interactions (i.e., neutral mine drainage). The weathering process is further enhanced in mine waste dumps, where the fractured and crushed rock has a higher surface area in contact with percolating meteoric waters.

In the investigated area, the use of the median concentration as the representative value for the MSs sampled more than once [22] involves discarding half of the data, and in particular the higher concentrations. This approach can be justified when wanting to exclude from the dataset those concentrations that could be partly affected by anthropogenic contamination. However, it is less suitable when wanting to evaluate the highest concentrations that can be naturally reached in a peculiar hydrogeochemical context such as the one investigated. A too conservative assessment of the NBLs would result in frequent overrunning of these values, initially causing unjustified alarm, and possibly leading to an underestimation of potential anthropogenic contamination processes or of the decline over time of groundwater quality. An alternative approach, once the normalisable distribution of the measurements at each station is verified, would be to use as representative the 95p instead of the median values. However, this approach cannot be adopted at our study site, as the MSs were sampled irregularly and the available measurements are not numerous enough.

Therefore, to estimate NBLs, we propose to use all the available measurements, as this dataset is more representative of the high concentrations that can naturally occur at the study site. More specifically, we propose to adopt:

- the 95p values for Cr, Ni and Mn, where Ni and Mi will exceed the TVs for groundwater quality [22];
- the 90p values for Fe and Zn, which will not exceed the TVs.

Finally, concerning Cr VI, the 95p can be used. However, it should be noted that the detectable concentrations are found in only six MSs, mostly located at the foothills of the mountain. Therefore, this NBL cannot be adopted for the full investigated area, that also includes the populated BPA.

These NBLs are summarised in Table 9. As the case study is classified as B (adequate spatial dimension, inadequate temporal dimension), with a number of stations of > 25 and a surface extension of > 700 m^2, these NBL estimates are considered to have a medium confidence level (except for Cr VI) [33].

Table 9. Proposed natural background levels (NBLs) for PTEs in the study area. These correspond to the 95p values calculated for all the available measurements, excluding the two MSs with low redox potential (for Cr, Cr VI, Ni and Mn), and to the 90p values for Fe and Zn. In bold, the NBLs exceeding the TVs [15]. For Cr VI, the NBL is not representative of the BPA.

	Percentile	Proposed NBL in µg/L
Cr	95p	39.3
Cr VI	95p	**38.1 ***
Co	-	ND
Ni	95p	**84**
Mn	95p	**71.36**
Fe	90p	58.4
Zn	90p	232.2

* For Cr VI, the NBL is not representative of the BPA.

7. Conclusions

In this study, we illustrated the procedure followed to derive PTE concentration values for Cr (total and hexavalent), Ni, Mn, Fe and Zn that can reasonably represent the NBLs for the former Balangero asbestos mine, a "Contaminated Site of National Interest". This area is characterised by the presence of ferromagnesian rocks, mainly serpentinite, and of sediments derived from their dismantling. In addition, metal ores (Fe, Ni and Co oxides, sulphides and alloys, also associated with other PTEs) are present, both disseminated and in veins. Therefore, the high groundwater PTE contents derive from both the weathering of ultramafic rocks, accounting especially for Cr, Cr VI and Ni, and the weathering of the metal ores, likely contributing Fe, Mn and Zn. This process is known as neutral mine drainage and is testified by increased sulphate contents and electrical conductivities, but with a circumneutral pH of the solutions.

The methodology used followed the Italian guidelines for NBL assessment, but some adjustments were required to take into account the complex (hydro)geology of the area and to better exploit all the available dataset. In particular, the guidelines recommend using the median as a representative concentration for each monitoring station. However, this involves discarding half of the measurements, and in particular the higher concentrations, thus resulting in too conservative estimates that are unable to represent this peculiar natural setting. Using, instead, all the available measurements and the recommended statistical evaluation, the derived NBLs were: Cr = 39.3, Cr VI = 38.1, Ni = 84, Mn = 71.36, Fe = 58.4, Zn = 232.2 µg/L. These values were compared with the concentrations reported in the literature for groundwater circulating in similar geological settings to support the conclusion on their natural origin.

The proposed NBLs were discussed with the authorities and were accepted as a preliminary assessment. To increase the confidence level of the estimates, in particular the temporal consistency of the dataset, further monitoring campaigns are foreseen in the coming years, targeting those MSs that exceed the TVs for Cr, Cr VI, Ni, Fe and Mn. The availability of a larger number of measurements will allow a better evaluation of the possible presence of trends in time and for finding more statistically sound NBL values. These will be again compared with data from the literature, and possibly be used for geochemical modelling of the water–rock interaction process.

Despite the positive outcome of this investigation, the results highlight the need to rethink the method indicated in the guidelines to define the MS representative concentrations. For example, different percentiles than the median could be recommended based on both the distribution and relative standard deviation of the measurements. Further discussion is needed in the scientific community about this topic, based on case studies such as the one presented, for the definition of more representative NBLs in naturally enriched environmental settings.

Supplementary Materials: The following are available online at https://www.mdpi.com/2073-4441/13/5/735/s1, Figure S1: Stiff diagrams. Colours of the polygons refer to the hydrogeological formation: red = SL; green = SERP; yellow = FGD; blue = BPA, Figure S2: Boxplots and normal Q–Q plots of the MS representative values: (a) total Cr; (b) Cr VI; (c) Ni; (d) Mn; (e) Fe; (f) Zn, Figure S3: Frequency histograms of the MS representative values (in μg/L), Figure S4: Frequency histograms of the total available measurements, including ND. Concentrations are in μg/L, Table S1: Physico-chemical parameters measured in the field, analytical results for major ions and charge balance error (%). GW = groundwater; SW = spring water; SL = Sesia–Lanzo; SERP = serpentinites; FGD = fluvioglacial deposits; BPA = Balangero Plain aquifer, Table S2: Minor and trace elements. Values in bold exceed the regulatory guidelines. GW = groundwater; SW = spring water; SL = Sesia–Lanzo; SERP = serpentinites; FGD = fluvioglacial deposits; BPA = Balangero Plain aquifer, Table S3: Total number of measurements of PTE concentrations available for the evaluation of the NBLs. Values in bold exceed the regulatory limits for groundwater quality, Table S4: Results of the trend analysis conducted for the MSs APP126 (a) and APE031 (b), Table S5: Results of the Shapiro–Wilk tests to evaluate the normal or lognormal distribution of the representative PTE values for each MS, Table S6: Results of the statistical tests to evaluate the presence of normal or lognormal distribution of the total number of PTE measurements.

Author Contributions: Conceptualisation, E.S. and M.B.; Data curation, E.L. and E.P.; Formal analysis, A.M. and J.-R.M.; Funding acquisition, M.B.; Investigation, E.S. and E.P.; Methodology, E.S., A.M. and J.-R.M.; Supervision, M.B.; Visualization, E.L. and E.P.; Writing—original draft, E.S.; Writing—review and editing, M.B., A.M. and J.-R.M. All authors have read and agreed to the published version of the manuscript.

Funding: This research was supported by R.S.A. srl with funds from the Italian Ministry of the Environment, Land and Sea.

Institutional Review Board Statement: Not applicable.

Informed Consent Statement: Not applicable

Data Availability Statement: The data presented in this study are available in the article and in the supplementary material.

Acknowledgments: Authors wish to acknowledge the constructive comments and the collaborative attitude of the ARPA, ISPRA and Piedmont region officers during the characterisation procedure, which contributed to achieving the goals of this research.

Conflicts of Interest: The authors declare no conflict of interest.

References

1. Pourret, O.; Hursthouse, A. It's time to replace the term "heavy metals" with "potentially toxic elements" when reporting environmental research. *Int. J. Environ. Res. Public Health* **2019**, *16*, 4446. [CrossRef] [PubMed]
2. Bradl, H.; Kim, C.; Kramar, U.; StÜben, D. *Heavy Metals in the Environment: Origin, Interaction and Remediation*; Elsevier: Amsterdam, The Netherlands, 2005; ISBN 9780120883813.
3. Meybeck, M. Heavy metal contamination in rivers across the globe: An indicator of complex interactions between societies and catchments. *IAHS-AISH Proc. Rep.* **2013**, *361*, 3–16.
4. Zheng, S.; Zheng, X.; Chen, C. Leaching Behavior of Heavy Metals and Transformation of Their Speciation in Polluted Soil Receiving Simulated Acid Rain. *PLoS ONE* **2012**, *7*, 1–7. [CrossRef] [PubMed]
5. World Health Organisation. Volume 1 Recommendations. In *Guidelines for Drinking Water Quality*, 2nd ed.; WHO: Geneva, Switzerland, 1993; ISBN 924154460.
6. BRIDGE Background cRiteria for the IDentification of Groundwater Thresholds. Available online: http://nfp-at.eionet.europa.eu/irc/eionet-circle/bridge/info/data/en/index.htm (accessed on 1 March 2021).
7. Vivona, R.; Preziosi, E.; Madé, B.; Giuliano, G. Occurrence of minor toxic elements in volcanic-sedimentary aquifers: A case study in central Italy. *Hydrogeol. J.* **2007**, *15*, 1183–1196. [CrossRef]
8. Viaroli, S.; Cuoco, E.; Mazza, R.; Tedesco, D. Dynamics of natural contamination by aluminium and iron rich colloids in the volcanic aquifers of Central Italy. *Environ. Sci. Pollut. Res.* **2016**, *23*, 19958–19977. [CrossRef]
9. Kazakis, N.; Kantiranis, N.; Voudouris, K.S.; Mitrakas, M.; Kaprara, E.; Pavlou, A. Geogenic Cr oxidation on the surface of mafic minerals and the hydrogeological conditions influencing hexavalent chromium concentrations in groundwater. *Sci. Total Environ.* **2015**, *514*, 224–238. [CrossRef]

10. Rotiroti, M.; Bonomi, T.; Sacchi, E.; McArthur, J.M.; Jakobsen, R.; Sciarra, A.; Etiope, G.; Zanotti, C.; Nava, V.; Fumagalli, L.; et al. Overlapping redox zones control arsenic pollution in Pleistocene multi-layer aquifers, the Po Plain (Italy). *Sci. Total Environ.* **2021**, *758*, 143646. [CrossRef] [PubMed]

11. Megremi, I.; Vasilatos, C.; Vassilakis, E.; Economou-Eliopoulos, M. Spatial diversity of Cr distribution in soil and groundwater sites in relation with land use management in a Mediterranean region: The case of C. Evia and Assopos-Thiva Basins, Greece. *Sci. Total Environ.* **2019**, *651*, 656–667. [CrossRef] [PubMed]

12. Barceloux, D.G.; Barceloux, D. Chromium. *J. Toxicol. Clin. Toxicol.* **1999**, *37*, 173–194. [CrossRef]

13. GWD Directive 2006/118/EC of the European Parliament and of the Council of 12 December 2006 on the protection of groundwater against pollution and deterioration. *Off. J. Eur. Union* **2006**, *L372*, 19–31.

14. WFD Directive 2000/60/EC of the European Parliament and of the Council of 23 October 2000 establishing a framework for Community action in the field of water policy. *Off. J. Eur. Union* **2000**, *L327*, 1–73.

15. D. Lgs n. 152/2006. Norme in materia ambientale. *Gazz. Uff. Ser. Gen.* **2006**, *88*, 592. Available online: https://www.minambiente. it/sites/default/files/dlgs_03_04_2006_152.pdf (accessed on 1 March 2021).

16. Rotiroti, M.; Di Mauro, B.; Fumagalli, L.; Bonomi, T. COMPSEC, a new tool to derive natural background levels by the component separation approach: Application in two different hydrogeological contexts in northern Italy. *J. Geochemical Explor.* **2015**, *158*, 44–54. [CrossRef]

17. Preziosi, E.; Giuliano, G.; Vivona, R. Natural background levels and threshold values derivation for naturally As, V and F rich groundwater bodies: A methodological case study in Central Italy. *Environ. Earth Sci.* **2010**, *61*, 885–897. [CrossRef]

18. De Caro, M.; Crosta, G.B.; Frattini, P. Hydrogeochemical characterization and Natural Background Levels in urbanized areas: Milan Metropolitan area (Northern Italy). *J. Hydrol.* **2017**, *547*, 455–473. [CrossRef]

19. Güler, C.; Thyne, G.D.; Tala, H.; Yjldjrjm, Ü. Processes governing alkaline groundwater chemistry within a fractured rock (Ophiolitic mélange) aquifer underlying a seasonally inhabited headwater area in the aladallar range (Adana, Turkey). *Geofluids* **2017**, *2017*, 13–15. [CrossRef]

20. Heikkinen, P.M.; Korkka-Niemi, K.; Lahti, M.; Salonen, V.P. Groundwater and surface water contamination in the area of the Hitura nickel mine, Western Finland. *Environ. Geol.* **2002**, *42*, 313–329.

21. Blowes, D.W.; Ptacek, C.J.; Jambor, J.L.; Weisener, C.G. The Geochemistry of Acid Mine Drainage. In *Treatise on Geochemistry*; Holland, H.H., Turekian, K.K., Eds.; Elsevier: Amsterdam, The Netherlands, 2003; pp. 149–204. ISBN 9780080437514.

22. ISPRA. Linee guida per la determinazione dei valori di fondo per i suoli e le acque sotterranee, Delibera del Consiglio SNPA. In *Seduta del 14.11.2017. Doc. n. 20/2017*; ISPRA: Rome, Italy, Manuali e Linee Guida 174/2018; 2018; ISBN 978-88-448-0880-8. Available online: https://www.isprambiente.gov.it/files2018/pubblicazioni/manuali-linee-guida/MLG_174_18.pdf (accessed on 1 March 2021).

23. Piccardo, G.B. The Lanzo peridotite massif, Italian Western Alps: Jurassic rifting of the Ligurian Tethys. *Geol. Soc. London Spec. Publ.* **2010**, *337*, 47–69. [CrossRef]

24. Piccardo, G.B.; Padovano, M.; Guarnieri, L. The Ligurian Tethys: Mantle processes and geodynamics. *Earth Sci. Rev.* **2014**, *138*, 409–434. [CrossRef]

25. Ferraris, C.; Compagnoni, R. Metamorphic evolution and significance of a serpentinized peridotite slice within the Eclogitic Micaschist Complex of the Sesia-Zone (Western Alps, Italy). *Schweizerische Mineralogische und Petrographische Mitteilungen* **2003**, *83*, 3–13.

26. Comune di Balangero Piano Regolatore Comunale Generale. Available online: http://www.comune.balangero.to.it/Home/Menu?IDDettaglio=135816 (accessed on 19 February 2021).

27. Rossetti, P.; Zucchetti, S. Early-alpine ore parageneses in the serpentinites from the Balangero asbestos mine and Lanzo Massif (Internal Western Alps). *Rend. Della Soc. Ital. Mineral. Petrol.* **1988**, *43*, 139–149.

28. Rossetti, P.; Zucchetti, S. Occurrence of native iron, Fe-Co and Ni-Fe alloys in the serpentinite from the Balangero asbestos mine (Western Italian Alps). *Ofioliti* **1988**, *13*, 43–56.

29. Castelli, D.; Rossetti, P. Sulphide-arsenide mineralization in rodingite from the Balangero mine (Italian Western Alps): Hydrothermalism related to metamorphic fluids. In Proceedings of the Giornata di Studio in Ricordo Del Prof. Stefano Zucchetti, Turin, Italy, 12 May 1994; pp. 103–110.

30. Castelli, D.; Rolfo, F.; Rossetti, P. Petrology of ore-bearing rodingite veins from the Balangero asbestos mine (Western Alps). *Boll. Mus. Regionale Sci. Nat. Torino* **1995**, *13*, 153–189.

31. Groppo, C.; Compagnoni, R. Metamorphic veins from the serpentinites of the Piemonte Zone, western Alps, Italy: A review. *Period. Mineral.* **2007**, *76*, 127–153.

32. Zucchetti, S.; Mastrangelo, F.; Rossetti, P.; Sandrone, R. Serpentinization and metamorphism: Their relationships with metallogeny in some ophiolitic ultramafics from the Alps. In Proceedings of the Zuffar' Days Symposium, Cagliari, Italy, 10–15 October 1988; pp. 137–159.

33. Ferrante, D.; Mirabelli, D.; Silvestri, S.; Azzolina, D.; Giovannini, A.; Tribaudino, P.; Magnani, C. Mortality and mesothelioma incidence among chrysotile asbestos miners in Balangero, Italy: A cohort study. *Am. J. Ind. Med.* **2020**, *63*, 135–145. [CrossRef] [PubMed]

34. Paglietti, F.; Gennari, F.; Giangrasso, M. The asbestos extraction in the Balangero mine: Environmental consequences. In Proceedings of the Goldschmidt Conference Abstracts, Davos, Switzerland, 21–26 June 2009; p. A985. Available online: https://goldschmidtabstracts.info/abstracts/abstractView?id=2009001620 (accessed on 1 March 2021).

35. Parrone, D.; Frollini, E.; Preziosi, E.; Ghergo, S. eNaBLe, an On-Line Tool to Evaluate Natural Background Levels in Groundwater Bodies. *Water* **2020**, *13*, 74. [CrossRef]

36. APAT; CNR-IRSA. *Metodi Analitici per le Acque*; Manuali e Linee Guida 29/2003; APAT: Rome, Italy, 2003; ISBN 88-448-0083-7.

37. Pourbaix, M. *Atlas of Electrochemical Equilibria in Aqueous Solution*, 1st ed.; Pergamon Press: Bristol, UK, 1966.

38. Edmunds, W.M.; Shand, P. (Eds.) *Natural Groundwater Quality*; Blackwell Publishing: Hoboken, NJ, USA, 2008; ISBN 978-14051-5675-2.

39. Piper, A.M. A graphic procedure in geochemical interpretation of water analyses. *Trans. Am. Geophys. Union* **1944**, *25*, 914–923. [CrossRef]

40. Stiff, H.A.J. The interpretation of chemical water analysis by means of patterns. *J. Pet. Technol.* **1951**, *3*, 15–16. [CrossRef]

41. Vitale Brovarone, A.; Martinez, I.; Elmaleh, A.; Compagnoni, R.; Chaduteau, C.; Ferraris, C.; Esteve, I. Massive production of abiotic methane during subduction evidenced in metamorphosed ophicarbonates from the Italian Alps. *Nat. Commun.* **2017**, *8*, 14134. [CrossRef] [PubMed]

42. Mann, H. Nonparametric tests against trend. *Econometrica* **1945**, *13*, 245–259. [CrossRef]

43. Kendall, M.G. *Rank Correlation Methods*, 3rd ed.; Hafner: New York, NY, USA, 1962.

44. Meals, D.W.; Spooner, J.; Dressing, S.A.; Harcum, J.B. *Statistical Analysis for Monotonic Trends*; Tetra Tech, Inc.: Fairfax, VA, USA, 2011. Available online: https://www.epa.gov/sites/production/files/2016-05/documents/tech_notes_6_dec2013_trend.pdf (accessed on 1 March 2021).

45. Sen, P.K. Estimates of the regression coefficient based on Kendall's tau. *J. Am. Stat. Assoc.* **1968**, *63*, 1379–1389. [CrossRef]

46. Shapiro, S.S.; Wilk, M.B. An analysis of variance test for normality (complete samples). *Biometrika* **1965**, *52*, 591–611. [CrossRef]

47. D'Agostino, R.B.; Belanger, A.; D'Agostino, R.B. A Suggestion for Using Powerful and Informative Tests of Normality. *Am. Stat.* **1990**, *44*, 316–321.

48. De Vos, W.; Tarvainen, T.; Salminen, R.; Reeder, S.; De Vivo, B.; Demetriades, A.; Pirc, S.; Batista, M.J.; Marsina, K.; Ottesen, R.-T.; et al. *Geochemical Atlas of Europe. Part 2—Interpretation of Geochemical Maps, Additional Tables, Figures, Maps, and Related Publications*; De Vos, W., Tarvainen, T., Eds.; Geological Survey of Finland, Espoo: Espoo, Finland, 2006; ISBN 951-690-956-6.

49. ARPA Piemonte. *Definizione dei Valori di Fondo Naturale per i Metalli Nelle Acque Sotterranee Come Previsto Dalla Direttiva 2006/118/CE e dal Decreto Legislativo 16 Marzo 2009 n.30*; ARPA Piemonte: Turin, Italy, 2012; Available online: https://www.arpa.piemonte.it/approfondimenti/temi-ambientali/acqua/acque-sotterranee/Relazionevaloridifondosotterranee.pdf (accessed on 1 March 2021).

50. Langone, A.; Baneschi, I.; Boschi, C.; Dini, A.; Guidi, M.; Cavallo, A. Serpentinite-water interaction and chromium(VI) release in spring waters: Examples from Tuscan ophiolites. *Ofioliti* **2013**, *38*, 41–57.

51. Apollaro, C.; Fuoco, I.; Brozzo, G.; De Rosa, R. Release and fate of Cr(VI) in the ophiolitic aquifers of Italy: The role of Fe(III) as a potential oxidant of Cr(III) supported by reaction path modelling. *Sci. Total Environ.* **2019**, *660*, 1459–1471. [CrossRef]

52. McClain, C.N.; Fendorf, S.; Webb, S.M.; Maher, K. Quantifying Cr(VI) Production and Export from Serpentine Soil of the California Coast Range. *Environ. Sci. Technol.* **2017**, *51*, 141–149. [CrossRef] [PubMed]

53. Izbicki, J.A.; Wright, M.T.; Seymour, W.A.; McCleskey, R.B.; Fram, M.S.; Belitz, K.; Esser, B.K. Cr(VI) occurrence and geochemistry in water from public-supply wells in California. *Appl. Geochem.* **2015**, *63*, 203–217. [CrossRef]

54. Iribar, V. Origin of neutral mine water in flooded underground mines: An appraisal using geochemical and hydrogeological methodologies. In Proceedings of the Mine Water 2004-Proceedings International Mine Water Association Symposium, Newcastle upon Tyne, UK, 20–25 September 2004; Volume 1, pp. 169–178. Available online: https://www.imwa.info/docs/imwa_2004/IMWA2004_22_Iribar.pdf (accessed on 1 March 2021).

55. Frau, F.; Medas, D.; Da Pelo, S.; Wanty, R.B.; Cidu, R. Environmental effects on the aquatic system and metal discharge to the mediterranean sea from a near-neutral zinc-ferrous sulfate mine drainage. *Water. Air. Soil Pollut.* **2015**, *226*, 1–17. [CrossRef]

Article

An Integrated Use of GIS, Geostatistical and Map Overlay Techniques for Spatio-Temporal Variability Analysis of Groundwater Quality and Level in the Punjab Province of Pakistan, South Asia

Huzaifa Shahzad [1], Hafiz Umar Farid [1], Zahid Mahmood Khan [1], Muhammad Naveed Anjum [2], Ijaz Ahmad [3], Xi Chen [4], Perviaz Sakindar [5], Muhammad Mubeen [6], Matlob Ahmad [7] and Aminjon Gulakhmadov [4,8,9,*]

1. Department of Agricultural Engineering, Bahauddin Zakariya University, Multan 60800, Pakistan; huzaifa.shahzad@ymail.com (H.S.); hufarid@bzu.edu.pk (H.U.F.); zahidmk@bzu.edu.pk (Z.M.K.)
2. Department of Land and Water Conservation Engineering, Faculty of Agricultural Engineering & Technology, PMAS Arid Agriculture University, Rawalpindi 46000, Pakistan; naveedwre@uaar.edu.pk
3. Centre of Excellence in Water Resources Engineering, University of Engineering &Technology, Lahore 54890, Pakistan; dr.ijaz@uet.edu.pk
4. Research Center for Ecology and Environment of Central Asia, Xinjiang Institute of Ecology and Geography, Chinese Academy of Sciences, Urumqi 830011, China; chenxi@ms.xjb.ac.cn
5. Agricultural Engineering Department, (Field Wing) Rawalpindi 46000, Pakistan; mhamzas@yahoo.com
6. Department of Environmental Sciences, COMSATS University Islamabad, Vehari Campus 61100, Pakistan; muhammadmubeen@ciitvehari.edu.pk
7. Department of Agricultural Engineering and Technology, Ghazi University, Dera Ghazi Khan 32200, Pakistan; mahmad@gudgk.edu.pk
8. State Key Laboratory of Desert and Oasis Ecology, Xinjiang Institute of Ecology and Geography, Chinese Academy of Sciences, Urumqi 830011, China
9. Institute of Water Problems, Hydropower and Ecology of the Academy of Sciences of the Republic of Tajikistan, Dushanbe 734042, Tajikistan
* Correspondence: aminjon@ms.xjb.ac.cn; Tel.: +86-991-782-3131

Received: 24 October 2020; Accepted: 14 December 2020; Published: 17 December 2020

Abstract: The rapidly changing climatic scenario is demanding periodic evaluation of groundwater quality at the temporal and spatial scale in any region for its effectual management. The statistical, geographic information system (GIS), geostatistical, and map overlay approaches were applied for investigating the spatio-temporal variation in groundwater quality and level data of 242 monitoring wells in Punjab, Pakistan during pre-monsoon and post-monsoon seasons of the years 2015 and 2016. The analysis indicated the higher variation in data for both the seasons (pre-monsoon and post-monsoon) as coefficient of variation (CV) values were found in the range of 84–175% for groundwater quality parameters. Based on the *t*-test values, the marginal improvement in groundwater electrical conductivity (EC), sodium absorption ratio (SAR) and residual sodium carbonate (RSC) and decrease in groundwater level (GWL) were observed in 2016 as compared to 2015 ($p = 0.05$). The spatial distribution analysis of groundwater EC, SAR and RSC indicated that the groundwater quality was unfit for irrigation in the lower south-east part of the study area. The groundwater level (GWL) was also higher in that part of the study area during the pre-monsoon and post-monsoon seasons in 2015 and 2016. The overlay analysis also indicated that the groundwater EC, RSC and GWL values were higher in south-east parts of the study area during pre-monsoon and post-monsoon seasons of 2015 and 2016. Hence, there is an instant need to apply groundwater management practices in the rest of the region (especially in the lower south-east part) to overcome the future degradation of groundwater quality.

Keywords: groundwater quality; groundwater level; geostatistics; *t*-test; spatial distribution modeling

---

1. Introduction

Groundwater is becoming a basic requirement for human utilization and crop production [1,2]. Approximately 2.5% of water worldwide is available as freshwater from all the global water resources. The amount of water accessible to humans is available in the form of rivers, lakes, reservoirs, and underground water [3]. Groundwater's function is gaining more importance because a crisis is arising regarding surface water resources with time [4]. In Pakistan, about 50% of water is supplied for irrigation through groundwater sources [5]. Presently, the annual abstraction from groundwater is 60 billion cubic meters [6]. Consequently, the groundwater quality has deteriorated, while the water table has increased in many irrigated areas [7]. The groundwater quality is also rapidly declining worldwide, particularly in developing countries, due to larger dependency on groundwater resources for irrigation [2,8]. Additionally, the deterioration of groundwater quality due to annual and seasonal climate variation may impose pressure on hydrologic and hydrogeologic systems. These seasonal changes were attributed to groundwater quality as a result of monsoonal-driven surface–groundwater interaction [9]. The variation in groundwater quality parameters at a site was mostly related to local conditions and hazards. Similarly, the groundwater level also increased after the precipitation events and then decreased gradually with evaporation. The groundwater level varied within the period from wet to dry seasons and showed seasonal variations because of the seasonal distribution of precipitation and evaporation. Estimation of seasonal flooding impacts on groundwater quality and level due to substantial rainfall is critical to the management of this precious resource [10]. Therefore, it is essential to know the seasonal variation in groundwater quality [11].

Moreover, a key role is played by the various natural processes and anthropogenic activities in degrading groundwater quality [12–14]. To ensure the sustainable safe use of these resources, it is very important to access the groundwater quality as well as its other resources [3]. The groundwater quality is affected by various factors such as regional topography, characteristics of soil, discharge and flow, groundwater circulation through different types of rocks, groundwater recharge, saline water intrusion and hydro-meteorological surroundings of the area [8]. An understanding of the spatial and temporal variation in groundwater quality is essential for sustainable water supplies under changing climate and local environmental pressure [2,15]. For monitoring water quality, traditional approaches have been found to be unreliable due to errors in sampling. Various researchers have developed different graphical and statistical techniques to assess the trends in groundwater quality which vary from simple linear regression to more advanced parametric and non-parametric methods [16–19]. To address the potential groundwater quality problems, a geographic information system (GIS) and trend detection techniques would be useful for examining the long-term water quality variations [20]. The non-parametric methods such as Mann–Kendall (MK), Spearman's rho (SR), Sen's slope estimator (SSE), innovative trend analysis (ITA), the Theil Sen approach and sequential MK have extensively used to detect trends in time series environmental data [21,22]. Similarly, Agca [23] used the *t*-test for the temporal analysis of groundwater quality parameters in Amik plain (South Turkey), which indicated the increasing or decreasing trends of groundwater quality parameters. The management of natural resources can also be achieved using the geographic information system (GIS) at temporal and spatial scale [20,24]. Many authors investigated the GIS contribution in analyzing the spatial distribution of groundwater [25,26].

Therefore, assessment of groundwater spatial distribution is an easy way to analyze the groundwater quality in a matter of its suitability for irrigation [24,27]. The distribution of concentration over space and time that derives from the relation between sample points can be assessed by geostatistical techniques [28,29]. The weighted overlay approach in GIS has also been used to identify the groundwater potential zones in Killinochi, northern Sri Lanka [30]. However, such past studies did not report the integrated use of statistical, map overlay, and geostatistical techniques for the spatio-temporal variation of groundwater quality and level. The aims of the present research are to use the combine approaches (statistical, geostatistical, and map overlay analysis) for identifying and investigating the spatio-temporal variation in groundwater quality and level in the Haveli Canal Circle (HCC) Command area.

2. Materials and Methods

2.1. Appearances of Study Area

The present study was conducted in the region of Haveli Canal Circle, Multan irrigation zone (MIZ), Punjab, Pakistan. The region lies between the longitude of 71.12° and 72.19° and latitude of 29.51° and 31.65° (Figure 1). The MIZ is situated in an arid and semi-arid regions where the climate is hot in summer and cold in the winter [31]. The highest recorded temperature in the region was 54 °C and the lowest recorded temperature was approximately −1 °C [27]. It has a flat topography and is suitable for agriculture purposes but receives very little rainfall throughout the year. The average annual rainfall in the study region varies from 100 to 300 mm. The rainfall is unreliable and can be distributed in two seasons (pre-monsoon and post-monsoon). About 60% of the total annual rainfall occurs during the summer season (monsoon rain) and the remaining rainfall is received during the rest of the year [32].

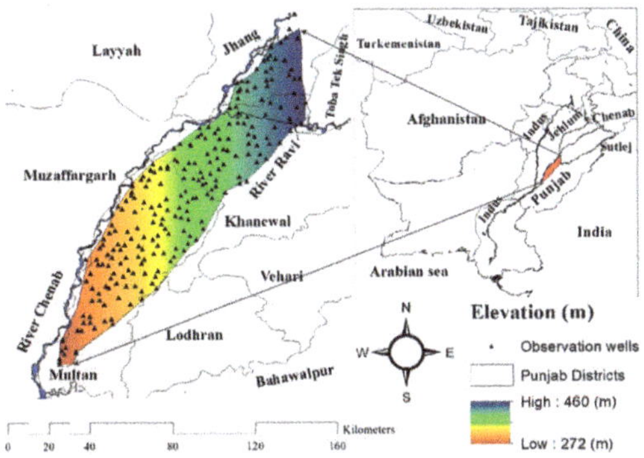

Figure 1. Location of study area and sampling points.

Wheat, cotton, sugarcane, and corn are the dominant crops grown in MIZ. The availability of irrigation water from surface sources is a crucial factor for the growth of crops. As the availability of surface water is highly variable, groundwater is pumped to fulfil crop water needs during prolonged dry periods. However, the quality of groundwater is not suitable for irrigation use in many parts of the study area [27]. The aquifers of the study region mostly consist of alluvial deposits which were transported by rivers from the Himalayan mountainous ranges. The alluvium generally contains sands and gravels, or mixtures of sands and gravels. Groundwater in the area is recharged through infiltration from precipitation and seepage from the Chenab River and its associated canal network.

2.2. Data Collection and Analysis

Two years (2015–2016) of groundwater quality data for 242 monitoring sites (wells) of Haveli Canal Circle, Multan Irrigation Zone (MIZ), Punjab for the pre-monsoon and post-monsoon seasons were collected from the Land Reclamation Department of Multan, Govt. of Punjab, Pakistan. Groundwater samples from every well installed in the study area were collected for pre-monsoon (May–June) and post-monsoon (October-November) meeting the standard protocols. Each sample was examined chemically to identify the groundwater properties of electrical conductivity (EC), sodium adsorption ratio (SAR), and residual sodium carbonate (RSC). The groundwater EC, SAR, and RSC are commonly used to assess the suitability of groundwater for irrigation in Pakistan [33]. Furthermore, the Punjab Irrigation Department also collected data on groundwater EC, SAR, and RSC for groundwater quality monitoring in the study area. The groundwater EC was measured with a portable multi-meter which was calibrated before its use. The SAR and RSC were calculated using the standard laboratory protocol as reported in the literature [34,35]. Similarly, the groundwater level (GWL) data were also recorded using the water level recorder.

The Shapiro–Wilk normality test was conducted to check the data distribution in the Statistical Package for Social Sciences (SPSS) software package. To visualize the normality of the data, normal Q-Q plots were used. The p values were used to confirm the parameters that showed normal distribution. The normality test showed that the data of all groundwater parameters did not conform to a normal distribution ($p = 0.05$). For exploratory analysis of the data, statistical parameters, i.e., minimum, maximum, mean, median, coefficient of variation (CV) and skewness, were determined using the Statistix 10.0 (Analytical Software, Tallahassee, FL, USA). Spearman's rank correlation analysis was determined in both the seasons for a better assessment of the relationship among groundwater parameters. The Spearman's rank correlation method is a non-parametric test method which can express the level of association between two groundwater parameters [36]. For the temporal evaluation of the 242 datasets during the sampling periods, an independent sample t-test was also carried out [23].

2.3. Spatial Variability Analysis

Spatial variability analysis of groundwater parameters was carried out using the ordinary kriging interpolation technique. Ordinary Kriging is a one of the geostatistical methods that can be used to interpolate a random variable at an unknown location considering its value at the nearby location [37]. Before applying the ordinary kriging interpolation technique, the data were log-transformed to conform to the necessary assumptions for ordinary kriging. The histogram and normal Q-Q plot were drawn using the log-transformed data in the SPSS software package to assess the normality of data and to ensure that the data conformed to the normal distribution. Then, that central tool of geostatistical methods such as semi-variance was applied to quantify the spatial autocorrelation of groundwater parameters based on the log-transformed data [38,39]. All the geostatistical analyses were carried using the GS+ software (RockWare Inc., Golden, CO, USA). Using the binned values fitness method, the best fit model was selected before applying the ordinary kriging interpolations based on the coefficient of determination (R^2) and, for scatter plots, each groundwater parameter. Moreover, the ratio of ((C0)/(C0 + C1)) was used to determine the spatial variation of groundwater EC, SAR, RSC, and GWL values. For the given parameters of groundwater, the ratio of ((C0)/(C0 + C1)), i.e., (<25%), (25–75%) and (>75%), indicates strong, moderate, and weak spatial dependence, respectively [40]. The leave one-out cross validation was performed by hiding the values of groundwater samples and estimating its values from the remaining data set. The same procedure was repeated for different semi-variogram models and found the statistics of cross validation error for selecting optimal model. After selecting the suitable model, the kriging method was found to be suitable for the interpolation and management of groundwater parameters using either GS+ software or ArcGIS software package. A study suggested that using the interpolation methods of GS+ software and ArcGIS is efficient in the prediction of unsampled data and gave almost the same result [41]. However, the ArcGIS 10.1 software package was used to interpolate groundwater EC, SAR, RSC, and GWL for both the pre-monsoon and post-monsoon seasons in each year based on the variogram parameters calculated using the GS+ software. A similar approach has also been used for geostatistical and interpolation of top soil properties by Zhang et al. [42]. All groundwater parameters were classified using the manual classification technique based on the groundwater quality standards for irrigation.

3. Results and Discussions

The minimum, maximum, mean, median, standard deviation (SD), coefficient of variance (CV) and skewness values of groundwater EC, SAR, RSC, GWL for pre-monsoon and post-monsoon seasons in 2015 and 2016 are shown in Tables 1 and 2, respectively. The variability in data can be addressed by the value of CV. A value of CV less than 10% indicates low variability, moderate variability is indicated when $10\% \leq CV \leq 100\%$, and high variability is indicated when $CV > 100\%$, respectively [21]. The values of CV for EC during the pre-monsoon and post-monsoon seasons in 2015 were found to be 115.77% and 135.34%, respectively, which indicated the high variability in the data. Similarly, the values of CV for RSC during the pre-monsoon and post-monsoon seasons in 2015 were found as 170.43% and 175.22%, respectively, which also indicated the high variability in the RSC data. Moreover, the values of CV for SAR and GWL during the pre-monsoon and post-monsoon seasons were found to be 84.97%, 92.97%, 79.74%, and 98.29%, respectively, which indicated the moderate variability in the data for both the seasons in 2015 (Table 1). The values of CV for EC during the pre-monsoon and post-monsoon seasons in 2016 were found to be 115.75 and 110.71%, respectively, which also indicated the high variability in the data for groundwater EC. Similarly, the values of CV for RSC and GWL during pre-monsoon and post-monsoon seasons in 2016 were found to be 174.80, 224.19%, 103.35% and 112.47%, respectively, which indicated the high variability in the data for groundwater RSC and GWL (Table 2).

Table 1. The statistical summary of groundwater properties in 2015 ($n = 242$).

Seasons	Parameters	Units	Min	Max	Mean	Median	CV (%)	Skewness
Pre-monsoon	EC Pre	dS/m	0.31	9.10	1.46	0.85	115.77	3.19
	SAR Pre	-	0.02	24.11	4.91	3.88	84.97	1.89
	RSC Pre	meq/l	0.00	7.80	0.76	0.00	170.43	2.00
	GWL Pre	m	1.58	77.08	22.06	29.91	79.74	0.30
Post-monsoon	EC Post	dS/m	0.37	17.50	1.59	0.95	135.34	4.46
	SAR Post	-	0.15	23.65	4.46	3.51	92.97	2.11
	RSC Post	meq/l	0.00	4.60	0.68	0.00	175.22	1.69
	GWL Post	m	1.58	78.00	21.14	30.75	98.29	0.28

SD standard deviation, CV coefficient of variation.

Table 2. The statistical summary of groundwater properties in 2016 ($n = 242$).

Seasons	Parameters	Units	Min	Max	Mean	Median	CV (%)	Skewness
Pre-monsoon	EC	dS/m	0.36	10.8	1.50	0.94	115.75	3.19
	SAR	-	0.19	28.88	5.50	4.42	86.98	1.98
	RSC	meq/l	0.00	7.00	0.70	0.00	174.80	1.96
	GWL	m	1.75	75.75	19.12	28.25	103.35	0.33
Post-monsoon	EC	dS/m	0.29	8.80	1.41	0.85	110.71	2.73
	SAR	-	0.02	40.00	4.10	2.50	118.36	3.52
	RSC	meq/l	0.00	7.30	0.48	0.00	224.19	3.36
	GWL	m	1.66	76.58	17.82	26.79	112.47	0.43

SD standard deviation, CV coefficient of variation.

The paired *t*-test was performed to analyze the difference in mean values of groundwater parameters for pre-monsoon and post-monsoon seasons [43]. The mean EC of 1.49 dS/m during the pre-monsoon season was significantly lower than that of mean EC of 1.65 dS/m during the post-monsoon season in 2015 ($p = 0.05$). The mean values for SAR, RSC, and GWL during the pre-monsoon season were higher than those of post-monsoon season in 2015 (Table 3). Similarly, significant negative difference was determined for groundwater EC between the pre-monsoon and post-monsoon seasons in 2015, having a t-state of −0.887. On the other hand, significant positive difference was observed between the pre-monsoon and post-monsoon seasons for groundwater SAR,

RSC, and GWL in 2015. However, the significant positive difference was observed for groundwater EC between the pre-monsoon and post-monsoon seasons, having a t-state value of 0.901 and the significant negative difference was observed for GWL between the pre-monsoon and post-monsoon seasons in 2016 ($p = 0.05$). The difference in results for both the years (2015 and 2016) for seasonal analysis may be due to the variation in rainfall.

Table 3. Inequality analysis of means for groundwater quality and level.

Parameters	Seasonal Comparison (2015)		Seasonal Comparison (2016)		Parameters	Annual Comparison	
	Mean	t-Stat	Mean	t-Stat		Mean	t-Stat
Pre-EC	1.49	−0.887 *	1.57	0.901 *	2015-EC	1.56	−0.206 *
Post-EC	1.65		1.43		2016-EC	1.59	
Pre-SAR	5.01	1.032 *	5.72	3.577	2015-SAR	4.80	0.291 *
Post-SAR	4.62		4.14		2016-SAR	4.69	
Pre-RSC	0.78	0.623 *	0.73	2.324	2015-RSC	0.75	1.324 *
Post-RSC	0.71		0.49		2016-RSC	0.62	
Pre-GWL	22.60	0.165 *	19.51	−0.128 *	2015-GWL	22.10	0.504 *
Post-GWL	21.02		20.86		2016-GWL	20.68	

* Significant at $p = 0.05$; electrical conductivity (EC) (dS/m); residual sodium carbonate (RSC) (meq/L); groundwater level (GWL) (m); pre (pre-monsoon season); post (post-monsoon season).

The seasonal variation in groundwater level mainly depends on the annual cycle of rainfall and river water levels [9–11]. The GWL increases due to over-pumping in the pre-monsoon season, which tends to degrade groundwater quality [44]. Annual comparison of the mean value for all groundwater parameters was also observed for both the years (2015 and 2016), which indicated positive significant difference with t-values of 0.291, 1.324, and 0.504 for annual groundwater SAR, RSC and GWL, respectively, at 0.05 significant level (Table 3). The marginal improvement in groundwater SAR and RSC and decrease in GWL were also observed in 2016 as compared to 2015.

3.1. Spearman's Rank Corelation Coefficient

The Spearman's rank correlation coefficients (r_s) among all groundwater parameters were calculated for correlation analysis in both the seasons. The values of (r_s) for each groundwater parameter were shown in Table 4. Interpretation of correlation analysis indicates a quick quality monitoring for groundwater parameters [45]. A significant positive relationship was observed between boring depth and discharge and boring depth and screen length, with $r_s = 0.372$ and 0.473, respectively, at 0.05 significant level. The analysis for groundwater EC in the pre-monsoon season of 2015 indicated the significant positive relationship with groundwater EC and SAR in the post-monsoon season with $r_s = 0.930$ and 0.729, respectively. Similarly, the analysis of groundwater SAR in pre-monsoon season of 2015 indicated a significant positive relationship with groundwater EC, SAR, and RSC for the post-monsoon season, with $r_s = 0.744$, 0.810, and 0.360, respectively. Moreover, groundwater RSC has a significant positive connection in pre-monsoon and post-monsoon seasons with $r_s = 0.842$. The significant positive connection between groundwater EC and SAR was also found, with r_s values varying from 0.297 to 0.783 for both the seasons (pre-monsoon and post-monsoon) in 2015 and 2016. These correlations may indicate some of the impacts of agricultural activities in the study area. This may also be related to the general process governing the groundwater formation [36,46].

Table 4. Spearman's rank correlation matrix for groundwater quality parameters.

	Boring Depth	Discharge	Screen Length	Pre-EC	Pre-SAR	Pre-RSC	Post-EC	Post-SAR	Post-RSC
Boring depth	1.000								
Discharge	0.372 **	1.000							
Screen length	0.473 **	−0.066	1.000						
Pre-EC	0.072	0.043	0.114	1.000					
Pre-SAR	0.090	0.036	0.105	0.783 **	1.000				
Pre-RSC	−0.043	−0.029	−0.031	0.117	0.416 **	1.000			
Post-EC	0.177 **	0.101	0.165 *	0.930 **	0.744 **	0.042	1.000		
Post-SAR	0.136 *	0.051	0.108	0.729 **	0.810 **	0.395 **	0.711 **	1.000	
Post-RSC	−0.046	−0.066	−0.015	0.079	0.360 **	0.842 **	0.052	0.451 **	1.000
Boring depth	1.000								
Discharge	0.372 **	1.000							
Screen length	0.473 **	−0.066	1.000						
Pre-EC	0.140 *	0.082	0.115	1.000					
Pre-SAR	0.059	0.052	0.032	0.768 **	1.000				
Pre-RSC	−0.081	−0.078	−0.034	0.082	0.457 **	1.000			
Post-EC	0.235 **	0.117	0.195 **	0.403 **	0.297 **	−0.003	1.000		
Post-SAR	0.142 *	0.063	0.148 *	0.363 **	0.261 **	−0.062	0.783 **	1.000	
Post-RSC	−0.039	0.058	0.014	0.029	0.036	0.109	−0.069	0.242 **	1.000

EC (dS/m); RSC (meq/L); GWL (m); ** Significant at = 0.05; * Significant at = 0.01; pre (pre-monsoon season); post (post-monsoon season).

3.2. Spatial Modelling of Groundwater Parameters

The spatial dependence of each groundwater parameter for pre-monsoon and the post-monsoon seasons was determined with the help of semi-variograms models for both the years [27]. The ratio of nugget variance (C_0) to sill variance ($C_0 + C1$) was used to determine the spatial variation of groundwater parameters i.e., EC, RSC, SAR and GWL.

The data are considered "strongly" spatially dependent if the ratio is less than or equal to 25%, "moderately" spatially dependent if the ratio ranges from 25 to 75%, and "weakly" spatially dependent if the ratio is greater than 75% [40]. The results indicate that the data of groundwater quality parameters were spatially autocorrelated over the years. From the analysis of the developed semi-variogram, the best fit model was selected based on the coefficient of determination (R^2) before the kriging interpolations to analyze the spatial distribution of groundwater parameters (Table 5). The nugget to sill ratios for EC, SAR, RSC, and GWL during the post-monsoon season in 2015 were found to be 13.86%, 12.52%, 15.57%, 13.40% and 8.10%, respectively, which indicated the strong spatial dependence. The nugget to sill ratios during the post-monsoon season in 2016 for EC, SAR, RSC and GWL were found to be 0.03%, 21.60%, 6.84%, 6.33%, 6.33% and 6.38%, respectively, which also indicated the strong spatial dependence. The moderate spatial dependence was analyzed for EC during the pre-monsoon season with nugget to sill ratios of 49.98% (in 2015) and 43.09% (in 2016) for both years. The spatial variability of groundwater properties was affected due to the naturally occurring factors such as topography, regional climate, groundwater flow pattern, groundwater runoff and hydrological conditions. When these naturally occurring factors have greater influence on groundwater, then the chances of strong spatial dependence become maximum for a given area [47]. Similarly, the detailed outcomes about the fitness and selection of different models for the interpolation of groundwater quality parameters are given in Table 5.

Table 5. Fitted semi-variogram model of groundwater properties in 2015 and 2016.

Years	Seasons	Parameter	Model	Range	C0	C0 + C	C0/(C0 + C) × 100	R^2
2015	Pre-monsoon	EC	Spherical	0.354	1.527	3.055	49.98	0.793
		SAR	Spherical	0.267	8.160	17.910	45.56	0.441
		RSC	Exponential	0.147	0.265	1.701	15.57	0.537
		GWL	Gaussian	1.186	65.00	842.50	7.71	0.971
	Post-monsoon	EC	Exponential	0.177	0.680	4.903	13.86	0.592
		SAR	Exponential	0.183	2.220	17.720	12.52	0.422
		RSC	Exponential	0.132	0.190	1.417	13.40	0.800
		GWL	Gaussian	1.172	72.00	888.60	8.10	0.973
2016	Pre-monsoon	EC	Spherical	0.289	1.372	3.184	43.09	0.771
		SAR	Exponential	0.222	5.180	23.980	21.60	0.689
		RSC	Exponential	0.579	0.844	1.689	49.97	0.938
		GWL	Gaussian	1.086	51.00	771.00	6.61	0.954
	Post-monsoon	EC	Exponential	0.339	0.001	2.690	0.03	0.858
		SAR	Exponential	0.354	1.760	25.700	6.84	0.782
		RSC	Spherical	0.114	0.076	1.200	6.33	0.571
		GWL	Gaussian	0.978	48.00	752.00	6.38	0.938

C0 Nugget Variance; C0 + C Sill Variance; EC (dS/m); RSC (meq/L); GWL (m).

3.3. Spatial Distribution of Groundwater Quality Parameters

The spatial distribution of groundwater EC for pre-monsoon season in 2015 indicated that the concentration of EC was higher in the south-east zone of the study area (Figure 2a). The values of EC in this zone were ranged from 2.50 to 4.20 dS/m. A higher concentration of groundwater EC was also found in the northern side for pre-monsoon season in 2015. Similarly, the groundwater EC for the post-monsoon season in 2015 indicated that the concentration of EC was higher in the lower south-east and upper north-west zones of the study area. The lower concentration of EC was observed in the south-west, central south-east, north-west, and north-east parts of the study area and the values ranged from 0 to 1.50 dS/m, respectively (Figure 2b). The groundwater EC for the pre-monsoon season in 2016 also indicated that the concentration of EC was higher in the lower south-east zone and the northern part of the study area (Figure 2c). The groundwater EC concentration showed minor changes in the area, as a higher concentration of groundwater EC was observed in the lower south-east and upper north-east zones (Figure 2d). However, the lower south-east part of the study area continuously showed higher concentration of groundwater EC for both seasons and years. The higher concentration of groundwater EC in that part may be due to the over-abstraction of groundwater. As a result of excessive groundwater extraction, fractured rocks and shales dissolve in the water, which increases the salinity level [48]. Furthermore, the salts tend to accumulate due to strong evaporation in arid and semi-arid areas and consequently, high salinity groundwater forms so that the roots of plants cannot absorb enough water to meet their metabolic requirements [49–51].

Apart from the spatial distribution analysis, overlay analysis was also conducted to analyze the combined effect of the groundwater properties for both seasons. In 2015, the overlay analysis for EC indicated that the lower portion of the south-east zone had a higher concentration of EC in the pre-monsoon and post monsoon seasons. The growth of the high regulation spatial model, GIS, and remote sensing has increased the need for map comparison. The importance of map comparison methods has been recognized and has stimulated growing interest among different researchers [52–54].

The value of EC in this zone was greater than 4.20 dS/m for both the seasons of 2015. The lower concentration of groundwater EC was analyzed in the central south-east and south-west parts of the study area for both seasons of 2015 (Figure 3a). In 2016, the overlay analysis for EC indicated that the lower portion in the south-east and south-west sides had a higher concentration of EC in both seasons (Figure 3b). The area of lower concentration of EC slightly increased in the post-monsoon season as compared to the pre-monsoon season. This indicated that the recharge of freshwater increases in the post-monsoon season from the Chenab river flows and precipitation because more than 60% of the annual rainfall occurred during the post-monsoon season in this region [27,55].

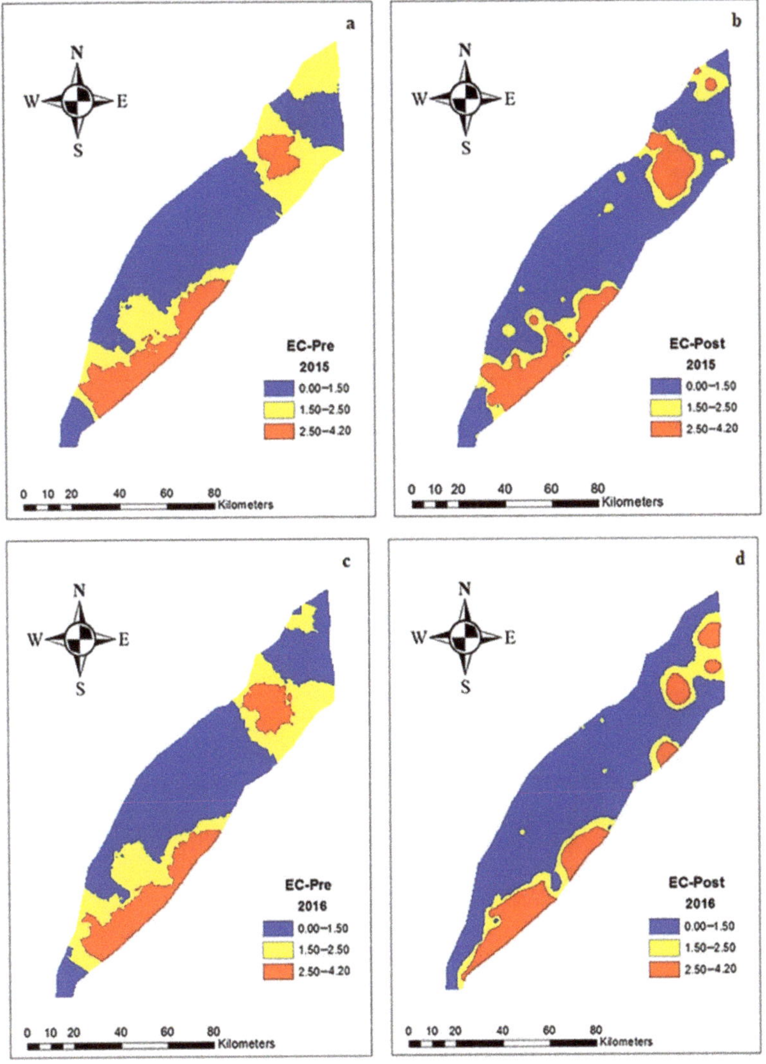

Figure 2. Spatial distribution of groundwater EC (dS/m) (**a**) for pre-monsoon season in 2015 (**b**) for post-monsoon season in 2015 (**c**) for pre-monsoon season in 2016 (**d**) for pre-monsoon season in 2016.

The spatial distribution of groundwater RSC for pre-monsoon season in 2015 indicated that the concentration of RSC was higher in the upper north-east zone of the study area (Figure 4a). Similarly, the groundwater RSC for the post-monsoon season in 2015 indicated that concentration of RSC was higher in the upper north-east and central south-east and south-west parts of the study area (Figure 4b). The groundwater RSC for pre-monsoon season in 2016 also indicated that the concentration of RSC in groundwater was higher in the upper north-east and central south-east zones of the study area (Figure 4c). Similarly, the groundwater RSC for the post-monsoon season in 2016 indicated that the concentration of RSC was higher in the upper north-east and central south-east parts of the study area and the values of RSC in that zone ranged from 1.50 to 4.20 meq/l (Figure 4d). In 2015, the overlay analysis for RSC indicated that the upper part of the north-east zone had a higher concentration of RSC in the pre-monsoon and post-monsoon season. The value of RSC in this zone was greater than 4.20 meq/l for both seasons (Figure 5a). Similarly, in 2016, the overlay analysis for RSC indicated that the upper north-east zone had a higher concentration of RSC in both seasons (Figure 5b).

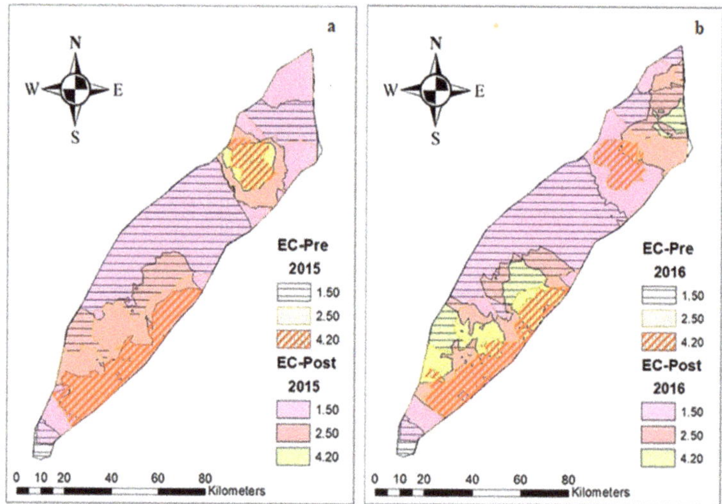

Figure 3. Overlay analysis of groundwater EC in pre-monsoon and post-monsoon season. (**a**) EC for 2015 and (**b**) EC for 2016.

The SAR in groundwater evaluates the effect of the sodium hazard with calcium and magnesium concentrations [56]. The spatial distribution of groundwater SAR for the post-monsoon season in 2015 indicated that concentration of SAR, ranging from 5.0 to 10.0, was marginally higher in the upper north-east and lower south-east parts of the study area (Figure 6a). The spatial distribution of groundwater SAR for pre-monsoon season in 2015 indicated that concentration of SAR was lower in the central south-east and south-west parts of the study area and the values ranged from 0 to 5.00 (Figure 6b). It has been reported that seasonal variation affects the groundwater SAR, which can reduce the soil permeability and thus inhibit the absorption of water by crops [57].

The lower concentration of SAR was observed mainly in the central south-east and south-west parts of the study area (Figure 6a). The spatial distribution of groundwater SAR for the pre-monsoon season in 2016 also indicated that the concentration of SAR was higher in the upper north-east and lower south-east parts of the study area (Figure 6c). Higher SAR values in groundwater make soil unfit for plant growth due to a loss of soil permeability. Sodium reduces soil permeability and encourages hardening of the soil. On the other hand, the lower groundwater SAR concentration was observed for the post-monsoon season in 2016 in the upper north-west, north-west, and lower south-west parts of the study area (Figure 6d).

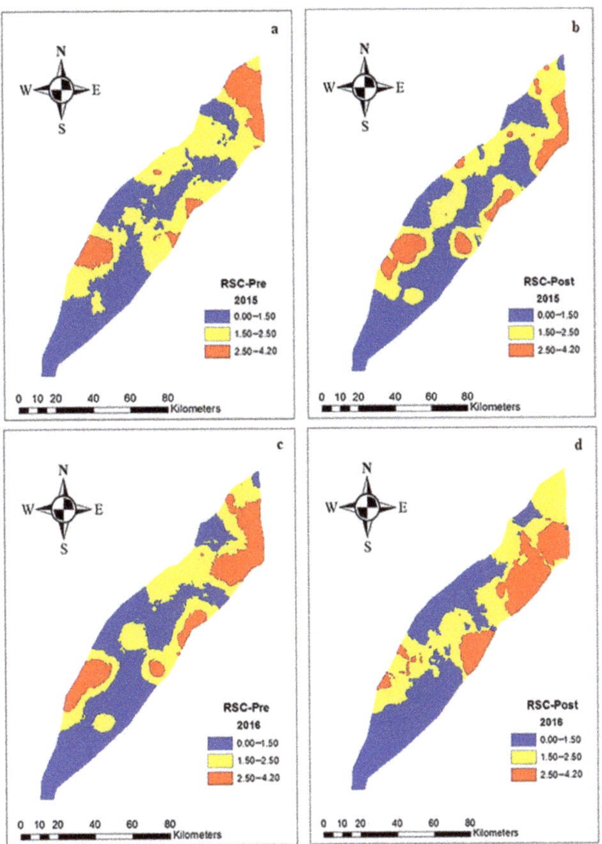

Figure 4. Spatial distribution of groundwater RSC (meql/l) (**a**) for pre-monsoon season in 2015 (**b**) for post-monsoon season in 2015 (**c**) for pre-monsoon season in 2016 (**d**) for pre-monsoon season in 2016.

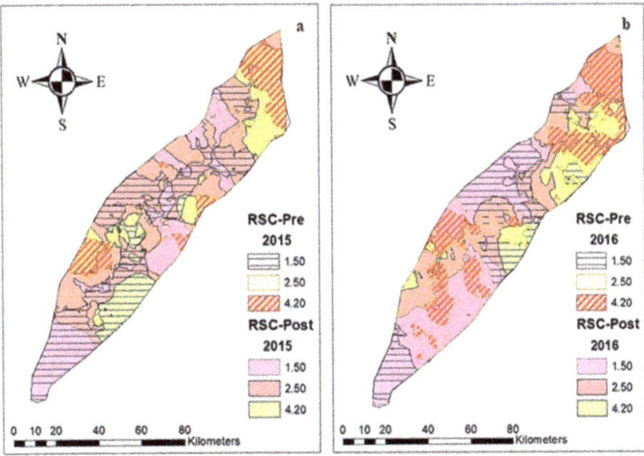

Figure 5. Overlay analysis of groundwater RSC. (**a**) RSC for 2015, (**b**) RSC for 2016.

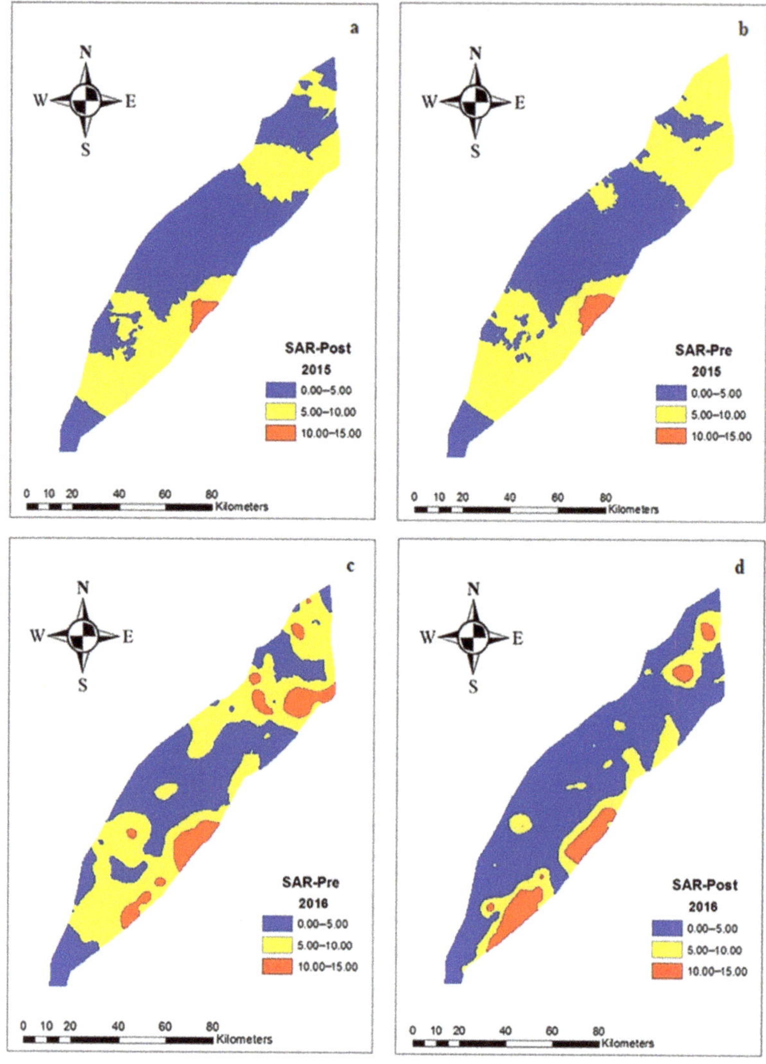

Figure 6. Spatial distribution of groundwater SAR (**a**) for post-monsoon season in 2015 (**b**) for pre-monsoon season in 2015 (**c**) for pre-monsoon season in 2016 (**d**) for pre-monsoon season in 2016.

In 2015, the overlay analysis for SAR indicated that the upper part of the north-east and lower south-east and south-west zones had a higher concentration of SAR mainly in pre-monsoon season (Figure 7a). In 2016, the overlay analysis for SAR indicated that the upper north-east zone had a higher concentration of SAR in both the seasons, and the value of SAR in this zone was greater than 15 (Figure 7b). The spatial distribution of groundwater level (GWL) for the pre-monsoon season in 2015 indicated that GWL was high over the lower south-east part of the study area (Figure 8a,b).

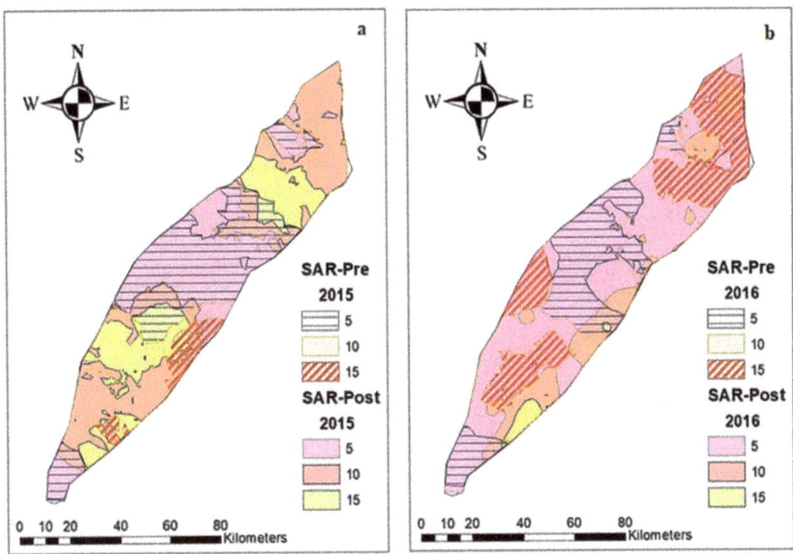

Figure 7. Overlay analysis of groundwater SAR. (**a**) SAR for 2015, (**b**) SAR for 2016.

It has been reported that with the increase in groundwater level, the osmotic and buoyancy pressure produced by the soil increases to a certain extent. This means that the soil bears a certain additional load within a certain range, resulting in the discharge of pore water from the soil, a reduction in pore water pressure, increase in effective stress, and consolidation and compaction of strata [58]. The GWL for pre-monsoon and post-monsoon seasons in 2016 also indicated that GWL was high in the lower south-east part of the study area. Moreover, the results for the upper north-east part of the study area also indicate that GWL was lower in this zone for both the seasons of 2016 (Figure 8c,d).

In 2015, the overlay analysis for GWL indicated that in the lower south-east part, the groundwater had greater depth in the pre-monsoon and post-monsoon seasons (Figure 9a). Similarly, the GWL results in 2016 indicated that in the lower south-east part, the groundwater had greater depth in pre-monsoon and post-monsoon seasons (Figure 9b). The overall analysis indicated that water quality in the lower south-east part of the study area is deteriorating and unfit for irrigation without possible treatment before its direct application to agricultural lands. The results indicate that potential management measures are needed to ameliorate these impacts, including decreasing groundwater pumping or implementing the activities of groundwater recharge in the region. In order to further reduce the deterioration of groundwater quality and to protect groundwater resources, urgent groundwater management practices are required in the rest of the region.

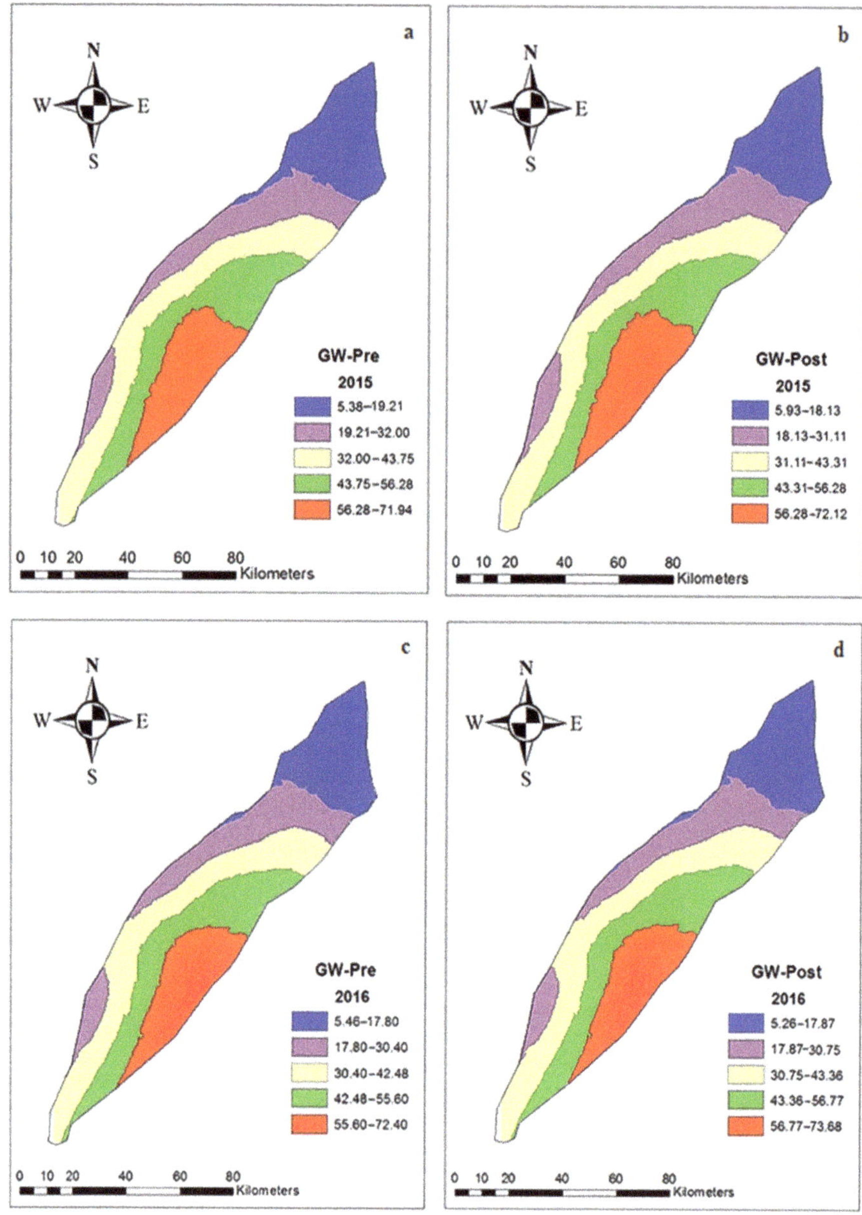

Figure 8. Spatial distribution of groundwater level (GWL) (**a**,**b**) in 2015 and (**c**,**d**) in 2016 for pre monsoon and post monsoon.

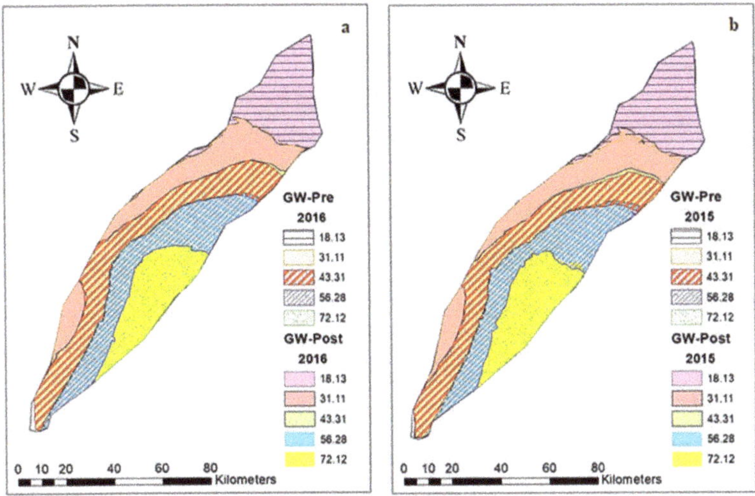

Figure 9. Overlay analysis of groundwater level. (**a**) GWL for 2015 (**b**) GWL for 2016.

4. Conclusions

The present study described the integrated use of statistical, GIS, geostatistical, and map overlay approaches for investigating the spatio-temporal variation in groundwater quality and level using the data of 242 monitoring wells in Punjab, Pakistan during the pre-monsoon and post-monsoon seasons of the years 2015 and 2016. The results show that CV values for all the groundwater parameters were greater than 84% for both the seasons (pre-monsoon and post-monsoon) and over the years. The annual comparison for groundwater parameters indicated that the groundwater EC followed a decreasing trend over the years with a 0.05 level of significance. However, groundwater RSC, SAR, and GWL indicated increasing trends over the year, with t-state values of 1.324, 0.291 and 0.504, respectively. However, marginal improvements in groundwater electrical conductivity (EC), sodium absorption ratio (SAR), and residual sodium carbonate (RSC), and a decrease in groundwater level (GWL), were observed in 2016 as compared to 2015 ($p = 0.05$). The results also indicate that the data of groundwater parameters were spatially autocorrelated over the years. The nugget to sill ratios in 2015 for post EC, post-SAR, pre-RSC, post-RSC, pre-GWL, and post-GWL were 13.86%, 12.52%, 15.57%, 13.40%, 7.71% and 8.10%, respectively, which indicates strong spatial dependence. The nugget to sill ratios in 2016 for post-EC, pre-SAR, post-SAR, post-RSC, pre-GWL and post-GWL were 0.03%, 21.60%, 6.84%, 6.33%, 6.61% and 6.38%, respectively, which also indicates strong spatial dependence. The spatial distribution analysis of groundwater EC, SAR, and RSC indicated that the groundwater quality was unfit for irrigation in the lower south-east part of the study area. The groundwater level (GWL) was also higher in that part of the study area during the pre-monsoon and post-monsoon seasons in 2015 and 2016. The overlay analysis also indicated that the groundwater EC, RSC and GWL values were higher in the south-east parts of the study area during pre-monsoon and post-monsoon seasons of 2015 and 2016. The higher values of these parameters in groundwater seem to be due to the simultaneous contribution of natural mineralization, the use of unpurified irrigation water, and anthropogenic inputs. Hence, there is an instant need to apply groundwater management practices in the rest of the region (especially in the lower south-east part) to overcome the future degradation of groundwater quality. The research findings of the present research study provide guidelines for the potential management of groundwater resources in relation to seasonal variation in other regions.

Author Contributions: Conceptualization, H.U.F., M.N.A., I.A., and A.G.; methodology, H.S., and H.U.F.; software, H.S., H.U.F., and M.N.A.; formal analysis, H.S., H.U.F., and M.N.A.; data curation, I.A., M.A., and P.S. writing—original draft preparation, H.S., and H.U.F.; writing—review and editing, Z.M.K., M.N.A., I.A., X.C.,

M.M., M.A., and A.G.; funding acquisition, A.G., and X.C. All authors have read and agreed to the published version of the manuscript.

Funding: This study was financially supported by the Strategic Priority Research Program of the Chinese -Academy of Sciences, the Pan-Third Pole Environment Study for a Green Silk Road (Grant No. XDA20060303), the National Natural Science Foundation of China (Grant No. 41950410575), the International Cooperation Project of National Natural Science Foundation of China (Grant No. 41761144079), the Xinjiang Tianchi Hundred Talents Program (Grant No.Y848041), the project of the research Center of Ecology and Environment in Central Asia (Grant No. Y934031), and the CAS PIFI fellowship (Grant No. 2021PC0002).

Acknowledgments: The authors are grateful to the Department of Land and Reclamation, Multan, for providing groundwater quality data to accomplish this research.

Conflicts of Interest: The authors declare no conflict of interest.

References

1. Vasanthavigar, M.; Srinivasamoorthy, K.; Vijayaragavan, K.; Rajiv Ganthi, R.; Chidambaram, S.; Anandhan, P.; Manivannan, R.; Vasudevan, S. Application of water quality index for groundwater quality assessment: Thirumanimuttar sub-basin, Tamilnadu, India. *Environ. Monit. Assess.* **2010**, *171*, 595–609. [CrossRef] [PubMed]

2. Tlili-Zrelli, B.; Gueddari, M.; Bouhlila, R. Spatial and temporal variations of water quality of mateur aquifer (Northeastern Tunisia): Suitability for irrigation and drinking purposes. *J. Chem.* **2018**. [CrossRef]

3. Ebrahimi, K.; Feiznia, B.S.; Jannat Rostami, C.M.; Ausati, K. *Assessing Temporal and Spatial Variations of Groundwater Quality (A Case Study: Kohpayeh-Segzi)*; IA University: Borujerd Branch, Iran, 2011; Volume 1.

4. Tatawat, R.K.; Chandel, C.P.S. A hydrochemical profile for assessing the groundwater quality of Jaipur City. *Environ. Monit. Assess.* **2008**, *143*, 337–343. [CrossRef]

5. Farid, H.U.; Mahmood-Khan, Z.; Ali, A.; Mubeen, M.; Anjum, M.N. Site-specific aquifer characterization and identification of potential groundwater areas in Pakistan. *Pol. J. Environ. Stud.* **2017**, *26*. [CrossRef]

6. Sharif, M.; Jabbar, A.; Niazi, M.A.; Mahr, A.B. Managing water availability and requirements in Pakistan: Challenges and way forward. *J. Agric. Res.* **2016**, *54*, 117–131.

7. Basharat, M. Spatial and temporal appraisal of groundwater depth and quality in LBDC command-issue. *Pak. J. Eng. Appl. Sci.* **2012**, *11*, 14–29.

8. Zhou, Y.; Wang, Y.; Li, Y.; Zwahlen, F.; Boillat, J. Hydrogeochemical characteristics of central Jianghan Plain, China. *Environ. Earth Sci.* **2013**, *68*, 765–778. [CrossRef]

9. Richards, L.A.; Magnone, D.; Sovann, C.; Kong, C.; Uhlemann, S.; Kuras, O.; van Dongen, B.E.; Ballentine, C.J.; Polya, D.A. High resolution profile of inorganic aqueous geochemistry and key redox zones in an arsenic bearing aquifer in Cambodia. *Sci. Total Environ.* **2017**, *590–591*, 540–553. [CrossRef]

10. Yan, S.F.; Yu, S.E.; Wu, Y.B.; Pan, D.F.; She, D.L.; Ji, J. Seasonal Variations in Groundwater Level and Salinity in Coastal Plain of Eastern China Influenced by Climate. *J. Chem.* **2015**, *2015*. [CrossRef]

11. Aladejana, J.A.; Kalin, R.M.; Sentenac, P.; Hassan, I. Assessing the impact of climate change on groundwater quality of the shallow coastal aquifer of eastern dahomey basin, southwestern Nigeria. *Water* **2020**, *12*, 224. [CrossRef]

12. Sauer, T.; Havlík, P.; Schneider, U.A.; Schmid, E.; Kindermann, G.; Obersteiner, M. Agriculture and resource availability in a changing world: The role of irrigation. *Water Resour. Res.* **2010**. [CrossRef]

13. Khatri, N.; Tyagi, S. Influences of natural and anthropogenic factors on surface and groundwater quality in rural and urban areas. *Front. Life Sci.* **2015**, *8*, 23–39. [CrossRef]

14. Nema, S.; Awasthi, M.K.; Nema, R.K. Spatial and temporal ground water responses to seasonal rainfall replenishment in an alluvial aquifer. *Biosci. Biotechnol. Res. Commun.* **2017**, *10*, 431–437. [CrossRef]

15. Tlili Zrelli, B.; Gueddari, M. Groundwater hydro-geochemistry of Mateur alluvial aquifer (Northern Tunisia). *J. Hydrogeol. Hydrol. Eng.* **2016**. [CrossRef]

16. Belkhiri, L.; Boudoukha, A.; Mouni, L. Amultivariate statistical analysis of groundwater chemistry data. *Int. J. Environ. Res.* **2011**, *5*, 537–544. [CrossRef]

17. Kim, J.H.; Kim, R.H.; Lee, J.; Cheong, T.J.; Yum, B.W.; Chang, H.W. Multivariate statistical analysis to identify the major factors governing groundwater quality in the coastal area of Kimje, South Korea. *Hydrol. Process.* **2005**. [CrossRef]

18. Mohammadi, Z. Assessing hydrochemical evolution of groundwater in limestone terrain via principal component analysis. *Environ. Earth Sci.* **2009**. [CrossRef]

19. Srivastava, P.K.; Han, D.; Gupta, M.; Mukherjee, S. Integrated framework for monitoring groundwater pollution using a geographical information system and multivariate analysis. *Hydrol. Sci. J.* **2012**. [CrossRef]

20. Liu, J.; Zhang, X.; Xia, J.; Wu, S.; She, D.; Zou, L. Characterizing and explaining spatio-temporal variation of water quality in a highly disturbed river by multi-statistical techniques. *Springerplus* **2016**. [CrossRef]

21. Zhou, Z.; Zhang, G.; Yan, M.; Wang, J. Spatial variability of the shallow groundwater level and its chemistry characteristics in the low plain around the Bohai sea, north China. *Environ. Monit. Assess.* **2012**, *184*, 3697–3710. [CrossRef]

22. Zhang, Q.; Singh, V.P.; Peng, J.; Chen, Y.D.; Li, J. Spatial-temporal changes of precipitation structure across the Pearl River basin, China. *J. Hydrol.* **2012**. [CrossRef]

23. Ağca, N. Spatial variability of groundwater quality and its suitability for drinking and irrigation in the Amik Plain (South Turkey). *Environ. Earth Sci.* **2014**. [CrossRef]

24. Omran, E.S.E. A proposed model to assess and map irrigation water well suitability using geospatial analysis. *Water* **2012**, *4*, 545–567. [CrossRef]

25. Srivastava, P.K.; Gupta, M.; Mukherjee, S. Mapping spatial distribution of pollutants in groundwater of a tropical area of India using remote sensing and GIS. *Appl. Geomat.* **2012**. [CrossRef]

26. Pacheco Castro, R.; Pacheco Ávila, J.; Ye, M.; Cabrera Sansores, A. Groundwater quality: Analysis of Its temporal and spatial variability in a karst aquifer. *Groundwater* **2018**. [CrossRef]

27. Farid, H.U.; Ahmad, I.; Anjum, M.N.; Khan, Z.M.; Iqbal, M.M.; Shakoor, A.; Mubeen, M. Assessing seasonal and long-term changes in groundwater quality due to over-abstraction using geostatistical techniques. *Environ. Earth Sci.* **2019**, *78*, 386. [CrossRef]

28. Yeh, H.F.; Lee, C.H.; Hsu, K.C.; Chang, P.H. GIS for the assessment of the groundwater recharge potential zone. *Environ. Geol.* **2009**. [CrossRef]

29. Piccini, C.; Marchetti, A.; Farina, R. Francaviglia, R. Application of indicator kriging to evaluate the probability of exceeding nitrate contamination thresholds. *Int. J. Environ. Res.* **2012**, *6*, 853–862.

30. Kumar, P.; Herath, S.; Avtar, R.; Takeuchi, K. Mapping of groundwater potential zones in Killinochi area, Sri Lanka, using GIS and remote sensing techniques. *Sustain. Water Resour. Manag.* **2016**. [CrossRef]

31. Iqbal, Z.; Abbas, F.; Ibrahim, M.; Ayyaz, M.M.; Ali, S.; Mahmood, A. Surveillance of heavy metals in maize grown with wastewater and their impacts on animal health in peri-urban areas of multan, Pakistan. *Pak. J. Agric. Sci.* **2019**. [CrossRef]

32. Abbas, F.; Ahmad, A.; Safeeq, M.; Ali, S.; Saleem, F.; Hammad, H.M.; Farhad, W. Changes in precipitation extremes over arid to semiarid and subhumid Punjab, Pakistan. *Theor. Appl. Climatol.* **2014**. [CrossRef]

33. Shakoor, A.; Arshad, M.; Bakhsh, A.; Ahmed, R. GIS based assessment and delineation of groundwater quality zones and its impact on agricultural productivity. *Pak. J. Agric. Sci.* **2015**, *52*, 837–843.

34. Singh, K.K.; Tewari, G.; Kumar, S.; Mancini, P.M. Evaluation of groundwater quality for suitability of irrigation purposes: A case study in the Udham Singh Nagar, Uttarakhand. *J. Chem.* **2020**. [CrossRef]

35. Frank, M. Eaton Significance of Carbonates in Irrigation Waters: Soil Science. Available online: https://journals.lww.com/soilsci/Citation/1950/02000/Significance_of_Carbonates_in_Irrigation_Waters.4.aspx (accessed on 29 November 2020).

36. Popugaeva, D.; Kreyman, K.; Ray, A.K. Assessment of Khibiny Alkaline Massif groundwater quality using statistical methods and water quality index. *Can. J. Chem. Eng.* **2020**, *98*, 205–212. [CrossRef]

37. Rossi, R.E.; Dungan, J.L.; Beck, L.R. Kriging in the shadows: Geostatistical interpolation for remote sensing. *Remote Sens. Environ.* **1994**. [CrossRef]

38. Chabala, L.M.; Mulolwa, A.; Lungu, O. Application of ordinary kriging in mapping soil organic carbon in Zambia. *Pedosphere* **2017**, *27*, 338–343. [CrossRef]

39. Gia Pham, T.; Kappas, M.; Van Huynh, C.; Hoang Khanh Nguyen, L. Application of ordinary kriging and regression kriging method for soil properties mapping in hilly region of central Vietnam. *ISPRS Int. J. Geo-Inf.* **2019**, *8*, 147. [CrossRef]

40. Mehrjardi, R.T.; Jahromi, M.Z.; Heidari, A. Spatial distribution of groundwater quality with geostatistics (Case study: Yazd-ardakan plain). *Appl. Sci.* **2008**, *4*, 9–17.

41. Al-Omran, A.M.; Aly, A.A.; Al-Wabel, M.I.; Al-Shayaa, M.S.; Sallam, A.S.; Nadeem, M.E. Geostatistical methods in evaluating spatial variability of groundwater quality in Al-Kharj Region, Saudi Arabia. *Appl. Water Sci.* **2017**, *7*, 4013–4023. [CrossRef]

42. Zhang, H.; Zhuang, S.; Qian, H.; Wang, F.; Ji, H. Spatial Variability of the Topsoil Organic Carbon in the Moso Bamboo Forests of Southern China in Association with Soil Properties. *PLoS ONE* **2015**, *10*, e0119175. [CrossRef]

43. Sharma, P.; Sood, S.; Duggirala, G. Seasonal Variation of Groundwater Quality in Rural Areas of Jaipur District, Rajasthan. *Indian J. Sci. Technol.* **2017**. [CrossRef]

44. Bui, D.D.; Kawamura, A.; Tong, T.N.; Amaguchi, H.; Nakagawa, N. Spatio-temporal analysis of recent groundwater-level trends in the Red River Delta, Vietnam. *Hydrogeol. J.* **2012**. [CrossRef]

45. Rehman, F.; Cheema, T.; Lisa, M.; Azeem, T.; Ali, N.A.; Khan, Z.; Rehman, F.; Rehman, S.U. Statistical analysis tools for the assessment of groundwater chemical variations in wadi bani malik area, Saudi Arabia. *Glob. Nest J.* **2018**. [CrossRef]

46. Srinivasamoorthy, K.; Vasanthavigar, M.; Vijayaraghavan, K.; Sarathidasan, R.; Gopinath, S. Hydrochemistry of groundwater in a coastal region of Cuddalore district, Tamilnadu, India: Implication for quality assessment. *Arab. J. Geosci.* **2013**, *6*, 441–454. [CrossRef]

47. Ploum, S.; Laudon, H.; Peralta-Tapia, A.; Kuglerová, L. Are hydrological pathways and variability in groundwater chemistry linked in the riparian boreal forest? *Hydrol. Earth Syst. Sci. Discuss.* **2019**. [CrossRef]

48. Mabrouk, M.; Jonoski, A.; Essink, G.H.P.O.; Uhlenbrook, S. Assessing the fresh-saline groundwater distribution in the Nile delta aquifer using a 3D variable-density groundwater flow model. *Water* **2019**, *11*, 1946. [CrossRef]

49. Chen, L.; Feng, Q. Geostatistical analysis of temporal and spatial variations in groundwater levels and quality in the Minqin oasis, Northwest China. *Environ. Earth Sci.* **2013**. [CrossRef]

50. Chen, J.; Huang, Q.; Lin, Y.; Fang, Y.; Qian, H.; Liu, R.; Ma, H. Hydrogeochemical characteristics and quality assessment of groundwater in an irrigated region, Northwest China. *Water* **2019**, *11*, 96. [CrossRef]

51. Xu, P.; Feng, W.; Qian, H.; Zhang, Q. Hydrogeochemical characterization and irrigation quality assessment of shallow groundwater in the central-Western Guanzhong basin, China. *Int. J. Environ. Res. Public Health* **2019**, *16*, 1492. [CrossRef]

52. Visser, H.; De Nijs, T. The map comparison kit. *Environ. Model. Softw.* **2006**, *21*, 346–358. [CrossRef]

53. Pontius, R.G.; Huffaker, D.; Denman, K. Useful techniques of validation for spatially explicit land-change models. *Ecol. Modell.* **2004**, *179*, 445–461. [CrossRef]

54. Levine, R.S.; Yorita, K.L.; Walsh, M.C.; Reynolds, M.G. A method for statistically comparing spatial distribution maps. *Int. J. Health Geogr.* **2009**, *8*, 7. [CrossRef]

55. Faisal, N.; Jameel, A.; Cheema, S.B.; Ghuffar, A.; Mahmood, A.; Rasul, G. Third successive active monsoon over Pakistan—An analysis and diagnostic study of monsoon 2012. *Pak. J. Meteorol.* **2013**, *9*, 73–84.

56. Li, H.; Lu, Y.; Zheng, C.; Zhang, X.; Zhou, B.; Wu, J. Seasonal and inter-annual variability of groundwater and their responses to climate change and human activities in arid and desert areas: A case study in Yaoba Oasis, Northwest China. *Water* **2020**, *12*, 303. [CrossRef]

57. Tahmasebi, P.; Mahmudy-Gharaie, M.H.; Ghassemzadeh, F.; Karimi Karouyeh, A. Assessment of groundwater suitability for irrigation in a gold mine surrounding area, NE Iran. *Environ. Earth Sci.* **2018**. [CrossRef]

58. Yang, J.; Cao, G.; Han, D.; Yuan, H.; Hu, Y.; Shi, P.; Chen, Y. Deformation of the aquifer system under groundwater level fluctuations and its implication for land subsidence control in the Tianjin coastal region. *Environ. Monit. Assess.* **2019**. [CrossRef]

Publisher's Note: MDPI stays neutral with regard to jurisdictional claims in published maps and institutional affiliations.

Article

eNaBLe, an On-Line Tool to Evaluate Natural Background Levels in Groundwater Bodies

Daniele Parrone , Eleonora Frollini * , Elisabetta Preziosi and Stefano Ghergo

IRSA-CNR, Water Research Institute—National Research Council, Via Salaria km 29.300, PB 10,
00015 Monterotondo (Rome), Italy; parrone@irsa.cnr.it (D.P.); preziosi@irsa.cnr.it (E.P.); ghergo@irsa.cnr.it (S.G.)
* Correspondence: frollini@irsa.cnr.it

Abstract: Inorganic compounds in groundwater may derive from both natural processes and anthropogenic activities. The assessment of natural background levels (NBLs) is often useful to distinguish these sources. The approaches for the NBLs assessment can be classified as geochemical (e.g., the well-known pre-selection method) or statistical, the latter involving the application of statistical procedures to separate natural and anthropogenic populations. National Guidelines for the NBLs assessment in groundwater have been published in Italy (ISPRA 155/2017), based mainly on the pre-selection method. The Guidelines propose different assessment paths according to the sample size in spatial/temporal dimension and the type of the distribution of the pre-selected dataset, taking also into account the redox conditions of the groundwater body. The obtained NBLs are labelled with a different confidence level in function of number of total observations/monitoring sites, extension of groundwater body and aquifer type (confined or unconfined). To support the implementation of the Guidelines, the on-line tool evaluation of natural background levels (eNaBLe), written in PHP and using MySQL as DBMS (DataBase Management System), has been developed. The main goal of this paper is to describe the functioning of eNaBLe and test the tool on a case study in central Italy. We calculated the NBLs of As, F, Fe and Mn in the southern portion of the Mounts Vulsini groundwater body, within the volcanic province of Latium (Central Italy), also separating the reducing and oxidizing facies. Specific results aside, this study allowed to verify the functioning and possible improvements of the online tool and to identify some criticalities in the procedure NBLs assessment at the groundwater body scale

Keywords: natural background; groundwater body; conceptual model; preselection; nitrates; confidence level

Citation: Parrone, D.; Frollini, E.; Preziosi, E.; Ghergo, S. eNaBLe, an On-Line Tool to Evaluate Natural Background Levels in Groundwater Bodies. *Water* **2021**, *13*, 74. https://doi.org/10.3390/w13010074

Received: 26 November 2020
Accepted: 27 December 2020
Published: 31 December 2020

Publisher's Note: MDPI stays neutral with regard to jurisdictional claims in published maps and institutional affiliations.

1. Introduction

The presence of inorganic potentially toxic elements in groundwater represents a significant problem in many parts of the world. They may derive from both natural processes and anthropogenic activities and natural background levels (NBLs) assessment is often used to distinguish these sources. The NBL has been defined in the Groundwater Directive (GWD) [1] as "the concentration of a substance or the value of an indicator in a groundwater body corresponding to no, or extremely limited, anthropogenic alterations, compared to unaltered conditions". The GWD requests the EU Member States to individuate appropriate threshold values (TVs) for various potentially harmful substances, taking into account NBLs when necessary, in order to evaluate the chemical status of groundwater bodies.

Currently, it is possible to distinguish a geochemical and a statistical approach for the NBLs assessment. The geochemical approach, originally called "pre-selection" (PS) within the BRIDGE (Background cRiteria for the Identification of Groundwater thrEsholds) project [2], requires the identification of groundwater with no or negligible human impact, using markers such as nitrates/ammonia in oxidizing/reducing environments, organic compounds and isotopes. Once the samples not affected by anthropogenic impact have been selected, a representative value (usually a chosen percentile) of the natural background is derived. Many examples of PS method applications, including comparisons between groundwater bodies of different European countries [3,4], can be found in the literature [5–12].

The statistical approach involves the separation of uninfluenced and influenced populations by means of statistical procedures. For this purpose, numerous statistical techniques have been developed and tested. Some of these methods point to eliminate the outliers, assuming that the remaining data belong to the natural background. In some cases, the same methods can be used to separate different data populations. The "mean + 2σ" [13], the Median Absolute Deviation (MAD) [14], the Box and Whisker plot [15], the component separation [11,16] and other parametric or non-parametric techniques, including graphical methods as probability plots or quantile–quantile plots, have been largely used for this purpose [17–20]. Please refer to Preziosi et al. [21] for a review of the most common approaches for NBLs assessment, including many statistical techniques and the pre-selection methods.

In 2017, national Guidelines for the NBLs assessment in groundwater were published in Italy [22]. Starting from these Guidelines, Frollini et al. [23] applied the procedure to define the NBLs at the groundwater body (GWB) scale, while Parrone et al. [24] tested a multi-methodological approach on two case studies, also suggesting new criteria for the choice of the nitrate threshold to be used for the pre-selection of non-contaminated samples. An automated approach implementing both the component separation and the pre-selection methods was recently proposed by Chidichimo et al. [25].

The procedure is based mainly on the pre-selection method but different assessment paths are proposed according to the redox conditions of the GWB, the sample size in both the spatial and temporal dimension and the type of the distribution, normal or not, of the pre-selected dataset. The complexity of the schema is essential because of the great heterogeneity of the different GWB monitoring data. For this reason, moreover, the obtained NBLs are labelled with a different confidence level in function of: The number of total observations or the number of total monitoring sites (MSs), extension of groundwater body and aquifer type (confined or unconfined). For GWBs characterized by NBLs with low confidence levels, further monitoring activities are requested.

To provide support to operators involved in the use of the Italian Guidelines and to achieve a harmonization of procedures between the different structures involved, evaluation of natural background levels (eNaBLe), an on-line tool implementing the sequence of operations for NBL assessment, was developed at Water Research Institute – National Research Council (IRSA-CNR) [26]. The procedure implemented in the eNaBLe tool, written in PHP and using MySQL as DBMS, is organized into three logical blocks:

- Selection of the calculation parameters;
- NBL calculation for all the chemical parameters that show concentration values exceeding the relative TV;
- Graphical output of the results.

The main goal of this work is to test the eNaBLe tool on a case study in Central Italy. We calculated the NBLs of As, F, Fe and Mn in the southern portion of the Mounts Vulsini GWB, within the volcanic province of Latium (Central Italy), also operating a separation of data based on the different redox conditions found in the study area.

2. Materials and Methods

This section is dedicated to the detailed description of the tool and of all the steps that lead to the calculation of the natural background values. A description of the case study to which the Guidelines were applied is also included.

2.1. The Italian Guidelines and the eNaBLe Tool

eNaBLe operates in compliance with the provisions of the Italian Guidelines for NBLs assessment, the general scheme of which is shown in Figure 1.

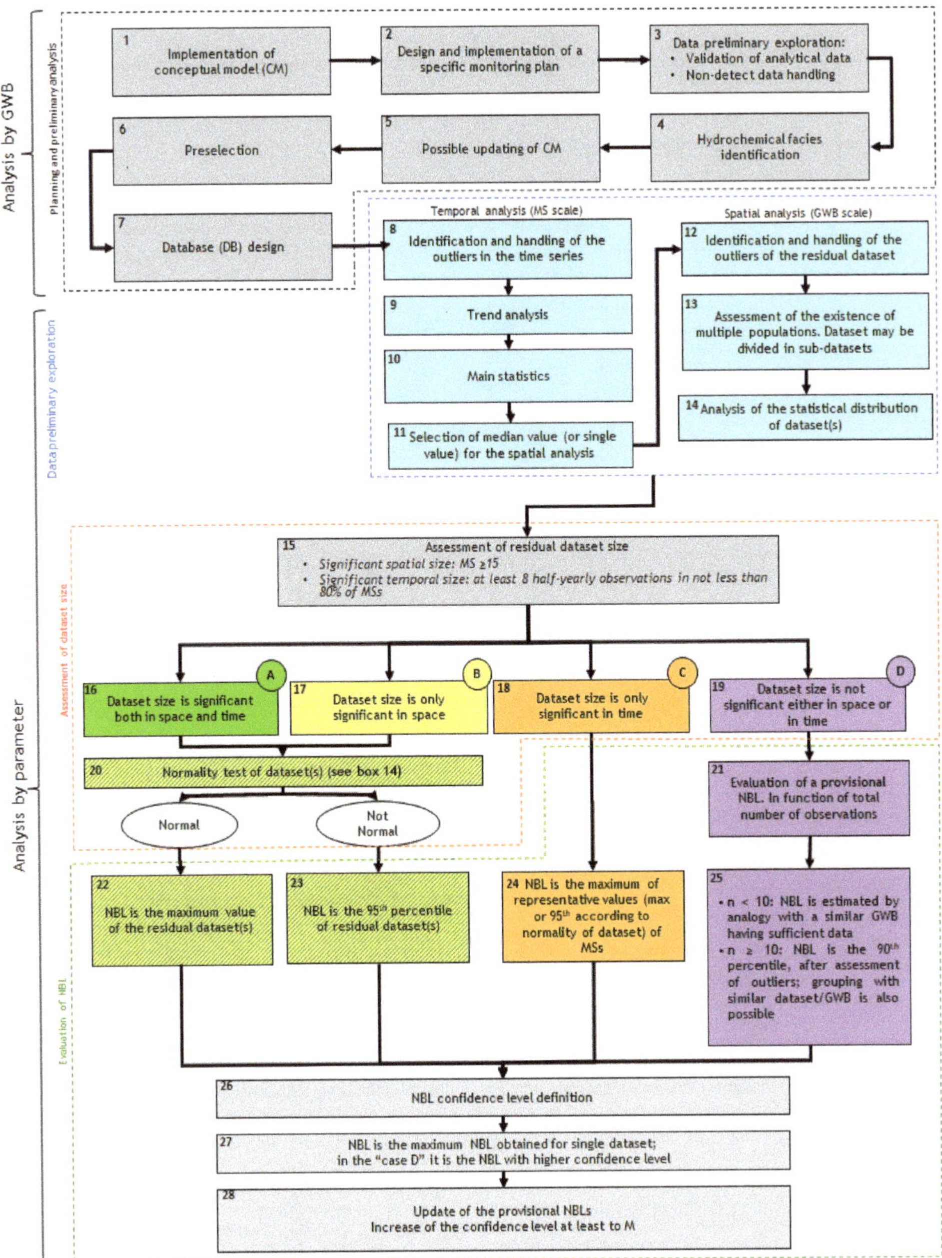

Figure 1. Italian Guidelines: Scheme of the procedure for the natural background levels (NBLs) assessment at the Groundwater Body (GWB) scale [23].

The general methodology adopted by the Guidelines is that of preselection which is accompanied by statistical evaluations in order to adapt the application of the procedure to the various groundwater bodies and allow the use of datasets of different numerical consistency. Four different evaluation paths (A, B, C, D) were therefore identified according to the spatial and temporal consistency of the available dataset. As a consequence of the great differences between the various GWB, both in terms of extension and number of monitoring networks, it was considered appropriate to associate also an index (confidence level) to the defined NBLs linked to the size of the statistical sample, to the extension and type of the aquifer (unconfined or confined) (Table 1). For more detailed explanation of the entire procedure, see [22].

Table 1. Confidence levels. The confidence level attributed to a NBL takes into account the characteristics of the dataset (case A, B, C, D), of the total number of observations and monitoring sites (MSs), also with reference to the extension of the GWB and the aquifer type (unconfined/confined). Confidence levels: H: High; M: Medium; L: Low; and LL: Very Low.

N. Total obs.	N. Total MS	Size of GWB of Portion of GWB Represented by Dataset (km^2) Aquifer Type (Unconfined/Confined)							
		<10 km^2		10–70 km^2		70–700 km^2		>700 km^2	
		Unconf.	Conf.	Unconf.	Conf.	Unconf.	Conf.	Unconf.	Conf.
	15–25	H	H	H	H	M	H	M	H
	>25	H	H	H	H	H	H	H	H
	15–25	M	H	M	H	L	M	L	M
	>25	H	H	H	H	M	H	M	M
≤15		M	M	L	M	LL	L	LL	LL
16–30		M	H	M	M	L	M	LL	L
>30		H	H	M	H	M	H	M	M
<10		L	L	LL	LL	LL	LL	LL	LL
≥10		L	L	L	L	LL	L	LL	LL

The complexity of the Guidelines framework advised the demand for an automated procedure which, while following the general formulation and the involved provisions, could facilitate the task of the operators. In 2018, a first automated approach was developed at IRSA-CNR. Later, the procedure was implemented as the eNaBLe on-line tool. The complete procedure, written in PHP [27] and using MySQL [28] as relational databases manager, is structured in modules, each one performing specific tasks on the input data, using specific configuration parameters. Some of these parameters are inherited from the Guidelines and cannot be changed; the others can be modified by the user. Each resulting dataset is then sent to the next module. The overall breakdown of eNable tool is shown in Figure 2. The modularity of eNaBLe has allowed us to easily integrate it into the Institute's Water Resources Database [29] which now is the source for the raw data for the NBL evaluation.

The user interface for the tool is divided into three logical phases which correspond to three different procedures and therefore to three distinct associated web pages: Selection of the analytical parameters for the evaluation of the NBLs and selection of the calculation options to be used; NBLs assessment and relative confidence levels; and output of results in graphical, tabular and report form. These procedures will now be described schematically.

2.1.1. Choice of the GWB, of the Parameters and the Calculation Options to Be Used for the NBLs Assessment.

Performing a query on the Water Resource Database [29], eNaBLe produces a table in which all the analytical parameters and their number are listed, proposing for the NBLs evaluation those parameters in which at least one exceedance of the relative TVs has been found. It is therefore possible to select the upper limits to be used in the validation, preselection and calculation procedures.

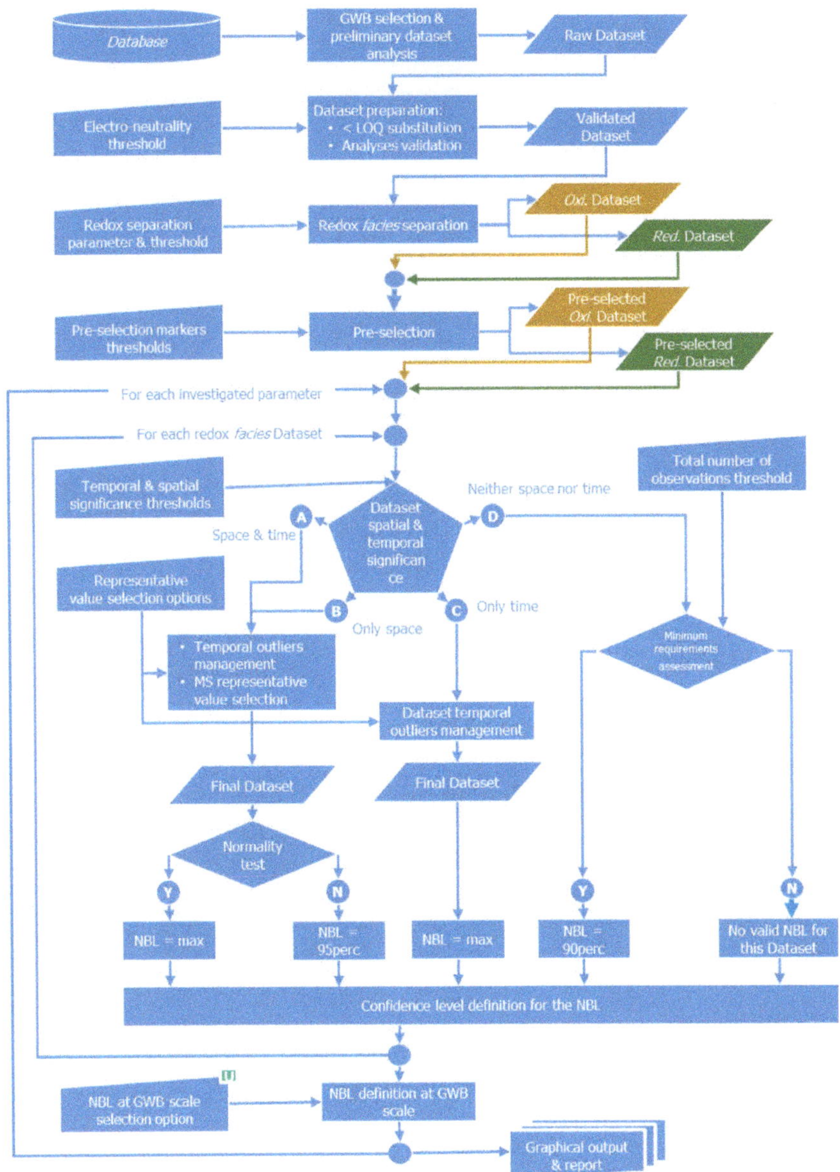

Figure 2. General scheme of the evaluation of natural background levels (eNaBLe) tool.

In particular, it is possible to define:

- The inclusion or not of those stations that, due to their overall characteristics, have been judged unsuitable for NBLs evaluation (checkbox "Use MPs exclusion list").
- A threshold for the control of the electrical balance, typically 5% [30] or 10% [2].
- The appropriate time interval for the data to be processed.
- Parameters to be used for the redox facies separation. The redox potential (ORP) or the dissolved oxygen concentration (DO) can be selected. The predefined limits proposed by the system for these two parameters (e.g., DO 3 mg/L or ORP 100 mV) can be

modified. It is also possible, by deselecting the checkbox, to completely disable the redox facies separation.

- The limits to be applied for the preselection process to nitrates or ammonia in relation to the oxidizing or reducing facies. The system proposes values equal to 75% of the expected quality standards.
- Methodology for the management of the time series in order to calculate the representative values for the monitoring stations. It is possible to select the option of the simple calculation of the median; else an analysis and elimination of the outliers and the choice of the maximum value after the elimination. In this last case, it is also possible to select the method for the identification of the outliers (Huber non-parametric test or boxplot) [31];
- Selection of the method for identifying and eliminating the outliers of the representative values of the MSs (Huber test or boxplot). It is also possible to disable the procedure of outliers elimination.
- Selection of the method of NBLs assessment at GWB scale. If the MSs have been classified based on the redox facies, it is possible to select the highest value or the one with the highest confidence level.

A screenshot of the configuration option selection window is shown in Figure 3.

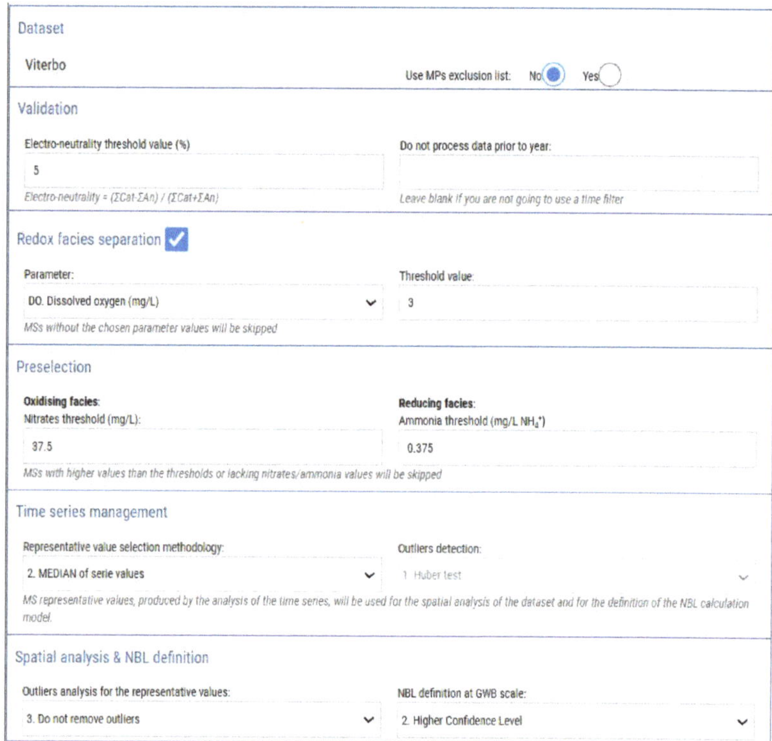

Figure 3. Selection of the calculation parameters within the eNaBLe tool.

2.1.2. NBLs Assessment and Relative Confidence Levels

The general NBLs assessment procedure can be summarized in the following operational phases:

1. Query to the database to retrieve data of the selected GWB, relative to the selected parameters and to the parameters linked to the facies separation.

2. MSs list filtering using the selected time interval and, possibly, the exclusion list.
3. Processing of the values below the limit of quantification (LOQ). All analytical values reported in the dataset as lower than the LOQ, are replaced with half of the LOQ value.
4. Validation of each sample analysis using the threshold for electrical balance entered in the configuration phase.
5. Redox facies separation using the parameter and threshold selected in the configuration phase.
6. Preselection. As regards the oxidizing facies, the monitoring stations with the median of the nitrate values higher than the selected limit in the configuration phase or with missing values are rejected. For the reducing facies, MSs with the median of ammonia values higher than the selected indicated in the configuration phase or with missing values are rejected. If redox facies separation has been disabled, stations with values greater than the respective limits of nitrates and/or ammonia or stations with missing data will be excluded from the dataset.
7. Analysis of the time series and calculation of the representative values for each MS using the methodology selected during the configuration phase, rejecting, if required, the outliers data.
8. Analysis of the representative values of the MSs using the methodology selected for the management of the outliers and verifying the normality of the resulting dataset distribution using the Shapiro–Wilk test [32].
9. Evaluation of the consistency of the dataset of representative values to identify the assessment path for NBL calculation. In order to distinguish different levels of spatial and temporal consistency of the data available for a given GWB, once the processing described in the previous sections has been concluded, the consistency of the individual datasets is definitively assessed. Therefore, 4 cases are identified:

   - Case A: Sample size adequate to describe the temporal and spatial variability of the parameter for the dataset in question;
   - Case B: Sample size adequate to describe the spatial variability of the parameter for the dataset in question, but not the temporal variability;
   - Case C: Sample size adequate to describe the temporal variability of the parameter for the dataset in question, but not spatial variability;
   - Case D: Sample size inadequate to describe the spatial and temporal variability of the parameter for the dataset in question.

   For sample size adequate to describe, from a purely statistical point of view, the spatial variability of the system in question, it means a minimum of 15 monitoring stations adequately distributed ($N \geq 15$). For sample size adequate to describe, from a purely statistical point of view, the temporal variability of the system in question, it means a minimum of 8 observations ($n \geq 8$) distributed regularly over at least 2 years for each station, over at least 80% of the monitoring stations. These requirements are those listed in the Guidelines and are not modifiable by the tool users.

10. Calculation of NBLs. The NBL is given for datasets of type A, B or C by the maximum value among the representative values of the MSs (in case the dataset shows a normal distribution) or by the 95th percentile of the representative values of the MSs (in case of non-normality of the dataset). For the type D datasets, if the number of total observations available is ≥ 10, the provisional NBL will be given by the 90th percentile of the total available observations. When the total number of observations is less than 10, the calculation will not be carried out and a provisional NBL can be obtained by analogy with other GWBs, or portions of GWBs, characterized by similar conditions in terms of geochemical facies, hydrogeological context and anthropic pressures.
11. Determination of the confidence level to be associated to the calculated NBLs. It was considered appropriate to associate an index (confidence level) to the NBLs, as a function of the size of the statistical sample on which the calculation was based, of

the dimensional (area) and typological (unconfined or confined) characteristics of the GWB (Table 1).

12. Assignment of a NBL at the groundwater body scale using the methodology selected in the configuration phase.

Steps 6 to 11 are performed by eNaBLe for the different redox facies and steps 7 to 12 for each of the analytical parameters selected.

2.1.3. Output of Results

At the end of the calculation procedures, eNaBLe produces a summary with the configuration options and the results (TV, validated data, minimum and maximum representative values, calculation model and normality of distribution) and the calculated NBLs with relative confidence levels for the investigated parameters, differentiated by redox facies. Finally, the system produces a table which shows the NBLs relative to the entire GWB. If during the configuration phase the separation of redox facies has been deactivated, only the NBLs calculated for the entire GWB are produced.

By the appropriate links contained in the results page, files in CVS format, containing the intermediate and final datasets produced during the data processing, are also accessible.

The tool will finally produce graphical reports of the selected parameters consisting of a table with the main statistical data (minimum and maximum value, mean, median, MAD, 95th percentile and normality of distribution), quantile-quantile plots and the georeferenced spatial distribution of the monitoring stations. A printable PDF files in which are summarized all the configuration parameters and the resulting NBLs, is also available.

2.2. *The Mounts Vulsini Groundwater Body*

The investigated area extends for about 60 km^2. It is located on the southern flank of the Mounts Vulsini groundwater body, an unconfined aquifer hosted in the Pleistocene volcanites of the Vulsini volcanic district (Central Italy). Groundwater in the study area flows from N to SW and ESE (Figure 4).

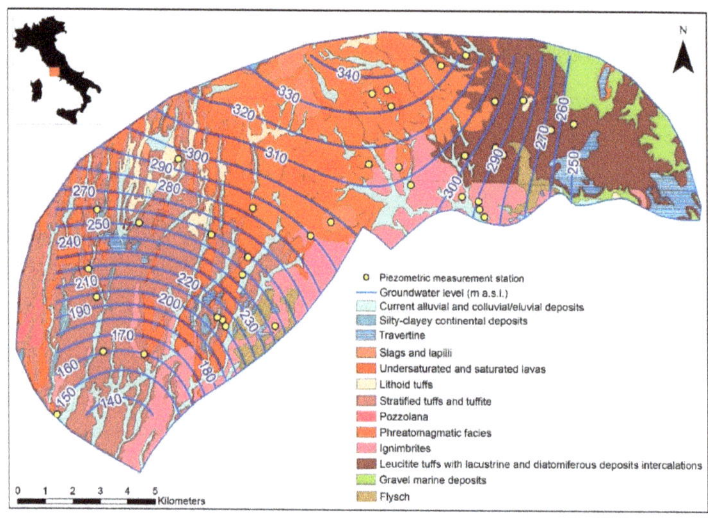

Figure 4. Groundwater level contour map of the southern Mounts Vulsini GWB and outcropping lithologies.

The main anthropogenic pressures are agriculture and animal husbandry. Urban areas and industrial sites including waste management facilities are also present in the area.

Groundwaters are mainly of the alkaline-earth bicarbonate and alkaline bicarbonate type; due to different natural phenomena (water–rock interaction, upwelling of geothermal fluids along fracture/fault systems and presence of mineral deposits), As and F are known to be widespread and co-present in the area [33–40], mainly in oxidizing conditions, with values often higher than the standards set for human consumption by WHO [41] and Directive 98/83/EC. Fe and Mn, on the other hand, are mainly linked to reducing conditions that are found locally, causing the reductive dissolution of their oxi-hydroxides and significant concentrations of the two elements in groundwater.

A total of 50 groundwater samples were collected from private wells in July 2017, 9 of which have been resampled twice more in January and July 2020 to increase the numerosity of the peculiar reducing facies existing in the area. Groundwater samples were collected following standardized sampling protocols [42,43]. Particular attention was paid to the in-line measurement of physical–chemical parameters, whose correct determination is crucial for the definition of the aquifer conceptual model and in the setting of the calculation parameters included in the procedure for NBLs assessment (e.g., DO and ORP accurate measurements for the redox facies separation).

Physical–chemical parameters and chemical data were used to build the starting database for the calculation of NBLs, operated by using the eNaBLe tool.

3. Results

3.1. Configuration and Preliminary Analysis on the Monitoring Stations

The first stage of the procedure includes a series of operations to be applied to the 50 monitoring stations (MSs), all belonging to the Mounts Vulsini groundwater body. All analytical values below the limit of quantification should be replaced with a value equal to half the LOQ. However, none of the parameters of interest for this study (As, F, Fe, Mn) showed values lower than the LOQ. As regards the data validation, the threshold of the electrical balance (5%) did not result in the elimination of any sample (maximum error = 3.8%).

The next step concerns the redox facies separation, for which a DO threshold of 3.0 mg/L was used. This value led to the identification of an oxidizing facies (41 water points), with DO > 3.0 mg/L, largely dominant and widely present throughout the study area, and a reducing facies (9 water points), with DO < 3.0 mg/L, not spatially diffused but rather linked to a few isolated points (Figure 5).

The third and last step of this phase of the procedure is represented by the preselection of the MSs, operated using two different markers of anthropogenic contamination, according to the redox conditions: NO_3^- (<37.5 mg/L) for the oxidizing facies and NH_4^+ (<0.375 mg/L) for the reducing facies. The selected concentration limits correspond to the 75% of the expected quality standards/TVs. As for the oxidizing facies, 15 samples were discarded, so the preselected dataset is composed of 26 MSs useful for the NBLs calculation (Figure 5). No points were discarded for the reducing facies, as all NH_4^+ values are below the chosen threshold. The preselected dataset is therefore composed of the 9 initial stations.

Before continuing with the next parameter-specific phase, it was decided to separately evaluate the correlation between the elements of interest and the redox parameters (DO and ORP), in order to evaluate their sensitivity to the redox conditions. The Pearson parametric and Spearman non-parametric correlation indexes (Table 2) show a significant negative correlation with the redox parameters for Fe and Mn (mutually proportional), while As and F are well correlated with each other but not redox sensitive elements.

Consequently, in the NBLs assessment we have decided to proceed in a diversified way, keeping the two redox facies separate for Fe and Mn and instead defining a single value, relative to the entire preselected dataset (35 water points), for As and F, not affected by the redox conditions.

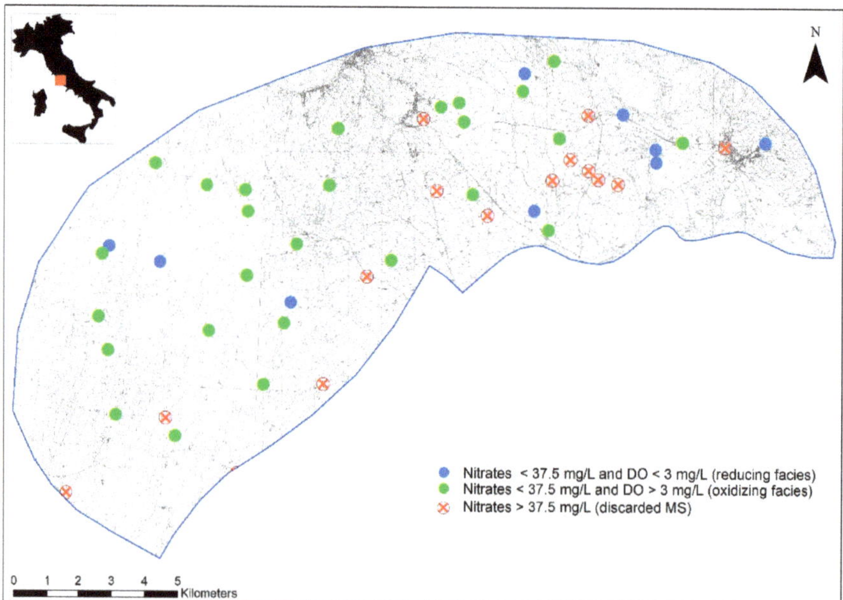

Figure 5. Redox facies subdivision based on dissolved oxygen concentration (DO) concentration and preselected MSs.

Table 2. Pearson (r) and Spearman (r_s) correlation coefficients for the redox parameters and the investigated elements. Statistically significant values in bold ($\alpha = 0.01$).

			Pearson (r)			
DO		0.58	**−0.25**	**−0.72**	**−0.69**	−0.15
ORP	**0.43**		−0.39	**−0.45**	**−0.42**	−0.21
F	−0.21	−0.29		0.18	0.21	**0.51**
Mn	**−0.66**	**−0.48**	0.16		**0.99**	0.13
Fe	**−0.53**	**−0.42**	0.28	**0.76**		0.11
As	−0.04	−0.09	**0.59**	−0.12	0.04	
	DO	**ORP**	**F**	**Mn**	**Fe**	**As**
			Spearman (r_s)			

3.2. NBLs Calculation for As and F

For the 26 stations of the oxidizing facies, the available data relate to a single sampling (2017), for which only the spatial analysis was carried out. On the other hand, as regards the 9 water points in reducing facies, the dataset also includes the samples of January and July 2020. For the two parameters, a temporal processing was therefore carried out, defining a representative value for each station, given by the median of the values measured in the three campaigns. Consequently, the As and F dataset used for the spatial analysis consists of 35 data, of which 26 individual values of the oxidizing facies and 9 median values of the reducing facies.

For both parameters, the presence of anomalous data was then analyzed using the Huber's non-parametric test, which identified two outliers for the F and one for the As. In the current state of knowledge, however, there are no valid scientific reasons to exclude these samples, which were therefore included in the final dataset for the definition of NBLs, consisting of 35 data. It is therefore the case B foreseen in the Guidelines, in which there is a significant spatial but not temporal dimension.

The Shapiro–Wilk test was then applied to verify the normality of data, showing that only F follows a Gaussian distribution. Consequently, according to the Guidelines, the NBL for F is given by the maximum of the statistical sample (3.56 mg/L), while for As it is equal to 95th percentile of the same statistical sample (20.5 µg/L).

Finally, considering the aquifer type (unconfined), its extension (60 km^2) and the number of total samples (35), for both parameters it was possible to associate a high confidence level to the defined NBLs.

3.3. NBLs Calculation for Fe and Mn (Oxidizing Facies)

As previously observed, for the 26 MSs in oxidizing facies, the available data belongs to a single sampling survey (2017), therefore only the spatial analysis was carried out. Huber's test found 2 outliers for Fe and 4 for Mn. However, also in this case there any scientific elements were identified to discard these data, hence the final dataset for the definition of NBLs remains the original one of 26 data. This is again the case B described in the Guidelines, in which there is a significant spatial but not a temporal dimension.

As expected, the subsequent Shapiro–Wilk test showed how the distributions of the two elements are far from normal, so in both cases the NBL should be set to the 95th percentile of the statistical sample (105.1 µg/L for Fe and 9.1 µg/L for Mn) and for both parameters a high confidence level was associated to the calculated NBL.

3.4. NBLs Calculation for Fe and Mn (Reducing Facies)

As regards the reducing facies, the dataset consists of only 9 water points, corresponding to the case D indicated in the Guidelines, the worst in terms of available information, in which neither a spatial nor a significant temporal dimension is reached. In these situations, the Guidelines suggest to estimate a provisional NBL, combining all available observations (in space and time). For Fe and Mn the total observations available are 26. Huber's test did not highlight possible outliers, so the 90th percentile of the observations for both parameters was calculated. The estimated NBLs are equal to 7364.7 µg/L for Fe and 805.0 µg/L for Mn. The associated confidence level is low and new monitoring observations are needed to improve the reliability of the dataset.

The calculated NBLs and the associated confidence levels are shown in Table 3. As indicated in the Guidelines, the NBL for each parameter was expressed with the same unit of measurement and rounded up with the same number of decimals as the relative limit set by Decree 152/2006 [44]. At present, as indicated by the Guidelines, for Fe and Mn the NBL for the groundwater body will correspond to that defined for the oxidizing facies, which has the highest confidence level.

Table 3. Calculated NBLs and associated confidence levels.

	F (mg/L)	Mn (µg/L)	Fe (µg/L)	As (µg/L)	Confidence Level
Preselected Dataset	3.6	-	-	21	High
Oxidizing Facies	-	9	105	-	High
Reducing Facies	-	805	7365	-	Low

4. Discussion

The study on the natural background values for the southern Mounts Vulsini groundwater body confirmed the presence of high concentrations of As and F, which show numerous exceedances of the TVs set at the national level by Decree 30/2009 (1.5 mg/L and 10 µg/L) [45] and are naturally present in groundwater. The relatively low range of concentration, the existence of single data populations and the normal (for fluoride) or close to normal (for arsenic) distributions suggest that the presence of the two elements in groundwater can be largely attributed to the water–rock interaction processes within the volcanic aquifer. The statistical analysis does not highlight further overlapping phenomena and this also translates into the definition of a particularly reliable NBL for the investigated groundwater body. Fe and Mn show a wide range of concentrations, depending on the redox conditions found in the aquifer. In the most common oxidizing conditions, the natural concentrations are generally low, while in the reducing conditions that can be found locally, the values are considerably higher and well above the limits set by Decree 152/2006 for the assessment of pollution while monitoring impacts on groundwater of e.g., industrial activities. The nature of these few peculiar points should be further investigated, in order to exclude any contribution of anthropogenic origin (currently not identifiable and quantifiable). Unlike the oxidizing facies, the few points belonging to the reducing facies do not show a particular spatial distribution within the GWB. This results in a geochemical population that can hardly be better characterized, due to the difficulty in finding other equally peculiar sampling points, which would make the dataset statistically more robust. The reliability of the NBL in this case can be improved mainly by monitoring these MSs, thus increasing the number of observations over time.

During the application of the Italian Guidelines for the NBLs assessment, some critical issues emerged, partly associated with the Guidelines themselves, others more specific of the online tool.

With regard to the first phase of the procedure, MSs specific, the software allows to perform the separation of the redox facies or alternatively to keep a single dataset. However, this choice cannot be applied in a different way depending on the parameters whose NBL is to be determined. Consequently, the procedure must be repeated twice, for the total dataset and for the separate facies.

Moreover, it is not clear in the Guidelines whether a temporal analysis of the pre-election markers (nitrates/ammonia) should be done. Currently, eNaBLe calculates the median of the temporal data and if this is < the selected limit (e.g., 37.5 mg/L for nitrates or 0.375 mg/L for ammonia), the MS is considered useful for the calculation of NBLs; otherwise, it is discarded. However, this entails the risk of considering stations that have exceedances of these limits. In addition, since an analysis of temporal trends is not envisaged, stations that show ascending temporal trends of the markers, suggesting contamination in progress, could be included in the preselected dataset. In this regard, an official Guideline including different statistical methods for trend analysis, estimating concentration scenarios and identification of trend reversal, have been published in Italy in 2017 [46]. The use of these techniques, currently to be applied externally to the tool, could be helpful to integrate the temporal analysis of data, in particular for substances of clear anthropogenic origin such as nitrates. About this, Frollini et al. [47] have recently applied a slightly modified version of the Guideline to a groundwater body in Northern Italy featuring nitrate pollution, discussing its advantages and limitations.

Furthermore, the software does not currently allow to evaluate the correlation between chemical elements and redox parameters or even other chemical-physical parameters; it is therefore not possible to statistically evaluate whether they are redox-sensitive (operation currently executable only externally to the tool) and make a justified choice of the facies separation.

As regards the second phase of the procedure, parameter-specific, the software evaluates the presence of data outliers through application of the non-parametric Huber test (or alternatively, extrapolating them from a simple boxplot). This is clearly a simplification of the procedure, as non-parametric methods do not require any knowledge or assumptions about the form of data distribution. These are robust techniques, as they are applicable to any situation. However, the tool could also provide for the possibility of applying parametric tests (e.g., Rosner test), that are statistical procedures based on the assumption of normality of the data, more powerful and preferable to non-parametric tests, in particular when the data follow the hypothesized distribution. In the case study, for example, the Huber test identifies 2 outliers for F. However, the distribution of data is normal, both including and excluding the outliers, and the Rosner test [48], recommended for Gaussian distributions with more than 25 data, applied externally to the tool, does not detect anomalous values. Hence in the definition of the NBL we have included the outliers, since their exclusion is not supported by evident scientific reasons. However, the adoption of a parametric test, in this case, would have simplified the path of definition of the NBL.

Furthermore, again regarding the statistical study, it is limited only to the evaluation of the normality of the data and the presence of anomalous values, which can be correctly discarded or not. The evaluation of the existence of multiple populations, which would lead to a subdivision of the dataset before calculating the NBLs as required by the Guidelines, should be conducted aside. eNaBLe shows the distribution of the data at the end of the procedure, through Q-Q plot, but any subdivision of the dataset before the calculation is not allowed. In this regard, the possibility of implementing further partitions of the dataset, based on the study of the conceptual model of the GWB and conducted outside the tool, is currently being evaluated.

About the temporal analysis of data, currently the software allows to calculate a representative value for each monitoring station through the calculation of the median or, alternatively, through the elimination of any outliers and using the maximum value of the residual distribution. Therefore, at the moment the evaluation of temporal trends (step foreseen in the Guidelines), which could lead to the elimination of MSs that do not show outliers but simply increasing trends for certain parameters, suggesting a possible contamination in progress, should be performed aside. In the direction of improving the tool in this part, the implementation of a statistical test for the estimation of the slope of ascending trends (e.g., Mann–Kendall test) [49,50] is planned.

Finally, for the cases in which there is neither spatial nor temporal significance (Case D), without further specification by the Guidelines, eNaBLe evaluates the outliers putting together all the observations. Indeed, Case D refers to a GWB with a small number of MSs (to the limit of one only MS); therefore, if we reduce the multiple time observations to one representative value (median) for each station, we could not perform a robust statistical evaluation of the outliers.

Following the Guidelines, for each parameter only one NBL can be defined even in presence of multiple datasets, e.g., when different geochemical facies or redox conditions have been recognized and evaluated separately in the GWB. The Guidelines suggest choosing the highest value among those of the single datasets, only in case D will the NBL with the highest confidence level be selected. At present, therefore, the NBLs for Fe and Mn assigned to the entire GWB would be those, rather low, specific to the oxidizing facies. However, it is apparent that the high concentrations of these metals are to be associated with the peculiar reducing conditions that are found locally within the GWB. In order to assign them a NBL more representative of the conditions which promote their presence in solution, it is necessary to continue the temporal monitoring until reaching case C (significant temporal dimension), or to increase the number of MSs until reaching case B (significant spatial dimension). Only when the confidence levels are all the same, it will be possible to select the NBL given by the maximum value among the different datasets, therefore the one (clearly higher) associated with the reducing facies.

5. Conclusions

eNaBLe is a versatile and user-friendly instrument, meant to facilitate the assessment of the NBLs following the national Guidelines. It performs automatically the sequence of operations as indicated by the Guidelines, easily allowing a very quick calculation of NBLs at the groundwater body scale, even in the presence of a limited amount of data. By setting a few configuration parameters, an assessment is rapidly reached which appears sufficiently representative of the natural state of groundwater. However, it should be noted that these settings must be made in a reasoned way, starting from a thorough knowledge of the conceptual model of the aquifer. In particular, the redox facies separation assumes that redox conditions of groundwater are known. Its determination is commonly based on physical-chemical indicators such as DO or ORP whose measure is often critical. Further, this information is frequently missing or scarcely reliable in datasets of groundwater quality monitoring. The validation of the data and their organization into coherent datasets are also fundamental to obtain an evaluation that is as significant as possible.

Some critical issues have emerged in the statistical analysis path, even in part deriving from the reference Guidelines, but the software can be improved and refined in every its part starting from the indications and criticalities deriving from these first applications. The tool is still under testing for a thorough verification of all the possible variants in dataset structures and GWB conceptual models, also taking into account the suggestions of the stakeholders.

Author Contributions: Conceptualization, D.P., E.F., E.P. and S.G.; methodology, D.P., E.P. and S.G.; software, S.G.; validation, D.P.; formal analysis, D.P.; investigation, D.P., E.F., E.P. and S.G.; data curation, S.G.; writing—original draft preparation, D.P.; writing—review and editing, D.P., E.F., E.P. and S.G.; visualization, D.P. and E.F.; supervision, E.P. and S.G. All authors have read and agreed to the published version of the manuscript.

Funding: The tool developing was not funded. Data used as case test were collected within a project funded by Regione Lazio (Area Ciclo Integrato Rifiuti), Italy (Det. G13172 18/10/2018) and Ecologia Viterbo s.r.l. (Contract 3269-2017 and 178/2020).

Data Availability Statement: Paper experimental data are not currently publicly available but can be supplied on request to the corresponding author.

Conflicts of Interest: The authors declare no conflict of interest. The funders had no role in the design of the study; in the collection, analyses, or interpretation of data; in the writing of the manuscript, or in the decision to publish the results.

References

1. GWD. *Groundwater Directive 2006/118/CE, Directive of the European Parliament and of the Council on the Protection of Groundwater Against Pollution and Deterioration*; Official Journal of European Union L372; European Union: Luxembourg, 2006; pp. 19–31.
2. Muller, D.; Blum, A.; Hart, A.; Hookey, J.; Kunkel, R.; Scheidleder, A.; Tomlin, C.; Wendland, F. Final proposal for a methodology to set up groundwater threshold values in Europe. In *Report to the EU project "BRIDGE"*; 2006; Deliverable D18. Available online: http://nfp-at.eionet.europa.eu/Public/irc/eionet-circle/bridge/library?l=/deliverables/bridge_groundw-205pdf/_EN_1.0_&a=d (accessed on 30 December 2020).
3. Hinsby, K.; de Melo, M.T.C.; Dahl, M. European case studies supporting the derivation of natural background levels and groundwater threshold values for the protection of dependent ecosystems and human health. *Sci. Total Environ.* **2008**, *401*, 1–20. [CrossRef] [PubMed]
4. Ducci, D.; de Melo, M.T.C.; Preziosi, E.; Sellerino, M.; Parrone, D.; Ribeiro, L. Combining natural background levels (NBLs) assessment with indicator kriging analysis to improve groundwater quality data interpretation and management. *Sci. Total Environ.* **2016**, *569–570*, 569–584. [CrossRef] [PubMed]
5. Griffioen, J.; Passier, H.F.; Klein, J. Comparison of selection methods to deduce natural background levels for groundwater units. *Environ. Sci. Technol.* **2008**, *42*, 4863–4869. [CrossRef] [PubMed]
6. Marandi, A.; Karro, E. Natural background levels and threshold values of monitored parameters in the Cambrian–Vendian groundwater body, Estonia. *Environ. Geol.* **2008**, *54*, 1217–1225. [CrossRef]
7. Wendland, F.; Berthold, G.; Blum, A.; Elsass, P.; Fritsche, J.G.; Kunkel, R.; Wolter, R. Derivation of natural background levels and threshold values for groundwater in the Upper Rhine Valley (France, Switzerland and Germany). *Desalination* **2008**, *226*, 160–168. [CrossRef]
8. European Commission. *Guidance on Groundwater Status and Trend Assessment, Guidance Document no 18*; Technical Report; European Communities: Luxembourg, 2009; ISBN 978-92-79-11374-1.
9. Gemitzi, A. Evaluating the anthropogenic impacts on groundwaters; a methodology based on the determination of natural background levels and threshold values. *Environ. Earth Sci.* **2012**, *67*, 2223–2237. [CrossRef]
10. Preziosi, E.; Giuliano, G.; Vivona, R. Natural background levels and threshold values derivation for naturally As, V and F rich groundwater bodies: A methodological case study in Central Italy. *Environ. Earth Sci.* **2010**, *61*, 885–897. [CrossRef]
11. Molinari, A.; Guadagnini, L.; Marcaccio, M.; Guadagnini, A. Natural background levels and threshold values of chemical species in three large-scale groundwater bodies in Northern Italy. *Sci. Total Environ.* **2012**, *425*, 9–19. [CrossRef]
12. Rotiroti, M.; Fumagalli, L. Derivation of preliminary natural background levels for naturally Mn, Fe, As and NH4 rich groundwater: The case study of Cremona area (Northern Italy). *Rend. Online Soc. Geol. Ital.* **2013**, *24*, 284–286.
13. Hawkes, H.E.; Webb, J.S. *Geochemistry in Mineral Exploration*; Harper & Row: New York, NY, USA, 1962; p. 415.
14. Tukey, J.W. *Exploratory Data Analysis*; Addison-Wesley Publishing Company: Boston, MA, USA, 1977; p. 688.
15. Reimann, C.; Filzmoser, P.; Garrett, G. Background and threshold: Critical comparison of methods of determination. *Sci. Total Environ.* **2005**, *346*, 1–16. [CrossRef]
16. Wendland, F.; Hannappel, S.; Kunkel, R.; Schenk, R.; Voigt, H.J.; Wolter, R. A procedure to define natural groundwater conditions of groundwater bodies in Germany. *Water Sci. Technol.* **2005**, *51*, 249–257. [CrossRef] [PubMed]

17. Edmunds, W.M.; Shand, P.; Hart, P.; Ward, R.S. The natural (baseline) quality of groundwater: A UK pilot study. *Sci. Total Environ.* **2003**, *310*, 25–35. [CrossRef]
18. Panno, S.V.; Kelly, W.R.; Martinsek, A.T.; Hackley, K.C. Estimating background and threshold nitrate concentrations using probability graphs. *Ground Water* **2006**, *44*, 697–709. [CrossRef] [PubMed]
19. Edmunds, W.M.; Shand, P. *Natural Groundwater Quality*; Wiley-Blackwell: Hoboken, NJ, USA, 2008; p. 488. ISBN 978-14051-5675-2.
20. Walter, T. Determining natural background values with probability plots. In Proceedings of the EU Groundwater Policy Developments Conference, UNESCO, Paris, France, 13–15 November 2008.
21. Preziosi, E.; Parrone, D.; Del Bon, A.; Ghergo, S. Natural background level assessment in groundwaters: Probability plot versus pre-selection method. *J. Geochem. Explor.* **2014**, *143*, 43–53. [CrossRef]
22. ISPRA—Istituto Superiore per la Protezione e la Ricerca Ambientale. *Linee Guida Recanti la Procedura da Seguire per il Calcolo dei Valori di Fondo Naturale per i Corpi Idrici Sotterranei (DM 6 luglio 2016)*; Manuali e Linee Guida 155/2017; ISPRA: Rome, Italy, 2017; ISBN 978-88-448-0830-3. Available online: https://www.isprambiente.gov.it/it/pubblicazioni/manuali-e-linee-guida/linee-guida-recanti-la-procedura-da-seguire-per-il-calcolo-dei-valori-di-fondo-per-i-corpi-idrici-sotterranei-dm-6-luglio-2016 (accessed on 26 November 2020).
23. Frollini, E.; Parrone, D.; Ghergo, S.; Preziosi, E. Natural background level assessment in groundwater: Application of the Italian national guideline; Geophysical Research Abstracts Vol.20, EGU2018-15674-1. In Proceedings of the EGU General Assembly 2018, Vienna, Austria, 8–13 April 2018. eISSN 1607-7962.
24. Parrone, D.; Ghergo, S.; Preziosi, E. A multi-method approach for the assessment of natural background levels in groundwater. *Sci. Total Environ.* **2019**, *659*, 884–894. [CrossRef] [PubMed]
25. Chidichimo, F.; De Biase, M.; Costabile, A.; Cuiuli, E.; Reillo, O.; Migliorino, C.; Treccosti, I.; Straface, S. GuEstNBL: The software for the guided estimation of the natural background levels of the aquifers. *Water* **2020**, *12*, 2728. [CrossRef]
26. Ghergo, S.; Parrone, D.; Frollini, E.; Preziosi, E. eNaBLe, an on-line tool to evaluate Natural Background Levels in groundwater following the Italian Guidelines. Geophysical Research Abstracts Vol.21, EGU2019-7443. In Proceedings of the EGU General Assembly 2019, Vienna, Austria, 7–12 April 2019. eISSN 1607-7962.
27. HP: Hypertext Preprocessor. Available online: https://www.php.net (accessed on 30 December 2020).
28. MySQL. Available online: https://www.mysql.com (accessed on 30 December 2020).
29. Water Resources Database. Available online: http://www.irsa.cnr.it/WRdB (accessed on 30 December 2020).
30. Appelo, C.A.J.; Postma, D. *Geochemistry, Groundwater and Pollution*, 2nd ed.; CRC Press: Boca Raton, FL, USA, 2005; p. 668.
31. Huber, P.J. *Robust Statistics*; John Wiley: New York, NY, USA, 1981; p. 320.
32. Shapiro, S.S.; Wilk, M.B. An analysis of variance test for normality (complete samples). *Biometrika* **1965**, *52*, 591–611. [CrossRef]
33. Angelone, M.; Cremisini, C.; Piscopo, V.; Proposito, M.; Spaziani, F. Influence of hydrostratigraphy and structural setting on the arsenic occurrence in groundwater of the Cimino-Vico volcanic area (Central Italy). *Hydrogeol. J.* **2009**, *17*, 901–914. [CrossRef]
34. Baiocchi, A.; Coletta, A.; Espositi, L.; Lotti, F.; Piscopo, V. Sustainable groundwater development in a naturally arsenic-contaminated aquifer: The case of the Cimino-Vico volcanic area (central Italy). *Ital. J. Eng. Geol. Environ.* **2013**, *1*, 5–18.
35. Dall'Aglio, M.; Giuliano, G.; Amicizia, D.; Andrenelli, M.C.; Cicioni, G.B.; Mastroianni, D.; Sepicacchi, L.; Tersigni, S. Assessing drinking water quality in Northern Latium by trace elements analysis. In *Proceedings of Water Rock Interaction (WRI-10), Villasimius, Italy, 10–15 June 2001*; Cidu, R., Ed.; A.A. Balkema: Exton, PA, USA, 2001; Volume 2, pp. 1063–1066.
36. De Rita, D.; Cremisini, C.; Cinnirella, A.; Spaziano, F. Fluorine in the rocks and sediments of volcanic area in central Italy: Total content, enrichment and leaching process and a hypothesis on the vulnerability of the related aquifers. *Environ. Monit. Assess.* **2012**, *184*, 5781–5796. [CrossRef]
37. Parrone, D.; Ghergo, S.; Frollini, E.; Rossi, D.; Preziosi, E. Arsenic-fluoride co-contamination in groundwater: Background and anomalies in a volcanic-sedimentary aquifer in central Italy. *J. Geochem. Explor.* **2020**, *217*, 106590. [CrossRef]
38. Preziosi, E.; Vivona, R.; Giuliano, G. Microinquinanti di origine naturale negli acquiferi vulcanici: Un approccio integrato quantitativo e qualitativo nel Lazio settentrionale [Micropollutants of natural origin in volcanic aquifers: An integrated quantitative and qualitative approach in northern Latium]. *IGEA* **2005**, *20*, 3–14.
39. Vivona, R.; Preziosi, E.; Giuliano, G.; Mastroianni, D.; Falconi, F.; Madé, B. Geochemical characterization of a volcanic-sedimentary aquifer in Central Italy. In *Proceedings of the 11th International Symposium on Water-Rock Interaction WRI-11*; Wanty, R., Seal, R.I.I., Eds.; Saratoga Springs: New York, NY, USA, 2004; pp. 513–517.
40. Vivona, R.; Preziosi, E.; Madé, B.; Giuliano, G. Occurrence of minor toxic elements in volcanic-sedimentary aquifers: A case study in central Italy. *Hydrogeol. J.* **2007**, *15*, 1183–1196. [CrossRef]
41. WHO. *Guidelines for Drinking-Water Quality*, 4th ed.; World Health Organization: Geneva, Switzerland, 2011; p. 564. ISBN 978-92-4-154815-1.
42. Preziosi, E.; Ghergo, S.; Frollini, E.; Parrone, D. Buone Pratiche per il Campionamento Delle Acque Sotterranee: Proposta di un Protocollo. (Best Practices for Groundwater Sampling: Proposal of a Protocol). Notiziario dei Metodi Analitici IRSA-CNR. 2017, Volume 1, pp. 23–36. Available online: http://www.irsa.cnr.it/images/docs/Notiz/notiz2017_1.pdf (accessed on 26 June 2018).
43. Frollini, E.; Rossi, D.; Rainaldi, M.; Parrone, D.; Ghergo, S.; Preziosi, E. A proposal for groundwater sampling guidelines: Application to a case study in southern Latium. Rend. *Online Soc. Geol. It.* **2019**, *47*, 46–51. [CrossRef]
44. *Decreto Legislativo 3 Aprile 2006, n. 152. Norme in Materia Ambientale. Gazzetta Ufficiale n. 88 del 14-04-2006*; Istituto Poligrafico e Zecca dello Stato s.p.a.: Rome, Italy, 2006.

45. *Decreto Legislativo 16 Marzo 2009, n. 30. Attuazione Della Direttiva 2006/118/CE, Relativa Alla Protezione Delle Acque Sotterranee Dall'inquinamento e dal Deterioramento. Gazzetta Ufficiale n. 79 del 4-4-2009*; Istituto Poligrafico e Zecca dello Stato s.p.a.: Rome, Italy, 2009.
46. ISPRA-CNR.IRSA. *Linee Guida per la Valutazione Delle Tendenze Ascendenti e D'inversione Degli Inquinanti nelle Acque Sotterranee (DM 6 Luglio 2016)*; Manuali e Linee Guida 161/2017; ISPRA: Rome, Italy, 2017; ISBN 978-88-448-0844-0.
47. Frollini, E.; Preziosi, E.; Calace, N.; Guerra, M.; Guyennon, N.; Marcaccio, M.; Menichetti, S.; Romano, E.; Ghergo, S. Groundwater quality trend and trend reversal assessment in the European Water Framework Directive context: An example with nitrates in Italy. *Environ. Sci. Pollut. Res.*. In press. [CrossRef]
48. Rosner, B. Percentage Points for a Generalized ESD Many-Outlier Procedure. *Technometrics* **1983**, *25*, 165–172. [CrossRef]
49. Mann, H.B. Nonparametric tests against trend. *Econometrica* **1945**, *13*, 245–259. [CrossRef]
50. Kendall, M.G. *Rank Correlation Methods*; Grin: London, UK, 1975.

Article

Ambient Background Values of Selected Chemical Substances in Four Groundwater Bodies in the Pannonian Region of Croatia

Zoran Nakić *, Zoran Kovač, Jelena Parlov and Dario Perković

Faculty of Mining, Geology and Petroleum Engineering, University of Zagreb, 10000 Zagreb, Croatia; zoran.kovac@rgn.unizg.hr (Z.K.); jelena.parlov@rgn.unizg.hr (J.P.); dario.perkovic@rgn.unizg.hr (D.P.)
* Correspondence: zoran.nakic@rgn.unizg.hr

Received: 25 August 2020; Accepted: 21 September 2020; Published: 24 September 2020

Abstract: Groundwater quality is a consequence of cumulative effects of natural and anthropogenic processes occurring in unsaturated and saturated zone, which, in certain conditions, can lead to elevated concentrations of chemical substances in groundwater. In this paper, the concept of determining the ambient background value of a chemical substance in groundwater was applied, because the long-term effects of human activity influence the increase in concentrations of substances in the environment. The upper limits of ranges of ambient background values were estimated for targeted chemical substances in four groundwater bodies in the Pannonian region of Croatia, according to the demands of the EU Groundwater Directive. The selected groundwater bodies are typical, according to the aquifer typology, for the Pannonian region of Croatia. Probability plot (PP), the modified Lepeltier method, as well as the simple pre-selection method, were used in this paper, depending on a number of chemical data in analysed data sets and in relation to the proportion of <limit of quantification (LOQ) values in a data set for each groundwater body. Estimates obtained by using PP and the modified Lepeltier method are comparable when data variability is low to moderate, otherwise differences between estimates are notable. These methods should not be used if the proportion of <LOQ values in a data set is higher than 30%; however, the integration of results of both methods can increase the confidence of estimation. If the proportion of <LOQ values is higher than 30%, it is recommended to use the robust pre-selection method with the adequate confidence level. For highly skewed data, the 90th percentile of the pre-selected data set is comparable with other methods and preferable over the 95th percentile. The estimates obtained for inert and mobile substances are comparable on different scales. For highly redox-sensitive substances, estimates may differ by one to two orders of magnitude, in relation to the observed heterogeneity of the aquifer systems. The critical issue in the estimation process is the determination of hydrogeological and geochemical homogeneous units within the heterogeneous aquifer system.

Keywords: ambient background values; probability plot; modified Lepeltier method; pre-selection method; LOQ; groundwater body; Croatia

---

1. Introduction

By definition, a geochemical background value of an element or compound in groundwater indicates the absence of anomalous, usually high, measured values of concentrations of substances that would indicate human influence. Matschullat et al. [1] define the geochemical background concentration as a relative measure for distinguishing between the natural and anthropogenic concentrations of an element or compound in a real sample set. The natural background concentrations of substances are due primarily to interactions between the rock matrix and water, i.e., dissolution of minerals and rocks, chemical and biological processes in the unsaturated and saturated zone, interactions between different groundwater bodies, groundwater residence time, and chemical composition of precipitation [2].

Due to the ubiquitous human impact, which is also reflected in the chemical composition of groundwater, the natural composition of groundwater, especially in shallow aquifers, is almost non-existent today. Accordingly, Reiman and Garrett [3] defined ambient background concentration as a background value under slightly changed conditions, when elevated concentrations of a substance in water result from a long-term human impact, such as agriculture, industry or urbanization, meaning that the measured values of concentrations of a substance cannot entirely reflect natural conditions. Other authors [4–6] take up this concept, recognizing the fact that for some substances in groundwater, e.g., nitrate, there are numerous natural and anthropogenic sources that could have influenced their concentrations.

When determining background values of substances in groundwater, it is necessary to take into considerations the concept of natural variability due to the heterogeneity of the aquifer system. Due to the geological variety of the different regions, some studies have shown the convenience of deriving local or regional background level [7]. It follows from this that the background value should not be presented as a single fixed value, since the background value thus defined does not provide information on the natural variability of the substance [1]. However, for the practical use of background values, in particular in the context of the application of threshold values to the requirements of the EU Groundwater Directive (2006/118/EC), the background value can be defined with a single value, as the upper limit of the range of background concentrations, with the adequate confidence level.

According to Molinari et al. [8], background value can be viably evaluated (i) from groundwater samples unaffected by human impact, including those samples taken from deep wells where no anthropogenic impacts from the surface are present, (ii) using multicomponent reactive transport modelling in real aquifer systems, in cases where discrepancies observed between reaction rates at laboratory and field scales, the problem of bridging across scales and the conceptual and parametric uncertainty can be fully addressed. Alternative to these two approaches is an estimation of background values by statistical analysis of a large set of monitoring data [7,8], with the aim to identify concentrations related to the contamination anomaly as opposed to those solely reflecting background processes.

At the EU level, there is currently no single set of criteria to ensure a standardized Europe-wide approach for defining natural background values [9]. The EU research project BaSeLiNe, Natural Baseline Quality in European Aquifers: A Basis for Aquifer Management, funded under the Fifth Framework Programme, has revealed that Median ± 2MAD and Mean ± 2SD rules have often been used as the main statistical parameters for initially defining original data distribution and background values in EU aquifers [10]. The problem arises from the fact that the use of these methods can give a wrong estimate of the location of the main body of data if data distribution is influenced by more than one process, resulting in multimodal distribution, which could be superimposed [11]. Reimann et al. [12] proposed the use of boxplot, both for determining extreme values and background concentrations, based on the results of comparative analysis, in which they compared the results of multiple statistical methods. They concluded that this method is appropriate if the number of extreme values is less than 10%, while the Median ± 2MAD method produces better results if the number of extreme values is greater than 15%. The EU research project BRIDGE, Background Criteria for Identification of Groundwater Thresholds [13], funded under the Sixth Framework Programme, resulted in a proposal of two methods to setting background values at European level. The component separation method is based on the separation of the relative frequency of concentration of a chemical substance into a natural and anthropogenic component, which are modelled with separate distributions, and is applied when a large amount of data is available for a chemical substance [14]. The pre-selection method is used in cases where a limited data set is available and when chemical samples do not show or show very little human influence [15–18]. It is clear that EU Member States apply very different approaches to determining background concentrations of substances in groundwater. Sweden estimates background concentrations by comparing groundwater chemical status with drinking water standards [19]. Buss et al. [20] state that the Irish Environmental Protection Agency determines concentrations of chemicals free from anthropogenic influences and calculates the upper and lower limits of the range of background concentrations from the extrapolation of the normal distribution curve of chemical concentrations. In Germany, the aforementioned component separation method is used [14]. In the Netherlands, several methods are applied, namely the historical method, the tritium method and the oxidation capacity method [17].

Nowadays, many researchers use the approach to determine background concentrations based on the use of probability plot (PP), triggered by research conducted by Sinclair in the early 1970s [21]. This approach is based on the assessment of one or more inflection points on a probability graph, which separate different populations within the distribution of all measured data for a substance. Kyoung et al. [22] strongly recommend this approach in cases where the distribution of measured data shows bimodality or multimodality. A complete data set from a statistical sample is subdivided into subgroups, which reflect relevant geochemical processes or pollution. A subgroup representing the background concentrations of an element or chemical compound has a characteristic probability density function that results from the cumulative influence of different processes in an aquifer. Such a subset of data can be approximated with a normal or lognormal distribution [23]. In many scientific papers, researchers more often use the lognormal distribution to show the distribution of natural background concentrations, while the normal distribution is mostly used to show the distribution of human-influenced data [8,14,21]. Similar to PP is the Lepeltier method [24] that visually evaluates cumulative sums in double-logarithmic scale graphs. The idea of the Lepeltier method is the assumption of a lognormal concentration distribution of the element for which the upper limit for background concentration is to be determined. Other methods that are intrinsically related to probability graph approach, by splitting the overall data distribution into distinct components, are the iterative 2-σ technique and the calculated distribution function. Both methods take a set of measured data and process the data, i.e., remove non-ambient values, until a normal distribution of ambient concentration range is obtained [1,5,6].

In this paper, three methods were applied to estimate the upper limit of the range of ambient background values (UL) of each targeted chemical substance in the context of assessing groundwater quality status of groundwater bodies in the Pannonian region of Croatia. PP and the modified Lepeltier method were used in accordance to results of a statistical simulation study, which evaluated the robustness and reliability of widely used methods for determining background values [25]. The pre-selection method was applied in accordance to recommendation stemming from the EU research project BRIDGE, in cases of limited data set availability and/or limited data quality.

As presented in Section 2, arsenic (As), sulphate (SO_4), chloride (Cl), and nitrate (NO_3) were considered, which are stipulated by EU and Croatian regulations as key substances for the assessment of the chemical status of groundwater bodies and should be taken into account when determining background concentrations. These substances occasionally occur in higher concentrations than reference values prescribed by EU and Croatian regulations. Following the recommendations of the EU research project BRIDGE, iron (Fe) was also included in the analysis. Iron is particularly sensitive to change in redox conditions due to human influence that can lead to an enrichment of dissolved iron in groundwater.

Four investigated groundwater bodies enabled comparison of UL estimates in similar hydrogeological settings. Selected groundwater bodies are typical, according to the aquifer typology, for the Pannonian region of Croatia and results are valid for unconsolidated gravel and sand aquifers.

The main aim of this research was to apply a robust methodology for the background estimation, applicable to data sets with moderate to high data variability and high percentage of limit of quantification (LOQ) values. To our knowledge, our work is a rare example of the application of the formal statistical procedures for an estimation of background values, which addresses the above-mentioned data quality issue.

2. Materials and Methods

2.1. Study area Description

Fifteen groundwater bodies were identified in the Pannonian part of the Republic of Croatia within the process of implementation of the Directive 2000/60/EC. In the River Basin Management Plan of the Republic of Croatia for the period 2016–2021 [26], each groundwater body was categorized as one, laterally and vertically, hydrogeological homogeneous unit. The four investigated groundwater bodies are located at the southern edge of the Pannonian basin, in the northern and the eastern part of Croatia (Figure 1).

Figure 1. Location of the investigated groundwater bodies in Croatia.

This area is characterized by vast plains of the Sava and Drava rivers, in which gravel and sand aquifers are formed at different depths. These aquifers are rich in water and represent the main water supply resource of the northern part of Croatia [26]. In the hilly area between Sava and Drava plains, small alluvial aquifers are contained in the catchment areas of large rivers tributaries, while spatially limited carbonate fissure and karst aquifers are found in the highest, mostly isolated parts of the hilly area. Table 1 lists the total thickness and average hydraulic conductivity of gravel and sand layers in each groundwater body as well as the area of four groundwater bodies analysed.

Table 1. Groundwater body characteristics. Thickness and hydraulic conductivity data refer to gravel and sand aquifers in groundwater bodies.

Groundwater Body	Area (km^2)	Total Thickness of Permeable Layers (m)	Average Hydraulic Conductivity (K) (m/day)
CDGI_19	402.1	80	210
CDGI_23	5010.9	120	30
CSGI_29	3329.4	50	110
CSGN_25	5188.1	40	50

Groundwater body CDGI_19 is in the western part of the Drava River plain (Figure 1). It is filled with thick gravel and sand layers separated by silt and clay interlayers and lenses. The groundwater body comprise two alluvial aquifers. The upper unconfined aquifer is of Quaternary age and is built of coarse-grained gravel and sand. The lower semi-confined aquifer is of Neogene age and has a higher amount of finer-grained sediments [27]. They are divided by a semipermeable silty to clayey layer of average thickness of several meters (Figure 2).

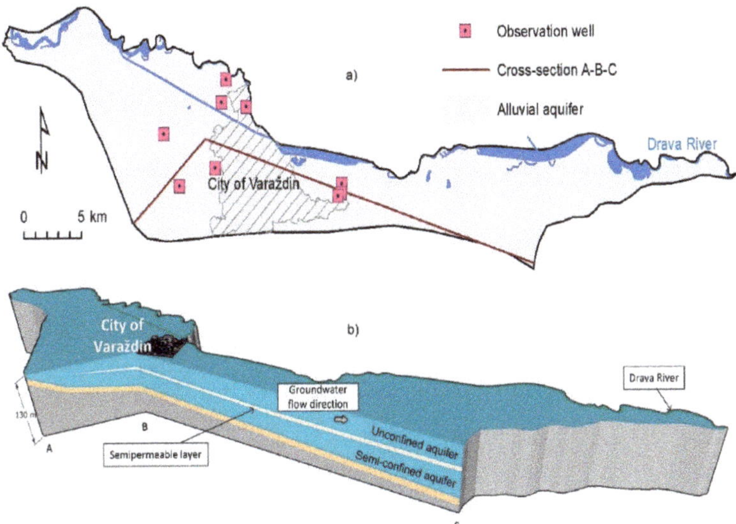

Figure 2. Conceptual model of the groundwater body CDGI_19: (**a**) Planar extent of the groundwater body; (**b**) Schematic 3D hydrogeological cross-section.

A severe degradation of groundwater quality, due to intensive agricultural activity, is recorded in the upper unconfined aquifer [28]. Before the construction of hydropower plants and reservoirs on the Drava River in 1970s, the groundwater quality was within the limits of drinking water standards. Afterwards, nitrate concentrations increased significantly in the groundwater in the shallow aquifer, in which the nitrate concentrations exceeded the maximum permissible concentration in drinking water. The probable cause of groundwater pollution was an increase in the flow of contaminated groundwater from areas with intensive agricultural activity, due to changes in boundary conditions after the construction of reservoirs and drainage channels.

Groundwater body CDGI_23 is in the eastern part of the Drava River plain (Figure 1). Sediments in the Drava River depression mostly originate from the Alp massif and to a lesser extent from Slavonian Mountains [29]. Here, coarse and fine-grained clastic sediments alternate laterally and vertically. The aquifer system is of Quaternary age and its thickness is more than 200 m, while thicknesses of single confined and semi-confined aquifers are from five to 50 m (Figure 3). Aquifers are predominantly composed of layers of medium to fine-grained sand in the western part of the groundwater body, while the fine-grained fraction prevails in the east. Sandy layers are separated by silt and clay interlayers and lenses. In the hilly area of the groundwater body, the rocks of the Middle Triassic carbonate complex, dolomites, dolomitic breccias and dolomitic limestones, form fissure and karst aquifers.

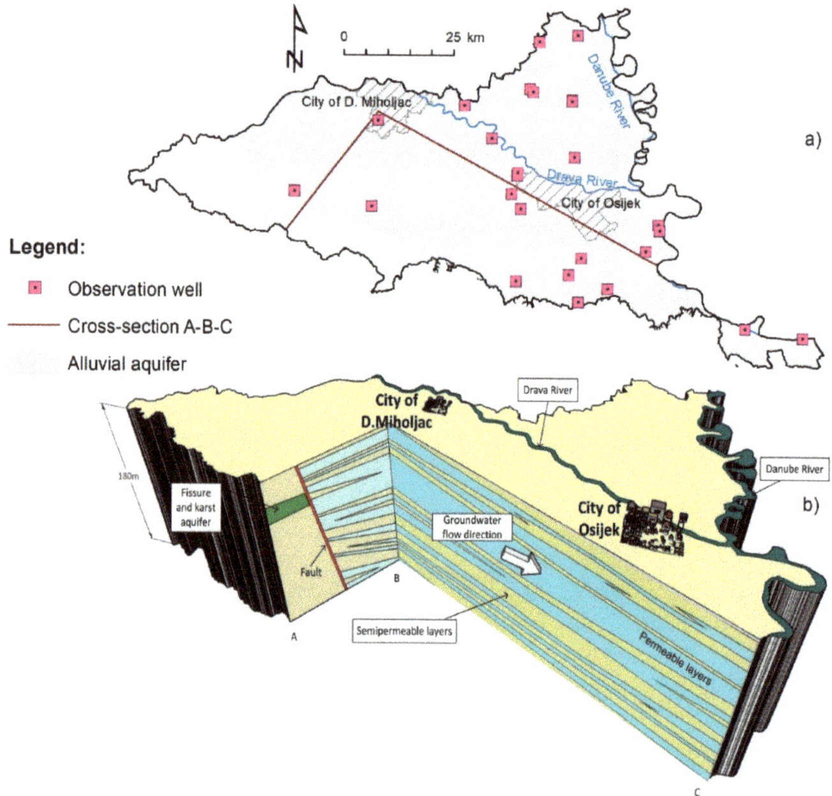

Figure 3. Conceptual model of the groundwater body CDGI_23: (**a**) Planar extent of the groundwater body; (**b**) Schematic 3D hydrogeological cross-section.

Groundwater body CSGI_29 is in the eastern part of the Sava River plain (Figure 1). The dominant factor on formation of the Sava River depression is transportation and deposition of eroded material by the River Sava tributaries. Rivers from Bosnia and Herzegovina deposited fan-shape coarse-grained clastic sediments in the peripheral areas of the depression. These sediments mostly originate from the Bosnian Mountains and to a lesser extent is of the Alpine provenance. The number of aquifers ranges from two to four, near the Sava River, to eleven in the northern part of the Sava depression (Figure 4). The thickness of the single aquifers is rarely greater than 30 m [29]. The aquifers near the Sava River, found from 20–70 m in depth, consist of gravel and sand layers, deposited during warm and humid interglacial periods, and alternate with fine sand, silt and clay, deposited during cold glacial periods in Pleistocene. Holocene deposits from 20 to 10 m in depth consist of silt, sand, and gravel, while uppermost deposits are fine grained, deposited by flooding of the Sava River and its tributaries and by erosion of loess plateau to the north. The aquifer thickness decreases from the Sava River to the north and the content of fine sediments, namely sand and silt, increases. Based on the measurement of tritium contents at locations far from the Sava River to the north, it is found that the relative mean residence time of groundwater is prior to the 1950s [30].

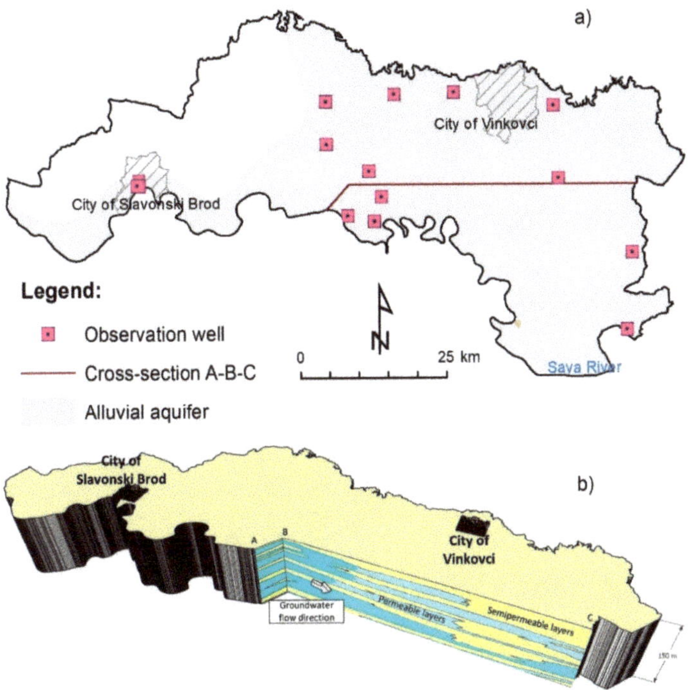

Figure 4. Conceptual model of the groundwater body CSGI_29: (**a**) Planar extent of the groundwater body; (**b**) Schematic 3D hydrogeological cross-section.

Chemical properties of deep groundwater in the eastern parts of the Drava and Sava plains are mainly controlled by natural geochemical processes, mostly by dissolution of carbonate and weathering of silicate minerals that form the particles of alluvial deposits [31]. Cation exchange processes, promoted by long groundwater residence time, can also significantly contribute solutes to groundwater. High concentrations of arsenic of natural origin are typical for the groundwater of the Pannonian Basin [32], where the type and geochemical composition of groundwater is result of lithological, sedimentological and palaeographic factors [33]. High arsenic content in groundwater in the eastern part of Croatia is mostly caused by reductive desorption of arsenic from iron oxides and/or clay minerals, reductive dissolution of iron oxides or competition for the sorption sites with organic matter and phosphate [29]. However, the influence of anthropogenic activity, namely agriculture and urbanisation, cause the deterioration of groundwater quality. In groundwater samples taken from observation wells located close to the Sava River, high content of nitrogen, potassium and chloride was periodically detected [30].

Groundwater body CSGN_25 is in the hilly area between Sava and Drava plains (Figure 1). Low to medium permeability unconfined alluvial aquifers of Quaternary age are found in the lowland part of the groundwater body (Figure 5). These spatially limited water-bearing layers of small thicknesses alternate with loose and unsorted silt and sandy clay sediments of low permeability. The total thickness of the Quaternary deposits in the lowland part of the groundwater body is 40 to 130 m. To a lesser extent, conglomerate, breccia, dolomite and limestone fissure, and cavernous aquifers of Triassic age have been identified in the hilly part of the groundwater body [34]. A small depth to groundwater and low protection capability of discontinuous aquitards in the lowland area make the groundwater body moderately to highly vulnerable to agricultural pollution.

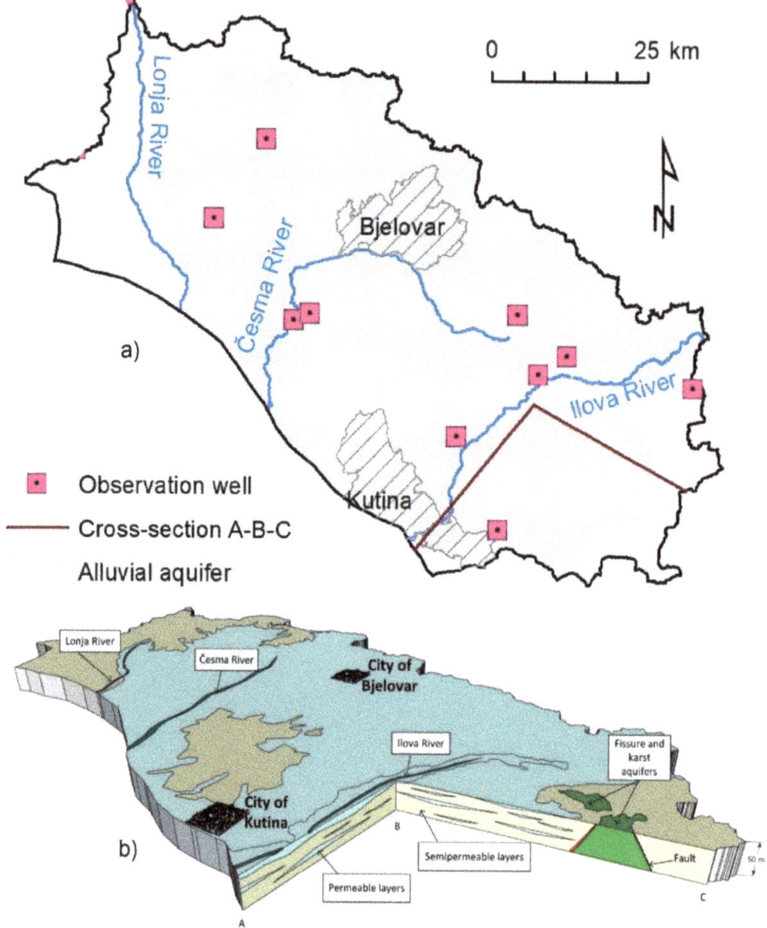

Figure 5. Conceptual model of the groundwater body CSGN_25: (**a**) Planar extent of the groundwater body; (**b**) Schematic 3D hydrogeological cross-section.

2.2. Available Data Set

The UL of each selected substance was estimated by groundwater chemistry data obtained from the national database of Croatian Waters: "National monitoring program of groundwater quality". Data from observations wells, included in the national monitoring program, have been used for this purpose. Nine observation wells have been used from the groundwater body CDGI_19, twenty-six from the groundwater body CDGI_23, fifteen from the groundwater body CSGI_29, and ten from the groundwater body CSGN_25 (Figures 2–5). All observation wells are attributed to unconsolidated sand and gravel aquifers at different depths in selected groundwater bodies. In the groundwater body CSGN_25, well screens are positioned only at shallow depths to monitor groundwater quality of unconfined alluvial aquifers.

Four data sets for chemical substances, one per groundwater body, are evaluated for the period from 2007 to 2017. Five chemical substances were chosen for further analysis: arsenic (As), sulphate (SO$_4$), chloride (Cl), nitrate (NO$_3$), and iron (Fe). The UL were estimated for substances that may be due to natural and anthropogenic conditions. The EU Groundwater Directive (2006/118/EC), in Annex II, specifically lists substances that occur naturally and under the human influence, as well as pollution indicators, which need to be considered when setting national groundwater quality standards. This group includes arsenic, sulphate and chloride, as well as nitrate, for which the Groundwater Directive sets Community criteria for the assessment of the chemical status of bodies of groundwater (50 mg NO_3/L). The EU research project "Background Criteria for Identification of Groundwater Thresholds (BRIDGE)", funded under the 6th Framework Program of the European Union, proposed that the determination of background concentrations of substances is carried out for important pollutants that occur as a result of natural conditions and for certain characteristic substances, such as iron, which may occur in elevated concentrations due to human activity. Hence, all selected substances are to be considered for the assessment of groundwater body chemical status, according to EU and Croatian regulations and guidelines.

Table 2 shows that arsenic, sulphate, and nitrate data contain a high proportion of <LOQ values. In addition, standard deviation and coefficient of variation show high to moderate data variability for selected substances.

Table 2. Main statistics for analysed substances.

Chemical Substance	Statistics	Groundwater Body			
		CDGI_19	CDGI_23	CSGI_29	CSGN_25
Arsenic (µg/L)	Arithmetic mean	-	37.3 (37.2 *)	14.3 (14.0 *)	5.5 (4.5 *)
	Standard deviation	-	67.3 (68.3 *)	19.8 (21.3 *)	6.5 (5.4 *)
	Coefficient of variation	-	1.8 (1.8 *)	1.4 (1.5 *)	1.2 (1.2 *)
	N of samples	342	400 (293 *)	227 (161 *)	135 (108 *)
	<LOQ	342	109 (93 *)	102 (87 *)	62 (54 *)
	Percentage of <LOQ	100.00	27.3 (31.7 *)	44.9 (54.0 *)	45.9 (50.0 *)
Iron (µg/L)	Arithmetic mean	21.6 (23.2 *)	1822.4	3402.2	219.4
	Standard deviation	67.1 (83.6 *)	1873.7	6101.2	382.6
	Coefficient of variation	3.1 (3.6 *)	1.0	1.8	1.7
	N of samples	284 (165 *)	404	227	135
	<LOQ	79 (52 *)	7	10	11
	Percentage of <LOQ	27.8 (31.5 *)	1.7	4.4	8.1
Sulphate (mg/L)	Arithmetic mean	34.8	14.8 (17.3 *)	7.7 (8.2 *)	15.9 (15.2 *)
	Standard deviation	31.7	35.2 (37.6 *)	10.5 (11.8 *)	28.7 (28.4 *)
	Coefficient of variation	0.9	2.4 (2.2 *)	1.4 (1.4 *)	1.8 (1.9 *)
	N of samples	342	454 (332 *)	255 (180 *)	151 (122 *)
	<LOQ	0	190 (128 *)	59 (35 *)	49 (34 *)
	Percentage of <LOQ	0.0	41.9 (38.6 *)	23.1 (19.4 *)	32.5 (27.9 *)
Chloride (mg/L)	Arithmetic mean	12.6	13.7	6.4	16.4
	Standard deviation	5.2	19.7	7.5	13.7
	Coefficient of variation	0.4	1.4	1.2	0.8
	N of samples	342	454	255	151
	<LOQ	0	1	0	0
	Percentage of <LOQ	0.0	0.2	0.0	0.0
Nitrate (mg/L)	Arithmetic mean	33.4	3.9 (3.2 *)	5.4 (6.2 *)	7.9 (8.0 *)
	Standard deviation	35.6	10.9 (4.7 *)	9.6 (9.2 *)	9.9 (10.0 *)
	Coefficient of variation	1.1	2.8 (1.5 *)	1.8 (1.5 *)	1.3 (1.2 *)
	N of samples	342	454 (332 *)	255 (180 *)	151 (122 *)
	<LOQ	13	349 (240 *)	167 (102 *)	74 (61 *)
	Percentage of <LOQ	3.8	76.9 (72.3 *)	65.5 (56.7 *)	49.0 (50.0 *)

Note: * Statistics for substances after exclusion of contaminated samples based on criteria for the pre-selection method.

It is worth noting that two selected substances frequently occur in higher concentrations than reference values stipulated by EU and Croatian regulations. From Table 2, it is evident that average values of arsenic and iron in some groundwater bodies exceed maximum permissible levels for drinking water (10 μg As/L; 200 μg Fe/L).

2.3. Description of Methods

The EU Groundwater Directive 2006/118/EC specifically defines background values as "concentration of a substance or the value of an indicator in a body of groundwater corresponding to no, or only very minor, anthropogenic alterations to undisturbed conditions". It further stipulates that the estimation of background values must be based on the characterization of groundwater bodies and on the results of groundwater monitoring. In cases where limited monitoring data are available, additional data need to be collected and background values must be determined by a simplified approach based on these limited monitoring data, considering geochemical reactions and processes in the groundwater system. Accordingly, the methodology for the estimation of ambient background values of substances must be as robust as possible to enable reliable evaluation of the chemical status and groundwater quality risk assessment, according to requirements of the Directive 2000/60/EC.

In this research, model-based objective methods were considered for the estimation of background values, which are based on the approach that a background population in a geological environment has a characteristic probability density function that results from the summation of natural processes that produce the background population. They differ from other methods, e.g., model-based subjective methods, in that the boundary between background and anomalous populations is defined by the data themselves rather than by an arbitrary decision of the researcher [23].

Well-known model-based objective methods have been selected for further testing, the probability plot (PP), the Lepeltier method, the iterative 2-σ technique, and the calculated distribution function. A statistical simulation study was conducted, which compared the reliability and robustness of these methods based on common criteria [25]. Since the lognormal distribution is frequently assumed for the background population [2,8,14,21], lognormal distribution parameters of hypothetical ambient and non-ambient populations were randomly selected using computer-generated values. The size of mixed population used in simulation study was predefined at 30, 100, 300, and 1000, assuming small, medium, or large number of groundwater samples. The mixing factor, i.e., the proportion of values belonging to the ambient and non-ambient distributions, was randomly selected from the uniform distribution, so that the percentages of ambient population vary from 30 to 70% of the mixed population. A proportion of <LOQ values in a hypothetical data set was set at 0%, 1%, 5%, 10%, 15%, 20%, 25%, and 30% and all <LOQ values were either discarded or were substituted with LOQ values or with randomly selected values from the uniform distribution from 0 to LOQ values. The simulation study revealed that PP and the Lepeltier method have the lowest relative and absolute error of the estimate of background values. The Lepeltier method gives better estimates than PP for small data sets with N < 100 and if the proportion of <LOQ values is between 20% and 30% [25].

The probability plot (PP) approach assumes that different processes generate data that have different probability distributions that can overlap, i.e., a certain part of the measurement range can be covered by multiple distributions, which can differ in parameters, while all distributions belong to the same distribution family. For example, it is possible that background and/or non-background processes result in data that can be described by normal or lognormal distribution, but with different parameters.

The aim of PP is to try to identify points that separate multiple distributions, i.e., the value to which the influence of one process is dominant and after that value the influence of another process grows stronger. If there is a partial overlap of background and non-background distributions, then a change in distributions can be seen on the probability graph as an inflection point, i.e., the point where the graph changes from concave to convex or vice versa. The concentration at the inflection point is defined as the threshold value (the upper limit of the range of natural concentrations), below which all measured values of the substance belong to background concentrations [6,23]. If background values follow a lognormal distribution, then the lognormal probability graph in its initial part should be very similar to a straight line. At the point of inflection, there is a visible deviation of the graph from the straight line and near that point a value is expected after which more and more influence has another process that generates non-background values.

In this paper, we have visually identified an inflection point below which all measured values of each analysed substance belong to ambient background concentrations. We have applied the following procedure to construct the probability graph using lognormal distribution:

1. Sort measured data $\{X_1, \ldots, X_n\}$ from the smallest to the largest. The label for sorted data is $\{X_{(1)}, \ldots, X_{(n)}\}$;
2. Calculate $p_i = (i - 0.5)/n$, where n is the number of data, and i denotes the index of the data in the sorted sequence from the step 1;
3. Calculate $t_i = \sqrt{-\ln(4p_i(1 - p_i))}$;
4. Calculate $z_i = sign(p_i - 0.5) \cdot 1,238 \cdot t_i(1 + 0.0262 \cdot t_i)$, where $sign(p_i - 0.5) = 1$ if $p_i - 0.5 > 0$, and $sign(p_i - 0.5) = -1$, if $p_i - 0.5 < 0$;
5. Take a natural logarithm of sorted data, i.e., to calculate $\{\ln(X_{(1)}), \ldots, \ln(X_{(n)})\}$;
6. The probability plot is obtained by displaying logarithmic values (from the step 5) on the x-axis and the corresponding values of z_i on the y-axis.

D'Agostino and Stephens consider that 25 is a minimum number of data for reliable use of PP [35]. Other researchers consider an approach where the lower limit is 100 data, below which the probability graph show significant deviation from normal distribution [4,23]. If a number of data (N) is less than 100, then the graph may look jagged or exhibit nonlinear behaviour, which can increase the likelihood of erroneous readings and making the wrong conclusion. In this paper, we have applied PP to determine the upper limit of ambient background concentrations if N > 100 and, based on the results of statistical simulation study [25], if the proportion of <LOQ values is ≤20%.

The advantage of the PP approach is that it enables the identification of multiple populations, which are determined on the graph by inflection points. In addition, each data value is observable on graph and extreme values can be clearly detected as single values [36]. A limitation in the application is that ambient and non-ambient distributions must be assumed a priori, most often lognormal, although researchers also use normal, gamma, and other distributions [4,8,14,37].

The Lepeltier method is a graphical method that analyzes the cumulative sum on a graph with logarithmic scales [24]. In his original work, Lepeltier suggested a "reverse procedure" for cumulative frequency calculation, i.e., cumulation from high to low values [24]. The aim was to overcome the problem of plotting the highest value at 100% on a probability scale and to compensate the analytical imprecision at the lower end where the cumulation starts, particularly if the proportion of <LOQ values is high in a data set [38]. Ashley and Keith [39] modified the Lepeltier approach by computing log estimators of the central value and dispersion rather than using unadjusted means and deviations obtained graphically.

The advantage of the Lepeltier method is that it enable the identification of ambient background values for relatively small data sets. A limitation in the application is that care must be taken to avoid the temptation to accept the visual deviations at the lower part of the curve as significant if values are close to the detection limit and on a highly magnified scale [38].

In this paper, cumulative relative frequencies have been graphically evaluated in double logarithmic scale graphs. A point is sought on a graph at which a significant visual alteration of the slope shows as a bend in a curve. In the case of finding such a value, for example the value of x, then all values less than or equal to x are further considered for the calculation of the upper limit of ambient background concentrations. Similar to Ashley and Keith [39], we have modified the Lepeltier method by computing both Mean + 2SD, following the approach described by Matschullat et al. [1], and Median + 2MAD, to compare results of different estimators of central value and spread of the data distribution. The advantage of the MAD estimator in calculating the upper background concentration, i.e., threshold value between ambient and anomalous populations, is that it is much less influenced by skewed data or outliers than other, less robust, spread estimators [36,40].

Although it is difficult to give a precise limit for the minimum number of data required for reliable analysis by this method, in his original work Lepeltier states that it is necessary to analyze a minimum of 50 data, with which meaningful results can be expected [24]. If the number of data is less than 50, then it can be difficult to visually determine the exact location on the graph where the distribution changes, i.e., where the graph has unexpectedly changed appearance, because the points in this case may be too spaced or very localized around one or more points. Accordingly, we have applied the modified Lepeltier method if N > 50 and if the proportion of <LOQ values is ≤30%.

The pre-selection method, developed under the EU research project BRIDGE [13], can be applied for estimation of background concentrations of substances in cases of limited data quality. Since 2008, numerous researchers have used this method [2,15–18]. It is based on the assumption that selected indicators can give a good insight into whether a sample is contaminated or not. In cases where concentrations of indicators exceed a predefined value, the sample is considered contaminated and excluded from the estimation of background concentration. In this paper, the following exclusion criteria have been applied: (a) ion balance error more than ±10%; (b) the sum of chloride and sodium higher than 1000 mg/L (salt or brackish water); (c) NO_3 > 50 mg/L or active substances in pesticides > 0.1 μg/L (>0.5 μg/L for total pesticides) or sum trichloroethylene and tetrachloroethylene > 10 μg/L, in line with the provision of the EU Groundwater Directive (2006/118/EC); and (d) anaerobic samples (DO < 1 mg/L), following the approach described in a previous study [18]. From the resulting data set per groundwater body (Table 2), the upper limit of background concentrations can be expressed as a 70th, 90th, or 95th percentile of the remaining data range, indicating appropriate confidence level, using the following procedure: (a) 95th percentile if N > 30; (b) 90th percentile if 20 < N < 30; and (c) 70th percentile if 10 < N < 20.

The advantage of the pre-selection method is that it can be applied if data does not allow for derivation of natural background levels by more advanced methods [18]. The limitation of this method is that the boundary between background and non-background population is defined by an arbitrary decision. It belongs to model-based subjective methods of background determination that include some type of formal statistical model to a set of selected geochemical values, making no assumptions about the form of the data distribution [11]. This characteristic makes the pre-selection method less sensitive to high proportion of LOQ values in a data set than model-based objective methods. In this paper, the pre-selection method was applied if other methods were not applicable and if, after exclusion of contaminated samples, the proportion of <LOQ values is ≤50%. The pre-selection method was also used in several cases to compare the results of the estimation of ambient background values obtained by other methods.

3. Results and Discussion

This section presents the results of an estimation of upper limits of ranges of ambient background values (UL) for selected substances in considered groundwater bodies. Methods for background estimation have been applied based on the criteria described in Section 2.3. As shown in Table 3, the modified Lepeltier method and the pre-selection method were more frequently used than the probability plot. The pre-selection method has been additionally used for arsenic (groundwater body CDGI_23), iron (groundwater body CDGI_19) and sulphate (groundwater body CSGI_29), to compare results with those obtained by the modified Lepeltier method. For As (groundwater bodies CDGI_19 and CSGI_29) and NO_3 (groundwater bodies CDGI_23 and CSGI_29), the proportion of <LOQ values was higher than 50%, hence UL were not estimated by either method.

Table 3. Methods selected for calculating the upper limits of ranges of ambient background concentrations of analysed substances based on criteria described in Section 2.3.

Chemical Substance	GW Body			
	CDGI_19	CDGI_23	CSGI_29	CSGN_25
Arsenic (µg/L)	none	modified Lepeltier method pre-selection method *	none	pre-selection method
Iron (µg/L)	modified Lepeltier method pre-selection method *	probability plot, modified Lepeltier method	probability plot, modified Lepeltier method	probability plot, modified Lepeltier method
Sulphate (mg/L)	probability plot, modified Lepeltier method	pre-selection method	modified Lepeltier method pre-selection method *	pre-selection method
Chloride (mg/L)	probability plot, modified Lepeltier method	probability plot, modified Lepeltier method	probability plot, modified Lepeltier method	probability plot, modified Lepeltier method
Nitrate (mg/L)	probability plot, modified Lepeltier method	none	none	pre-selection method

Note: * For comparison purpose only.

3.1. Arsenic

Table 4 shows the results obtained by applying the modified Lepeltier method and the pre-selection method to arsenic data representative for two groundwater bodies (CDGI_23 and CSGN_25). It can be seen that UL estimates for these two bodies, obtained by the 95th percentiles for the pre-selected data set, differ by an order of magnitude. High value of the UL estimate obtained for the body CDGI_23, which significantly exceeds the EU drinking water standard for arsenic, can be associated with the chemical composition of deep groundwater, which is controlled by natural geochemical processes that contribute high amount of solutes to groundwater [29]. It has been shown that arsenic occur naturally in very high concentrations in unconsolidated aquifers at high depth in the eastern part of the Drava River plain and can vary due to local hydrogeological and geochemical conditions in aquifers [41,42]. In the groundwater body CDGI_23, a high value of the coefficient of variation (1.8) for arsenic is noted (Table 2), indicating significant variability of the chemical composition in different parts of the groundwater body.

Table 4. Estimated Mean + 2SD and Median + 2MAD ranges (for modified Lepeltier method) and the upper limits of ranges of ambient background concentrations (UL) of arsenic (As) obtained by selected methods. EU Drinking Water Standard for As is 10 µg/L.

GW Body	Method	Mean + 2SD (µg/L)	Median + 2MAD (µg/L)	UL (µg/L)	
CDGI_23	modified Lepeltier method	29.5 + 96.6	6.1 + 10.2	126.1 (Mean + 2SD)	16.3 (Median + 2MAD)
	pre-selection method *		-	174.9	
CSGN_25	pre-selection method *		-	15.5	

Note: * UL expressed as 95th percentiles.

Figure 6 depicts the relative cumulative frequencies of arsenic for the groundwater body CDGI_23. High percentage of <LOQ values (27.3%) can be clearly seen on the graph at the lower end of data distribution, while a slight bend in the curve can be visually identified at the upper end of data distribution.

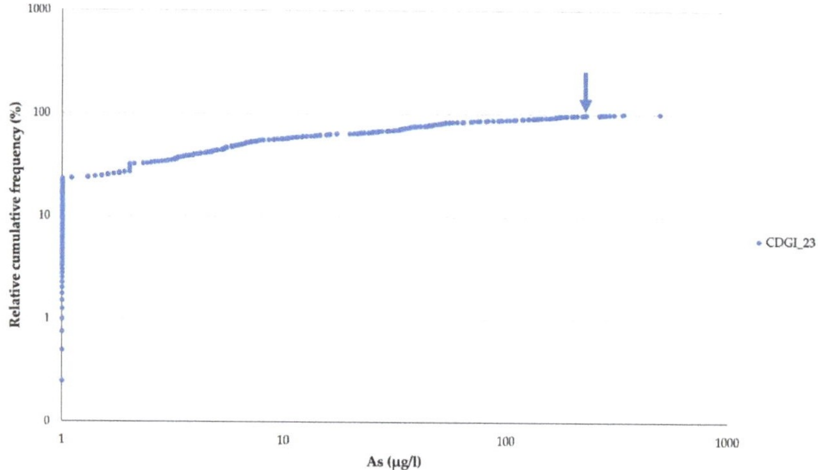

Figure 6. Cumulative relative frequencies of arsenic (As) for the groundwater body CDGI_23. Arrow indicates value **x** considered for the calculation of Mean + 2SD and Median + 2MAD ranges.

The UL estimate for the body CDGI_23 obtained by Mean + 2SD is significantly higher compared to the Median + 2MAD estimate. The Mean + 2SD estimate is lower, but comparable to the estimate obtained by the pre-selection method. The Median + 2MAD estimate is an order of magnitude lower than the UL estimate obtained by the pre-selection method (Table 4).

3.2. Iron

Iron is a highly redox-sensitive element that has low background concentrations in unconfined aquifers, where oxygen is present, but background concentrations of iron can be increased significantly across redox boundaries [9]. It is evident from Table 5 that estimated UL for analysed groundwater bodies vary over one to two orders of magnitude, which is consistent with findings of background concentration ranges of iron in groundwater across Europe [43]. The highest estimates obtained for bodies CDGI_23 and CSGI_29 are associated with the highest average depth of aquifers. The lowest values are noted for the unconfined aquifer within the body CDGI_19.

Very high UL estimates (Table 5) and mean values (Table 2), detected for groundwater bodies CDGI_23 and CSGI_29, by far exceed EU drinking water standard for iron and indicate fast and pronounced reductive dissolution of iron species in anoxic groundwater. It is well known that water-quality thresholds may be frequently breached for iron, which occur in groundwater by natural processes, such as the geochemical conditions existing in the aquifer or due to the specific geology of the area [44].

Table 5 shows that UL estimates for iron, obtained by the probability plot, are higher but comparable with those obtained by the Mean + 2SD. Median + 2MAD estimates are significantly lower in comparison with other methods. Exception is noted for the body CDGI_23, for which all estimates are the same order of magnitude, with Mean + 2SD estimate being the highest one. Tables 2 and 5 suggest that: (a) Mean + 2SD and probability plot estimates match well for highly variable data (coefficient of variation > 1), (b) Median + 2MAD and probability plot estimates are comparable for moderate to low data variability (coefficient of variation ≤ 1). It is noted that with higher data variability a difference between Mean + 2SD and Median + 2MAD estimates increases.

Table 5. Estimated Mean + 2SD and Median + 2MAD ranges (for modified Lepeltier method) and the upper limits of ranges of ambient background concentrations (UL) of iron (Fe) obtained by selected methods. EU Drinking Water Standard for Fe is 200 µg/L.

GW Body	Method	Mean + 2SD (µg/L)	Median + 2MAD (µg/L)	UL (µg/L)	
CDGI_19	modified Lepeltier method	9.5 + 16.2	5.6 + 6.6	25.7 (Mean + 2SD)	12.2 (Median + 2MAD)
	pre-selection method *	-		47.5	
CDGI_23	modified Lepeltier method	1156.4 + 2239.8	712.0 + 1421.4	3396.2 (Mean + 2SD)	2133.4 (Median + 2MAD)
	probability plot	-		1950.0	
CSGI_29	modified Lepeltier method	301.2 + 1141.8	39.0 + 74.0	1443.1 (Mean + 2SD)	113.0 (Median + 2MAD)
	probability plot	-		4270.0	
CSGN_25	modified Lepeltier method	83.2 + 178.8	30.1 + 56.2	262.0 (Mean + 2SD)	86.3 (Median + 2MAD)
	probability plot	-		292	

Note: * UL expressed as 95th percentiles.

Figure 7 depicts the cumulative relative frequencies of iron. Significant visual deviation at the lower end of curve from the main body of data is identified for groundwater bodies CDGI_19 and CSGN_25. These observed deviations are related to analytical imprecision due to high percentage of <LOQ values in data set, which is particularly evident for the body CDGI_19 (27.3%). Small alterations of the slope at the upper end of data distribution can be visually identified for all plots.

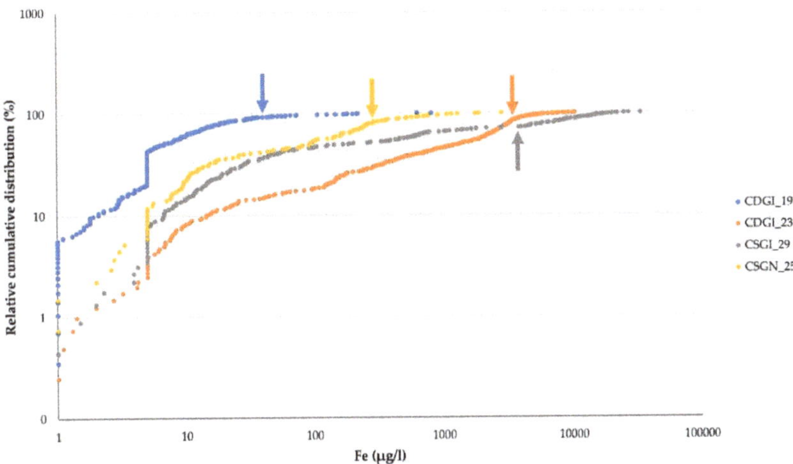

Figure 7. Cumulative relative frequencies of iron (Fe) for analysed groundwater bodies. Arrow indicates value x considered for the calculation of Mean + 2SD and Median + 2MAD ranges for each groundwater body.

Several inflection points can be identified on the lognormal plot for each groundwater body (Figure 8), which correspond to thresholds between different natural and/or anthropogenic populations. Due to high sensitivity of iron on abrupt changes across redox boundaries, it is difficult to conclude in this particular case the causes of the appearance of multiple inflection points on data plots, in terms of whether they are a consequence of natural processes or are the result of direct or indirect anthropogenic impacts, which may be reflected, e.g., through intensive water abstraction.

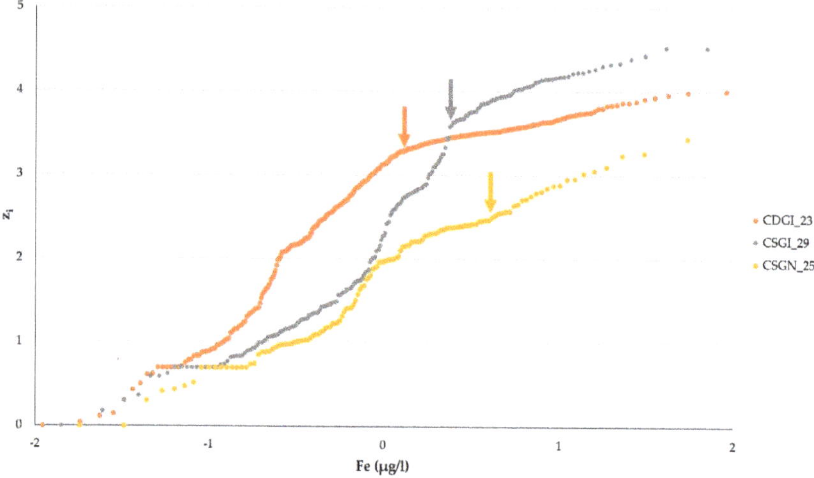

Figure 8. Lognormal probability plot of iron (Fe) for analysed groundwater bodies. Arrow indicates inflection point that separate ambient and non-ambient populations for each groundwater body.

3.3. Sulphate

Table 6 summarizes the main results of UL values for sulphate. An increase of the UL estimate is observed for deep confined aquifers in body CDGI_23, while significantly lower values are attributed to unconfined and semi-confined aquifers within water bodies CDGI_19 and CSGI_29. The highest value of the coefficient of variation (2.4) is noted for the water body CDGI_23 (Table 2), which indicates a high sensitivity of sulphate to changes of geochemical conditions within the aquifer system. It is well known that high concentrations of sulphate may be triggered by dissolution of minerals that control its natural abundance in water or by various land use [9]. EU Groundwater Directive (2006/118/EC) specifically states sulphate as indicator of the saline intrusion resulting from human activities.

Table 6. Estimated Mean + 2SD and Median + 2MAD ranges (for modified Lepeltier method) and the upper limits of ranges of ambient background concentrations (UL) of sulphate (SO_4) obtained by selected methods. EU Drinking Water Standard for SO_4 is 250 mg/L.

GW Body	Method	Mean + 2SD (mg/L)	Median + 2MAD (mg/L)	UL (mg/L)	
CDGI_19	modified Lepeltier method	24.8 + 14.8	27.9 + 5.6	39.6 (Mean + 2SD)	33.5 (Median + 2MAD)
	probability plot		-	33.6	
CDGI_23	pre-selection method *		-	121.4	
CSGI_29	modified Lepeltier method	5.8 + 11.0	2.9 + 5.4	16.8 (Mean + 2SD)	8.3 (Median + 2MAD)
	pre-selection method *		-	44.5	
CSGN_25	pre-selection method *		-	87.3	

Note: * UL expressed as 95th percentiles.

An increase of the sulphate concentration due to human impact is apparent in unconfined aquifers within water body CSGN_25. Estimated UL in bodies CSGI_29 and CSGN_25, obtained by the 95th percentiles for the pre-selected data set (Table 6), differ by two times, which can be associated with pronounced human impact from household diffuse pressure or water abstraction in the body CSGN_25.

Figure 9 depicts cumulative relative frequencies of sulphate for groundwater bodies CDGI_19 and CSGI_29. The jagged appearance of the cumulative relative frequency graph at the lower end of CSGI_29 data distribution can be attributed to the high percentage of <LOQ values (23.1%). A significant alteration of the slope at the upper end of CDGI_19 data distribution can be clearly identified on the cumulative relative frequency graph.

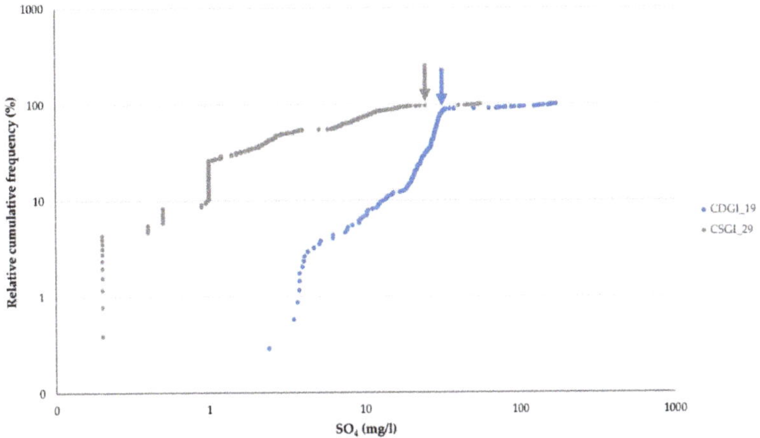

Figure 9. Cumulative relative frequencies of sulphate (SO_4) for groundwater bodies CDGI_19 and CSGI_29. Arrow indicates value **x** considered for the calculation of Mean + 2SD and Median + 2MAD ranges for each groundwater body.

In the lognormal plot (Figure 10), multiple inflection points can be identified at the middle part and at the upper end of CDGI_19 data distribution, which denote thresholds between several natural and/or anthropogenic populations in groundwater of this shallow alluvial aquifer system.

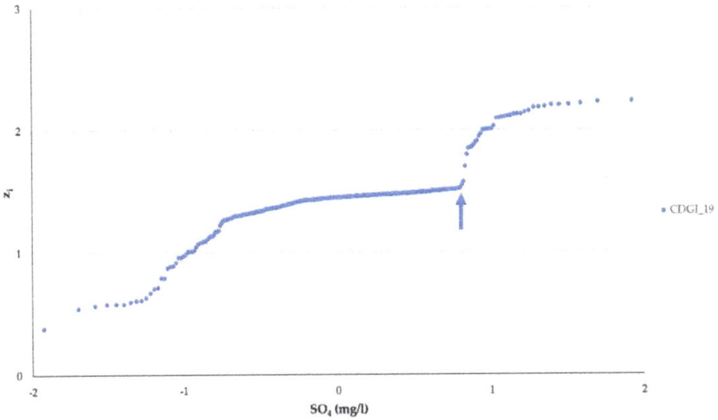

Figure 10. Lognormal probability plot of sulphate (SO_4) for the groundwater body CDGI_19. Arrow indicates inflection point that separate ambient and non-ambient populations.

An interesting observation is noted comparing results of the modified Lepeltier method and the probability plot for the groundwater body CDGI_19 (Table 6). Both Mean + 2SD and Median + 2MAD UL estimates are directly comparable, but also very similar to the UL estimate obtained with the probability plot. It appears that all UL estimates match well in the case of the low data variability (coefficient of variation < 1).

3.4. Chloride

Chloride is an inert and mobile compound, which natural amount depends on the geographical location, distance to sea, and amount of precipitation, but also on the regional influence of saline water inputs to the groundwater [9]. The average concentrations of chloride in groundwater of analysed water bodies (Table 2, 6.4 to 16.4 mg/L) are consistent with findings of Thunqvist [45], who stated that natural mean concentrations of chloride in groundwater vary between 10–15 mg/L. Multiple natural and anthropogenic sources of chlorides cause great variability of ambient background concentrations across Europe. The EU research project BRIDGE revealed that background values for chloride in the groundwater of Europe range from 6 to 548 mg/L [18].

Due to low percentage of <LOQ data, UL were estimated only by the modified Lepeltier method and probability plot. Close inspection of Table 7 indicates that UL estimates, obtained by different methods and across water bodies, are directly comparable. A slight increase of UL estimates for the shallow aquifers in the body CSGN_25 can be attributed to direct (salt used on roads to remove ice at low air temperatures, leakage from sewage, fertilizers), or indirect (water abstraction) anthropogenic impacts.

Table 7. Estimated Mean + 2SD and Median + 2MAD ranges (for modified Lepeltier method) and the upper limits of ranges of ambient background concentrations (UL) of chloride (Cl) obtained by selected methods. EU Drinking Water Standard for Cl is 250 mg/L.

GW Body	Method	Mean + 2SD (mg/L)	Median + 2MAD (mg/L)	UL (mg/L)	
CDGI_19	modified Lepeltier method	3.2 + 1.5	3.0 + 1.3	4.7 (Mean + 2SD)	4.3 (Median + 2MAD)
	probability plot	-		5.0	
CDGI_23	modified Lepeltier method	4.9 + 3.1	5.1 + 2.4	8.0 (Mean + 2SD)	7.5 (Median + 2MAD)
	probability plot	-		7.7	
CSGI_29	modified Lepeltier method	3.6 + 3.5	3.7 + 3.5	7.1 (Mean + 2SD)	7.2 (Median + 2MAD)
	probability plot	-		6.4	
CSGN_25	modified Lepeltier method	4.9 + 7.3	4.0 + 5.4	12.2 (Mean + 2SD)	9.4 (Median + 2MAD)
	probability plot	-		15.3	

Note: * UL expressed as 95th percentiles.

Figures 11 and 12 depict cumulative relative frequencies and lognormal plots of chlorides for analysed groundwater bodies. Cumulative relative frequency graph show distinct bend in a curve at the upper end of data distribution (Figure 11). Lognormal plots indicate curved data distributions, but visible changes at inflection points can be easily detected, indicating the existence of multiple natural populations as well as, in the case of shallow aquifers, anthropogenic populations at the upper end of data distribution (Figure 12).

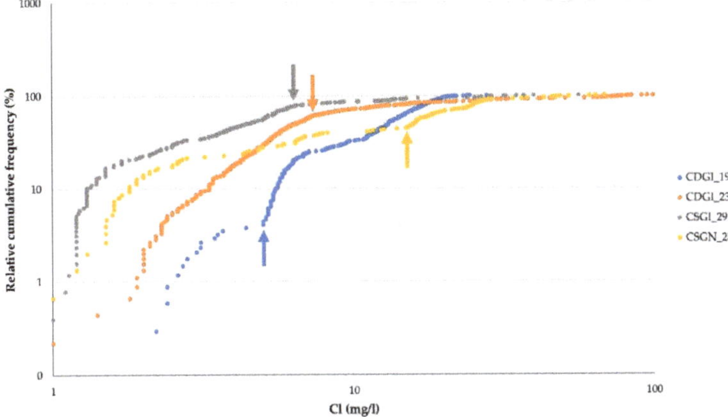

Figure 11. Cumulative relative frequencies of chloride (Cl) for analysed groundwater bodies. Arrow indicates value x considered for the calculation of Mean + 2SD and Median + 2MAD ranges for each groundwater body.

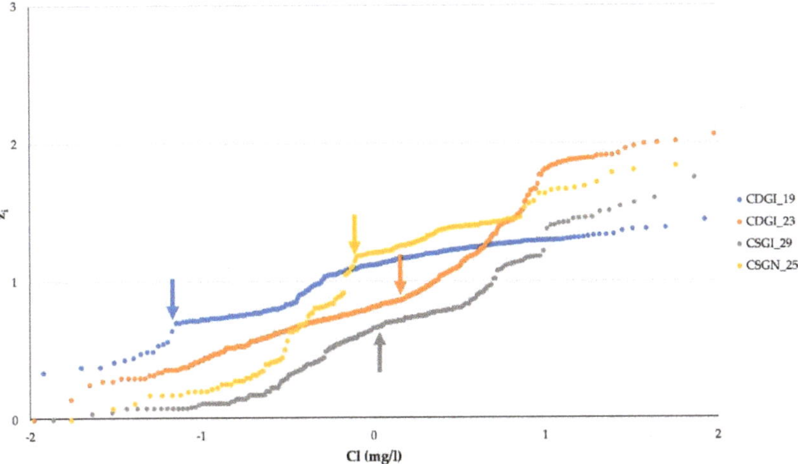

Figure 12. Lognormal probability plot of chloride (Cl) for analysed groundwater bodies. Arrow indicates inflection point that separates ambient and non-ambient populations for each groundwater body.

3.5. Nitrate

Nitrate is frequently detected in high concentrations in unconfined alluvial aquifers in Croatia [46], while its origin, in some cases, can be linked with the existence of elevated chloride and sulphate concentrations, particularly in urban areas [47]. Increased nitrate concentrations in groundwater body CDGI_19, which occasionally exceed EU drinking water standard (50 mg NO_3/L), were attributed to agricultural activities, sewage leaks and household discharges not connected to sewerage systems [26]. From Table 2 it is clear that the mean value of nitrate in the body CDGI_19 is significantly higher than mean values recorded in other bodies.

Table 8 shows results obtained by used methods to nitrate data representative for two groundwater bodies (CDGI_19 and CSGN_25). An increase in the UL estimates for the shallow aquifers in groundwater body CDGI_19, obtained by the modified Lepeltier method and probability plot, can be associated to deterioration of groundwater quality due to prolonged human influence.

Close inspection of Table 8 indicates that the UL estimate for the body CSGN_25, obtained by the 95th percentiles for the pre-selected data set, is even higher than estimates obtained for the body CDGI_19. This is not consistent with the average nitrate concentrations observed in these two groundwater bodies (Table 2). However, it is noted that 5% of samples from the pre-selected data set for the body CSGN_25, with high values of nitrate (>30 mg NO_3/L), significantly deviate from the majority of data characterized by the high percentage of <LOQ values in data set, thus having a strong influence on the calculation of the UL estimate.

Table 8. Estimated Mean + 2SD and Median + 2MAD ranges (for modified Lepeltier method) and the upper limits of ranges of ambient background concentrations (UL) of nitrates (NO_3) obtained by selected methods. EU Drinking Water Standard for NO_3 is 50 mg NO_3/L.

GW Body	Method	Mean + 2SD (mg/L)	Median + 2MAD (mg/L)	UL (mg/L)	
CDGI_19	modified Lepeltier method	8.6 + 9.6	8.7 + 15.6	18.2	24.3
	probability plot		-	19.0	
CSGN_25	pre-selection method *		-	29.1	

Note: * UL expressed as 95th percentiles.

Figures 13 and 14 depict relative cumulative frequencies and lognormal plot of nitrate for the groundwater body CDGI_19. Cumulative relative frequency graph (Figure 13) shows significant bend in the curve at the upper end of data distribution. In the lognormal plot (Figure 14), multiple inflection points can be identified at the middle part and at the upper end of data distribution, indicating the existence of natural and probably multiple anthropogenic populations, which is consistent with observed contribution of nitrate from multiple anthropogenic sources in the body CDGI_19.

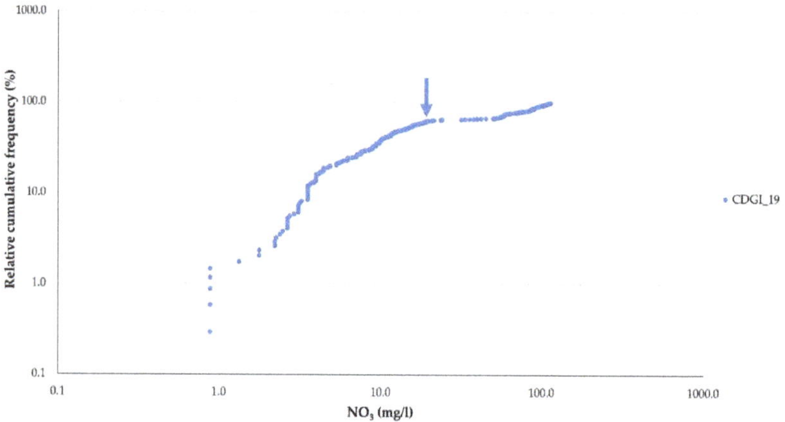

Figure 13. Cumulative relative frequencies of nitrate (NO_3) for the groundwater body CDGI_19. Arrow indicates value x considered for the calculation of Mean + 2SD and Median + 2MAD ranges.

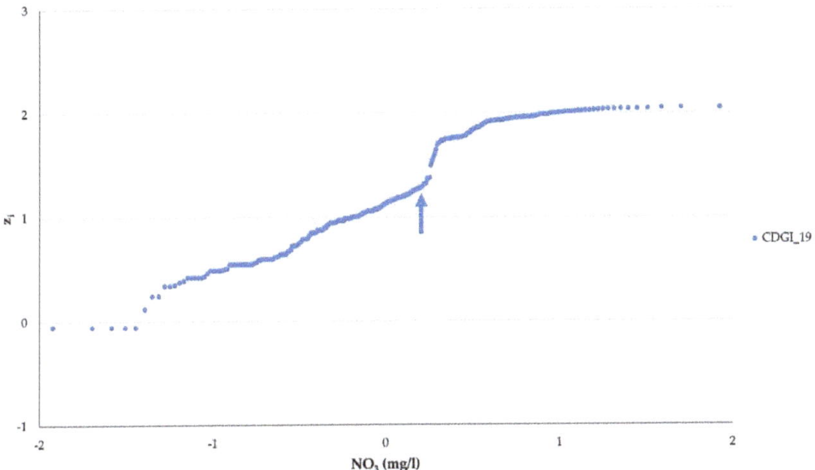

Figure 14. Lognormal probability plot of nitrate (NO₃) for the groundwater body CDGI_19. Arrow indicates inflection point that separates ambient and non-ambient populations.

3.6. Comparison of UL Estimates and Methods

Observed similarities and/or differences between estimated UL values can be attributed to local hydrogeology and geochemistry of analysed groundwater bodies and to characteristics of selected methods. For mobile and inert compound like chloride, UL estimates are comparable both between groundwater bodies and at the level of each groundwater body (Table 7). For highly redox-sensitive substances (arsenic, iron) UL estimates across groundwater bodies range over one to two orders of magnitude (Tables 4 and 5).

Comparing the model-based objective methods, the modified Lepeltier method and probability plot, differences between estimated UL are related to data variability and type of estimators used (for the modified Lepeltier method). When data variability is low to moderate, e.g., for chloride, Mean + 2SD and Median + 2MAD estimators give similar results and are directly comparable with probability plot estimates. Otherwise, for highly variable data, e.g., for iron, differences between UL estimates increase, particularly between Median + 2MAD and probability plot.

Median + 2MAD is a robust equivalent of Mean + 2SD, which is not affected by extreme values in data set and is highly resistant up to 50% of the data values being extreme [36]. However, by determining the point x on the cumulative relative frequency curve at which a bend in a curve is noted, the non-ambient population from the mixed data distribution is significantly reduced, so the occurrence of potential extreme values in the data set is also reduced. Given that Median + 2MAD estimates are systematically the lowest, especially in cases where the greatest variability of data was recorded, it appears that Median + 2MAD is excessively conservative and underestimate an actual UL value.

The choice of an inflection point to discriminate between ambient and non-ambient populations is a critical issue for model-based objective methods. These methods contain element of subjectivity in that the choice of the UL estimate (probability plot) or point x on the cumulative relative frequency curve (the modified Lepeltier method) is subject to visual detection and depends to great extent on the experience of researchers. As noted before, one should avoid temptation to detect a threshold between natural and anthropogenic populations at lower part of curve, close to LOQ values. A further limitation inherent to probability plot is existence of the multiple inflection points due to multiple natural and/or anthropogenic populations. Hence, to increase the confidence in the estimation process, it is necessary to detect non-linear behavior of data distribution, related to analytical imprecision or to erroneous data, and, whenever possible, to combine two or more model-based objective methods, particularly in cases of limited data set and/or limited data quality.

The UL estimates for arsenic (groundwater body CDGI_23), iron (groundwater body CDGI_19), and sulphate (groundwater body CSGI_29), obtained by the 95th percentile for the pre-selected data set, are systematically higher than those obtained by the modified Lepeltier method. The choice of an appropriate percentile, calculated from a data range remaining after the exclusion of contaminated samples, directly relates to corresponding confidence level of the UL estimate. It is influenced by a number of samples in data set [18] or by positive skewness of data distribution [2]. Since each data set contains more than hundred data, the choice of 95th percentile, according to "number of samples" criterion set by Hinsby et al. [18] is met. The Shapiro–Wilk normality test for all data sets (verified at 0.05 significance level) indicated very high positive skewness of all data distributions, probably due to high percentage of <LOQ values in corresponding data sets (Table 2). Hence, the 90th percentile criterion was additionally tested to examine UL estimates for different percentiles. Comparing UL estimates for arsenic, iron and sulphate (Table 9), differences are evident between the 90th and 95th percentiles, however, the 90th percentile estimates are comparable to the Mean + 2SD estimates, shown in Tables 4–6. Although the pre-selection method is less sensitive to limited data quality compared to model-based objective methods, the high proportion of <LOQ values in the data set significantly affects the UL estimate. In other words, the higher proportion of <LOQ values in data set, the higher degree of uncertainty in the UL estimate obtained with a high percentile.

Table 9. Comparison of 90th and 95th percentile UL estimates for pre-selected data sets of arsenic (As), iron (Fe), and sulphate (SO_4) for three groundwater bodies.

GW Body	Substance	UL Estimate	
		90th Percentile	95th Percentile
CDGI_19	iron (µg/L)	29.7	47.5
CDGI_23	arsenic (µg/L)	133.6	174.9
CSGI_29	sulphate (mg/L)	16.4	44.5

The UL estimates for redox-sensitive substances obtained by using the same statistical methods differ between groundwater bodies. Table 5 shows that UL estimates of iron for unconfined aquifers within groundwater bodies CDGI_19 and CSGN_25, obtained by the modified Lepeltier method, differ by an order of magnitude. Similarly, it is noted that UL estimates of iron for semi-confined and confined aquifers within groundwater body CDGI_23 are not comparable with UL estimates of iron for groundwater body CSGI_29 characterized by the same type of aquifer, although the same methods were used for the estimation. A related outcome can be observed comparing UL estimates of sulphate, obtained by the 95th percentiles for the pre-selected data set, between groundwater bodies CDGI_23 and CSGI_29 (Table 6).

Given the high data variability (coefficient of variation > 1) of redox-sensitive substances, these results point to the great heterogeneity of considered groundwater bodies. Due to variable dynamics of groundwater flow during the hydrological year, the movement of solutes due to geochemical and hydraulic gradient as well as geochemical barriers that cause retention of solutes on the rock matrix, natural variability is common even in lithologically homogeneous aquifers. As noted by Matschullat et al. [1], the geochemical background concentration must be determined for homogeneous units or areas within a natural system, primarily in relation to climatological, hydrogeological, lithological and pedological characteristics. This concept is taken by Preziosi et al. [48], who singled out representative aquifers as homogeneous units and determined background concentrations of substances in groundwater for each aquifer and for the whole groundwater body. Similarly, Molinari et al. [8] found that background concentrations of redox-sensitive substances increase with depth, showing that the geochemical and hydrogeological stratification can be an important factor in determining background concentrations and that it is desirable to carry out detailed characterization of aquifer system in order to determine hydrogeochemically homogeneous areas.

Assuming a relevance of determining background values of substances for each homogeneous unit, the concept of "regional" background concentration at the level of groundwater body, as followed in [26], needs to be re-examined, because the actual background value of a substance may differ between lithologically similar aquifers with different geochemical and hydrodynamic conditions. In compliance with the "sample size" and the "percentage of <LOQ values" criteria for the use of methods presented in Section 2.3, this changing paradigm inevitably requires balancing the need for extending the actual monitoring program in relation to the costs of drilling new boreholes and regular monitoring of parameters set by EU and Croatian regulations and guidelines.

4. Conclusions

Groundwater quality data from four groundwater bodies with similar hydrogeological settings, located in the northern and the eastern part of Croatia, were analysed in order to estimate the upper limits of ranges of background concentrations (UL) for targeted chemical substances. The concept of determining the ambient background value of a substance was applied, recognizing the fact that elevated concentrations of substances in groundwater are no entirely of natural origin and reflect long-term human impact on the chemical composition of groundwater. The UL were estimated using model-based objective methods, probability plot, and modified Lepeltier method, as well as the simple and robust the pre-selection method.

Model-based objective methods are sensitive to high variability of data and to high percentage of <LOQ values in the data set. UL estimates obtained by probability plot and modified Lepeltier method are comparable when data variability is low to moderate, otherwise differences between estimates are notable. It appears that an UL estimate calculated as a Median + 2MAD range of data from undisturbed (ambient) part of mixed data distribution obtained from the cumulative relative frequency curve, particularly underestimate an actual UL value. High percentage of <LOQ values in data set influences non-linear behavior of data distribution at lower part of a curve, thus affecting the detection of inflection point to discriminate between ambient and non-ambient populations. Results from this research indicate that both methods should not be used if data set contains more than 30% of <LOQ values. However, combining results of two or more model-based objective methods, particularly in cases of limited data set and/or limited data quality, can increase the confidence of estimation of threshold values.

The robust pre-selection method proved less sensitive to limited data quality or data availability compared to model-based objective methods. This method is preferable over the others if data set contains more than 30% of <LOQ values. For highly skewed data, the 90th percentile of the pre-selected data set is comparable with other methods and preferable over the 95th percentile estimate.

The estimated UL values for inert and mobile substances, e.g., chloride, are comparable on different scales. On the other hand, significant variability of UL estimates for redox-sensitive substances, e.g., arsenic and iron, can be attributed to change of chemical composition of groundwater across redox boundaries. Observed differences of the UL estimates for redox-sensitive substances between considered groundwater bodies are related to the heterogeneity of aquifer systems. This stresses the value of high-resolution conceptual model of groundwater bodies in the future update of the UL estimates. The critical issue in this process is the determination of hydrogeological and geochemical homogeneous units within the heterogeneous aquifer system.

Author Contributions: Z.N. made the conceptualization and wrote the draft of the manuscript. Z.N. and Z.K. developed the methodology. Z.K. made the data investigation and formal analysis, while Z.N. and Z.K. made most of the data interpretation and construction of tables. Z.K., J.P., and D.P. participated in the creation of figures, review and editing of the manuscript. All authors have read and agreed to the published version of the manuscript.

Funding: The publication process was supported by the Development Fund of the Faculty of Mining, Geology, and Petroleum Engineering of the University of Zagreb.

Acknowledgments: Authors would like to thank Croatian Waters for providing data for this research within the projects "Trend definition and groundwater status assessment in the Pannonian part of Croatia" and "Definition of criteria for the determination of background concentrations and threshold values of contaminants in the groundwater bodies in the Pannonian part of Croatia".

Conflicts of Interest: The authors declare no conflict of interest.

Abbreviation

Full Term Name	Abbreviation
European Union	EU
Natural Baseline Quality in European Aquifers: A Basis for Aquifer Management	BaSeLiNe
Probability plot	PP
Limit of quantification	LOQ
European Commission	EC
Median absolute deviation	MAD
Standard deviation	SD
Background Criteria for Identification of Groundwater Thresholds	BRIDGE
Upper limit of the range of ambient background values	UL
Number of data	N
Dissolved Oxygen	DO

References

1. Matschullat, J.; Ottenstein, R.; Reimann, C. Geochemical background—Can we calculate it? *Environ. Geol.* **2000**, *39*, 990–1000. [CrossRef]
2. Preziosi, E.; Parrone, D.; Del Bon, A.; Ghergo, S. Natural background level assessment in groundwaters: Probability plot versus pre-selection method. *J. Geochem. Explor.* **2014**, *143*, 43–53. [CrossRef]
3. Reimann, C.; Garrett, R.G. Geochemical background—Concept and reality. *Sci. Total Environ.* **2005**, *350*, 12–27. [CrossRef] [PubMed]
4. Panno, S.V.; Kelly, W.R.; Martinsek, A.T.; Hackley, K.C. Estimating background and threshold nitrate concentrations using probability graphs. *Groundwater* **2006**, *44*, 697–709. [CrossRef]
5. Nakić, Z.; Posavec, K.; Bačani, A. A visual basic spreadsheet macro for geochemical background analysis. *Groundwater* **2007**, *45*, 642–647. [CrossRef]
6. Nakić, Z.; Posavec, K.; Parlov, J. Model-based objective methods for the estimation of groundwater geochemical background. *Aqua Mundi* **2010**, *1*, 65–72.
7. Rodriguez, J.G.; Tueros, I.; Borja, A.; Belzunce, M.J.; Franco, J.; Solaun, O.; Valencia, V.; Zuazo, A. Maximum likelihood mixture estimation to determine metal background values in estuarine and coastal sediments within the European Water Framework Directive. *Sci. Total Environ.* **2006**, *370*, 278–293. [CrossRef]
8. Molinari, A.; Guadagnini, L.; Marcaccio, M.; Guadagnini, A. Natural background levels and threshold values of chemical species in three large scale groundwater bodies in Northern Italy. *Sci. Total Environ.* **2012**, *425*, 9–19. [CrossRef]
9. Balderacchi, M.; Benoit, P.; Cambier, P.; Eklo, O.M.; Gargini, A.; Gemitzi, A.; Gurel, M.; Klöve, B.; Nakić, Z.; Preda, E.; et al. Groundwater pollution and quality monitoring approaches at European-level. *Crit. Rev. Environ. Sci. Technol.* **2013**, *43*, 323–408. [CrossRef]
10. Edmunds, W.M.; Shand, P. Natural groundwater quality: Summary and significance for water resources management. In *Natural Groundwater Quality*; Edmunds, W.M., Shand, P., Eds.; Blackwell Publishing Ltd.: London, UK, 2008; pp. 441–462.
11. Reimann, C.; Filzmoser, P. Normal and lognormal data distribution in geochemistry: Death of a myth. Consequences for the statistical treatment of geochemical and environmental data. *Environ. Geol.* **2000**, *39*, 1001–1014. [CrossRef]
12. Reimann, C.; Filzmoser, P.; Garrett, R.G. Background and threshold: Critical comparison of methods of determination. *Sci. Total Environ.* **2005**, *346*, 1–16. [CrossRef] [PubMed]
13. BRIDGE, Background cRiteria for the IDentification of Groundwater thrEsholds. Available online: http://nfp-at.eionet.europa.eu/irc/eionet-circle/bridge/info/data/en/index.htm (accessed on 14 April 2020).

14. Waendland, F.; Hannapel, S.; Kunkel, R.; Schenk, R.; Voigt, H.J.; Wolter, R. A procedure to define natural groundwater conditions of groundwater bodies in Germany. *Water Sci. Technol.* **2005**, *51*, 249–257. [CrossRef]

15. Coetsiers, M.; Blaser, P.; Martens, K.; Walraevens, K. Natural background levels and threshold values for groundwater in fluvial Pleistocene and Tertiary marine aquifers in Flanders, Belgium. *Environ. Geol.* **2009**, *57*, 1155–1168. [CrossRef]

16. Gemitzi, A. Evaluating the anthropogenic impacts on groundwaters: A methodology based on the determination of natural background levels and threshold values. *Environ. Earth Sci.* **2012**, *67*, 2223–2237. [CrossRef]

17. Griffioen, J.; Passier, H.F.; Klein, J. Comparison of selection methods to deduce natural background levels for groundwater units. *Environ. Sci. Technol.* **2008**, *42*, 4863–4869. [CrossRef]

18. Hinsby, K.; Condesso de Melo, M.T.; Dahl, M. European case studies supporting the derivation of natural background levels and groundwater threshold values for the protection of dependent ecosystems and human health. *Sci. Total Environ.* **2008**, *401*, 1–20. [CrossRef]

19. Swedish Environmental Protection Agency. Environmental Quality Criteria—Groundwater. 2000. Available online: https://www.naturvardsverket.se/Documents/publikationer/620-6033-3.pdf (accessed on 16 April 2020).

20. WFD—Development of a Methodology for the Characterisation of Unpolluted Groundwater (2002-W-DS-7), Main Report. 2004. Available online: https://www.epa.ie/pubs/reports/research/water (accessed on 16 April 2020).

21. Sinclair, A.J. Selection of threshold values in geochemical data using probability graphs. *J. Geochem. Explor.* **1974**, *3*, 129–149. [CrossRef]

22. Kyoung-Ho, K.; Seong-Taek, Y.; Hyun-Koo, K.; Ji-Wook, K. Determinations of natural backgrounds and thresholds of nitrate in South Korean groundwater using model-based statistical approach. *J. Geochem. Explor.* **2015**, *148*, 196–205.

23. Sinclair, A.J. A fundamental approach to threshold estimation in exploration geochemistry: Probability plots revisited. *J. Geochem. Explor.* **1991**, *41*, 1–22. [CrossRef]

24. Lepeltier, C. A simplified treatment of geochemical data by graphical representation. *Econ. Geol.* **1969**, *64*, 538–550. [CrossRef]

25. Nakić, Z.; Ugrina, I.; Špoljarić, D.; Kovač, Z. Applicability of selected model-based objective methods for natural background level assessment. Unpublished work. 2020.

26. River Basin Management Plan of the Republic of Croatia 2016–2021. Official Gazette. 2016. Available online: https://narodne-novine.nn.hr/clanci/sluzbeni/dodatni/441070.pdf (accessed on 18 August 2020).

27. Brkić, Ž.; Briški, M. Hydrogeology of the western part of the Drava Basin in Croatia. *J. Maps* **2018**, *14*, 173–177. [CrossRef]

28. Urumović, K.; Duić, Ž.; Hlevnjak, B. Positive and negative influences of hydroelectric plants on Drava river to groundwater quality. In Proceedings of the XXIst Conference of the Danubian Countries on the Hydrological Forecasting and Hydrological Bases of Water Management, Bucharest, Romania, 22–26 September 2002; pp. 1–7.

29. Ujević, M.; Duić, Ž.; Casiot, C.; Sipos, L.; Santo, V.; Dadić, Ž.; Halamić, J. Occurrence and geochemistry of arsenic in the groundwater of Eastern Croatia. *Appl. Geochem.* **2010**, *25*, 1017–1029. [CrossRef]

30. Brkić, Ž.; Briški, M.; Marković, T. Use of hydrochemistry and isotopes for improving the knowledge of groundwater flow in a semiconfined aquifer system of the Eastern Slavonia. *Catena* **2016**, *142*, 153–165. (In Croatian) [CrossRef]

31. Mutić, R. Correlation of the Eastern Slavonian Quaternary deposits based on mineralogical and petrographical analyses (Eastern Croatia). *Acta Geol.* **1993**, *23*, 1–37. (In Croatian)

32. Oreščanin, V. Arsenic in the water—The origin of the toxic effect of the removal methods. *Hrvatske vode* **2013**, *21*, 7–16. (In Croatian)

33. Rowland, H.A.L.; Omoregie, E.O.; Millot, R.; Jimenez, C.; Mertens, J.; Baciu, C.; Hug, S.J.; Berg, M. Geochemistry and arsenic behaviour in groundwater resources of the Pannonian Basin (Hungary and Romania). *Appl. Geochem.* **2011**, *26*, 1–17. [CrossRef]

34. Hećimović, I. Tectonic Relationships of the Wider Kalnik Area. Ph.D. Thesis, University of Zagreb, Zagreb, Croatia, 1995. (In Croatian)

35. D'Agostino, R.B.; Stephens, M.A. *Goodness-of-Fit Techniques (Statistics: A Series of Textbooks and Monographs*, 1st ed.; Marcel Dekker Inc.: New York, NY, USA, 1986; p. 576.

36. Reimann, C.; Filzmoser, P.; Garrett, R.G.; Dutter, R. *Statistical Data Analysis Explained: Applied Environmental Statistics with R*, 1st ed.; John Wiley & Sons, Ltd.: West Sussex, UK, 2009; p. 343.

37. Jakeman, A.; Taylor, J. A hybrid ATDL-gamma distribution model for predicting urban area source acid gas concentration. *Atmos. Environ.* **1985**, *19*, 1959–1967. [CrossRef]

38. Parslow, G.R. Determination of background and threshold in exploration geochemistry. *J. Geochem. Explor.* **1974**, *3*, 319–336. [CrossRef]

39. Ashley, R.P.; Keith, W.J. *Distribution of Gold and Other Metals in Silicified Rocks of the Goldfield Mining District, Nevada*; Geological Survey Professional Paper 843-B; United States Government Printing Office: Washington, DC, USA, 1976; p. 17.

40. Reimann, C.; de Caritat, P. Establishing geochemical background variation and threshold values for 59 elements in Australian surface soil. *Sci. Total Environ.* **2017**, *578*, 633–648. [CrossRef]

41. Ujević-Bošnjak, M.; Fazinić, S.; Duić, Ž. Characterization of arsenic-contaminated aquifer sediments from eastern Croatia by ion microbeam, PIXE and ICP-OES techniques. *Nucl. Instrum. Methods Phys. Res. B* **2013**, *312*, 23–29. [CrossRef]

42. Ujević Bošnjak, M.; Casiot, C.; Duić, Ž.; Fazinić, S.; Halamić, J.; Sipos, L.; Santo, V.; Dadić, Ž. Sediment characterization and its implications for arsenic mobilization in deep aquifers of eastern Croatia. *J. Geochem. Explor.* **2013**, *126*, 55–66. [CrossRef]

43. Shand, P.; Edmunds, W.M. The baseline inorganic chemistry of European groundwaters. In *Natural Groundwater Quality*; Edmunds, W.M., Shand, P., Eds.; Blackwell Publishing Ltd.: London, UK, 2008; pp. 22–58.

44. Edmunds, W.M.; Shand, P. Groundwater baseline quality. In *Natural Groundwater Quality*; Edmunds, W.M., Shand, P., Eds.; Blackwell Publishing Ltd.: London, UK, 2008; pp. 1–21.

45. Thunqvist, E.L. Regional increase of mean chloride concentration in water due to the application of deicing salt. *Sci. Total Environ.* **2004**, *325*, 29–37. [CrossRef] [PubMed]

46. Kovač, Z.; Nakić, Z.; Pavlić, K. Influence of groundwater quality indicators on nitrate concentrations in the Zagreb aquifer system. *Geol. Croat.* **2017**, *70*, 93–103. [CrossRef]

47. Kovač, Z.; Nakić, Z.; Barešić, J.; Parlov, J. Nitrate origin in the zagreb aquifer system. *Geofluids* **2018**, *2018*, 2789691. [CrossRef]

48. Preziosi, E.; Giuliano, G.; Vivona, R. Natural background levels and threshold values derivation for naturally As, V and F rich groundwater bodies: A methodological case study in Central Italy. *Environ. Earth Sci.* **2010**, *61*, 885–897. [CrossRef]

 *water*

Article

GuEstNBL: The Software for the Guided Estimation of the Natural Background Levels of the Aquifers

Francesco Chidichimo [1,*], Michele De Biase [1], Alessandra Costabile [2], Enzo Cuiuli [3], Orsola Reillo [2], Clemente Migliorino [3], Ilario Treccosti [2] and Salvatore Straface [1]

[1]   Department of Environmental Engineering, University of Calabria, via P. Bucci 42B, 87036 Rende, Italy; michele.debiase@unical.it (M.D.B.); salvatore.straface@unical.it (S.S.)

[2]   Department of Environmental Policy, Calabria Region, Cittadella Regionale, Località Germaneto, 88100 Catanzaro, Italy; alessandra.costabile@regione.calabria.it (A.C.); o.reillo@regione.calabria.it (O.R.); i.treccosti@regione.calabria.it (I.T.)

[3]   Regional Agency for the Protection of the Environment—ARPACAL, via Lungomare (Loc Mosca) (Giovino), 88100 Catanzaro, Italy; e.cuiuli@arpacal.it (E.C.); c.migliorino@arpacal.it (C.M.)

*   Correspondence: francesco.chidichimo@unical.it

Received: 28 August 2020; Accepted: 28 September 2020; Published: 29 September 2020

Abstract: Natural background levels (NBLs) for targeted chemical elements characterize a specific groundwater body, the knowledge of which represents a fundamental information for environmental agencies responsible for the protection, management, and remediation of territory. the large number of areas subject to strong anthropogenic pressures of a different nature and magnitude makes the job of control authorities particularly difficult. the process to distinguish effective anthropogenic contamination from natural conditions and to define realistic environmental clean-up goals goes through the computation of several mutually dependent statistical methods, some of which have non-trivial resolution and interpretation. in this study, we presented a new tool designed to drive those working in the sector into an articulated path towards NBL assessment. the application software was developed in order to read environmental input data provided by a user-friendly web-based geographic information system (GIS) and to return the NBL estimate of a given chemical element following a wizard that allows for the implementation of two methodologies, i.e., component separation or pre-selection. the project was born from a collaboration between the Department of Environmental Engineering of the University of Calabria and the Department of Environmental Policies of the Calabria Region. the software was used to estimate NBLs in selected chemical species at potentially contaminated industrial sites located in Lamezia Terme, Italy. in the future, the developed calculation program will be the official evaluation tool of the Calabria Region for identifying groundwater thresholds.

Keywords: natural background levels; software implementation; parameters estimation; statistical methods; component separation method; pre-selection method

---

1. Introduction

Groundwater characterization activities may be marked by the observation of large concentration values related to an aquifer's petrographical composition rather than the pollution status of an investigated site [1]. the reliable quantification of the actual natural contribution to detected concentrations, at a field scale, is underlined by the European Union (EU) Water Framework Directive (WFD 2000/60/EC, article 17), which identified significant and feasible clean-up goals. Applying remediation strategies based on compliance levels guided by current regulations might lead, in some cases, to ineffective and unaffordable cleaning targets for areas where specific natural conditions take place. in such a context,

171

estimating natural background levels (NBLs) in groundwater gained large attention in the last decade. the Ground Water Daughter Directive (GWDD 2006/118/EC) defines the NBL as "the concentration of a substance or the value of an indicator in a body of groundwater corresponding to no, or only very minor, anthropogenic alterations to undisturbed conditions". Chemical and biological processes taking place in the water-rock system, as well as intakes from other water bodies and rainfall contribution, may cause local effects to give rise to the spread of background concentrations [2–6]. The correct distinction between anthropogenic and natural origin contamination is mandatory for the estimation of reliable NBLs. When this last condition is not met, the misleading classification of a site as potentially highly contaminated can be assessed by detecting large concentrations. the legal implications of such a conclusion have direct consequences, especially in cases where the possible negative effects of human activities occur in correspondence with polluting areas (industrial sites, landfills, intensive agriculture, farming systems, etc.). Optimizing remediation strategies occurs via proper NBL evaluation and the associated revision of the compliance values that can be locally modified to account for a specific context under investigation. Practical applications, based on the statistical analysis of monitored data also proposed by the EU research project BRIDGE (2007) (background criteria for the identification of groundwater thresholds [7]) have been frequently employed for NBL estimation and are available in the literature for several countries such as Germany, Italy, Spain, and South Korea [2–6,8–17].

Recently, the Italian Institute for Environmental Protection and Research, ISPRA, has promoted initiatives aimed at addressing the methodological aspects related to determining NBLs of soil and groundwater, as well as to draw up guidelines to be followed at the national level. This target was pursued by joining the experiences and skills of the Regional Environmental Protection Agencies (ARPA). the Resolution of the Council of the National System for the Protection of the Environment (SNPA), dated 14/11/2017, laid the foundations for drafting the aforementioned guidelines collected in the manual 174/2018 [18]. the procedural iter, required by this document, can be extremely complex, especially for operators without a high level of knowledge and experience in statistical calculation or iterative optimization processes.

In this study a software was presented to estimate the NBLs in aquifers, complying with indications provided by ISPRA guidelines, including an additional feature for the component separation methodology, which is not covered by the guidelines. This project was born from a collaboration between the Department of Environmental Engineering of the University of Calabria and the Department of Environmental Policies of the Calabria Region, and aimed at providing the public bodies responsible for environmental control with a tool to simplify and automate complex calculations and estimation processes. the software was able to manage environmental data (i.e., concentration of chemical elements were analyzed in the sampling points and collected into a user-friendly web-based geographic information system (GIS)) and to provide the estimate of the background values following different methodologies indicated by the guidelines. the wizard, implemented on Java Platform, was equipped with a step-by-step graphical interface that guided the operator by showing the results obtained for each elaboration, and by allowing him or her to make choices based on his/her experience. The data was analyzed after implementing two methodologies: component separation and pre-selection [7]. At the end of the analysis, performed for the single chemical element, the software printed a PDF report where the used environmental data, the adopted methods, the pursued decisions, the generated graphs/tables, and the estimated NBL were sequentially recorded. in the future, the developed calculation program will be the official evaluation tool of the Calabria Region to identify groundwater thresholds. the work also presented a case study in which the software was used to estimate the NBL of a selected chemical species in a potentially contaminated industrial site located in Lamezia Terme, Italy. To the best of our knowledge, the computer system presented in this work is the first of its kind. It could be easily exported and used in international contexts since the methodological approaches included therein are widely recognized by the whole scientific community.

2. Materials and Methods

This section is devoted to the detailed description of the internal structure of the software-based procedure, i.e., how it operates to reach the NBLs estimation starting from the environmental data, as well as an overview of the case study.

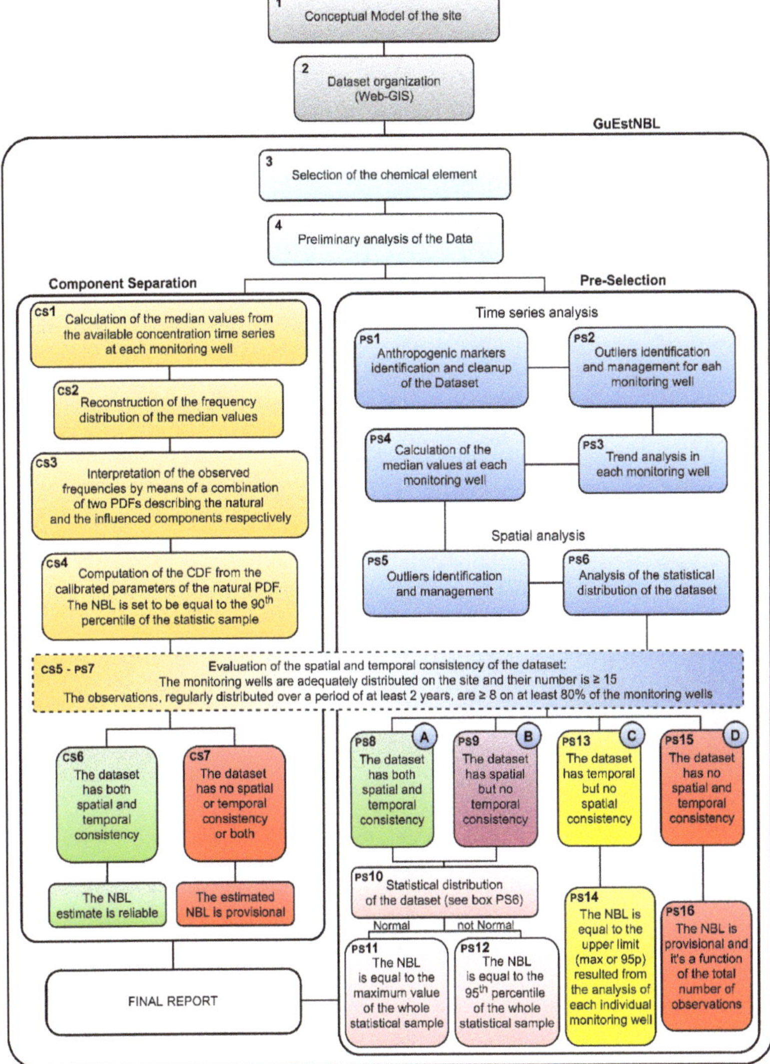

Figure 1. Scheme of the procedure implemented to estimate the natural background levels (NBLs) of aquifers. the blocks numbering (bold characters at the top right of each box) having the CS prefix refers to the procedures included into the component separation method, while the PS prefix is related to the pre-selection method. Other terms appearing in the scheme are: Web-GIS which stands for online geographical information system; PDF which stands for probability density function; CDF which stands for cumulative distribution function; bold letters A, B, C, D indicating four different cases in which the dataset can fall; 95p which stands for 95th percentile.

2.1. Structure of the Software-Based NBL Determination Procedure

Figure 1 shows the overall scheme of the procedure, which collects all aspects planned to guide the operator to determine the NBL for selected chemical species. the action depicted in each numbered block is expanded in the following dedicated subsections.

2.1.1. Data Scheduling and Acquisition

- Block 1 (Conceptual model of the site)

The definition of the conceptual model of the site is a fundamental step for any NBL estimation procedure. It is not an internal gear of the implemented procedure, since its definition is above the latter and it is included in the scheme for the sake of completeness. the conceptual model is aimed at identifying the factors (sources and processes) that determine the distribution, in space and time, of the parameters of interest. It constitutes the cognitive framework of the area. It contains interpretative and relational elements that allows for the understanding of the processes at play in the site in relation to the presence of the targeted substances. the conceptual model guides and supports some choices during the NBL determination procedure (e.g., the data grouping, the exclusion of specific observations, the identification of the most suitable statistical indicator, etc.). a good formulation of the conceptual model cannot exclude information on the geological, geochemical, and hydrogeological nature of the investigated environmental matrices, and on the anthropic pressures that, in the past or present, have impacted the study area. All these details are necessary to ensure that the analyzed data are from homogeneous environmental horizons.

- Block 2 (Dataset organization (Web-GIS))

This is the first step of the procedure falling within the guided system. It relies on a web-based geographic information system setup by the Department of Environmental Policies of the Calabria Region to contain the environmental data of the Water Protection Plan. This facility, providing a validated and constantly updated database, was equipped with specific additional accessories in order to select the data, follow appropriate screening criteria, and make them readily portable into the NBL estimation procedure. in particular, the system is designed to display the regional map of the monitoring wells, each of which contains the time series of the chemical analysis carried out on the collected samples. the system can be initially queried by entering specific spatial filters (e.g., provincial or municipal limits) in order to have a first rough delineation of the investigation area together with the included sampling stations. Subsequently, the knowledge gained from the conceptual model of the area, together with the use of the Web-GIS overlapping layers, containing the geological information of the site, or the land use map (industrial zones location, urban or agricultural areas, dumps position, etc.) allows to further circumscribe the monitoring points falling within portions of territory with homogeneous characteristics. This operation can be done by means of a "lasso tool" with which freehand selected areas are drawn on the map and the position of the inside monitoring wells, together with their data, are automatically downloaded into an Excel file. This file has a format suitable to be the input of the GuEstNBL software (implemented by the Department of Environmental Engineering—University of Calabria and owned by the Calabria Region, Italy). the tabular format of the file allows the operator to preliminary open it, to easily inspect and edit it if necessary. in this way, new data, not yet included in the GIS, can be added and analyzed together with the pre-existing ones, and those data, pertaining to chemical species having evident correlations with the chemical-physical characteristics of the sampled water, can be divided into homogeneous datasets according to these characteristics. This is particularly the case of redox-sensitive elements such as As, Fe, and Mg.

2.1.2. GuEstNBL

- Block 3 (Selection of the chemical element)

When the software is launched, a starting window appears on the PC desktop and a loading icon allows the operator to browse the computer and find the input file generated by the Web-GIS. Once the data are visualized into the GuEstNBL software, the graphical interface of the program shows a tabular representation of the concentrations of all the analyzed species. the operator can easily select the data associated with a chemical element just clicking on the specific name.

- Block 4 (Preliminary analysis of the data)

The concentrations of some parameters may be lower than the limit of quantification (LOQ) or the limit of detection (LOD) of the analytical method with which they were analyzed. the software automatically identifies these occurrences within the dataset and assigns to the <LOQ or <LOD observations a concentration value equal to LOQ and LOD, or $\frac{1}{2}$LOQ and $\frac{1}{2}$LOD, respectively [7,18]. a box shows the number of "non-detected" concentrations that are now provided with a value and can be part of the subsequent analysis. At the end of this step, the operator selects the methods for the NBL determination and the guided wizard shows the next interactive window. the numbering of the following blocks having the CS prefix refers to the procedures included into the component separation method, while the PS prefix is related to the pre-selection method.

2.1.3. Component Separation Method

The methodology follows the philosophy according to which the concentration of a chemical species in groundwater is due to the combination of natural and anthropogenic (where it exists) components [6]. the natural component is associated with the hydro-geochemical characteristics of the aquifer and to solid-water interaction processes. the anthropogenic component is due to the effects of human activities concerning specific substances whose detection have no causal relationship with the characteristics of a given site and the natural phenomena occurring in it [19–24].

- Blocks CS1 and CS2 (Calculation of the median values and construction of the observed frequency distribution)

The data related to the chosen chemical element are displayed in a table in which they appear chronologically and are associated with their own monitoring well. the software calculates the median values from the available concentration time series at each monitoring well and displays them into another table placed next to the previous one. the observed frequency distribution of the median values is automatically reconstructed and showed into a graph.

- Block CS3 (Interpretation of the observed frequencies)

The frequency distribution of the observed median concentration is interpreted by means of the following frequency distributions combinations:

$$f_{obs}(c) = f_{nat}(c) + f_{inf}(c) = k \cdot \left(A \cdot pdf_{LogNorm} + (1-A)t_N \cdot pdf_{Norm} \right), \tag{1}$$

where f_{obs} are the observed frequencies and c is the concentration of a given environmental parameter. According to Wendland et al., (2005) [6], the first term of the sum is associated with the natural component (f_{nat}) described by a log-normal frequency distribution ($pdf_{LogNorm}$), while the second term, represented by a normal distribution (pdf_{Norm}), constitutes the influenced component (f_{inf}). Moreover, k is the average width of the classes associated with the observed frequencies, A is a weight coefficient, and t_N is a truncation factor (equal to 0.5).

The two probability density functions (PDFs) are both characterized by a standard deviation and a mean that are estimated by imposing an optimization criterion (nonlinear least square method) within a calibration procedure automatically performed by the software. the best fitting of the observed data is showed as a graph.

- Block CS4 (NBL calculation)

Parameter calibration is then followed by the automatic computation of the range of concentrations comprised between the 10th (NBL10) and the 90th (NBL90) percentile of the identified log-normal PDF (a graph shows the cumulative distribution function (CDF)), then the NBL is set to be equal to NBL_{90} [6,7,18], and its value is displayed in a box next to the CDF graph.

- Blocks CS5-CS7 (NBL reliability)

NBL determination must take into account the spatial and temporal variability of the hydro-chemical characteristics of aquifers. the reliability of an estimated NBL depends on the coverage ratio of the available data: a reasonable sample size, suitable to describe the spatial variability of the system under examination, should have a minimum of 15 adequately distributed monitoring points; a reasonable sample size, adequate to describe the temporal variability of the system under examination, should have a minimum of 8 observations, regularly distributed over a period of 2 years on the 80% of the monitoring points. If the dataset meets both requirements, a high level of confidence is attributed to the determined NBL; otherwise, it is considered as provisional pending further data collection [6,7,18].

2.1.4. Pre-Selection Method

The pre-selection global statistical method, outlined in the BRIDGE project just like the component separation method, is based on the assumption that the concentration of specific indicator substances, detected into the analyzed samples, is strictly related to anthropogenic influence. If the concentration of these species is higher than well-defined values, the involved samples are excluded from the NBLs estimation procedure, as they are affected by human impact. the opportunity of excluding the data, because they are considered as not representative of the natural system under investigation, is based on the following criteria: the presence of concentrations of organic contaminants, or of other substances related to anthropogenic activity, greater than 75% of the threshold value provided by the current regulations; the presence of nitric or ammonia nitrogen concentrations whose values exceed 10.0 mg/L for nitrates (NO_3^-) and 0.1 mg/L for ammonia (NH_4^+) [7,18].

- Block PS1 (Dataset Pre-selection)

Based on what was said above, the software displays an interactive window in which the anthropogenic markers can be selected to exclude the samples with values higher than the established thresholds. the program is provided with a list containing the anthropogenic species together with their control values, and automatically displays those included in the dataset. When a marker is selected and confirmed, the dataset is filtered by those samples not meeting the above requirements [7,18].

- Block PS2 (Temporal outlier identification and management)

In this step, the temporal analysis of the data is carried out to identify potential outliers. the outliers represent concentration values that strongly differ from those within the temporal series of each monitoring well. the challenge is to recognize their nature according to the phenomenon under investigation. Concentration spikes could be caused by anthropic contamination or by the strong presence of a specific element in the minerals of a portion of the aquifer solid matrix. in the first case, the high values are certainly outside the objective of the study, while in the second case they could be considered representative of the natural background. Therefore, the removal of potential outliers must be carefully evaluated. These extreme concentration values, frequently occurring in environmental data, are highlighted by the software through graphic methods or through specific statistical tests.

In particular, when the outliers analysis is run, the program allows to evaluate, simultaneously or well by well, the results of the most used graphical methods (normal quantile–quantile (Q-Q) plots and box plots) and those of the best known statistical methods such as the Discordance [24], Huber [25], Walsh [26], Dixon [27], and Rosner [28] tests. Graphical and statistical results must be evaluated in the light of the outcomes of the conceptual model, so that the operator can decide to keep or discard the single detected outlier, providing a justification for the second operation.

- Block PS3 (Trend analysis)

The concentration time series collected in each monitoring well are analyzed, at this point, looking for the occurrence of a trend in the data. the software estimates the slope of a possible trend line through the implementation of the Mann–Kendall test [29,30]. Depending on the obtained results, the following actions are proposed: the analysis does not show, for the single monitoring well, a significant trend in the observed time window, so the fluctuation in the data can be attributed to seasonal variations under natural conditions; the analysis shows a significant trend in the time series of a specific monitoring well, so the investigated element is suspected of being subject to non-natural control factors. in the first case, the monitoring well is suitable to move forward in the guided procedure. in the second case, an assessment of whether to exclude the monitoring well from the NBL estimation must be done, also considering the indications coming from the conceptual model. the trend analysis window displays the results in the form of graphs, showing the chronological distribution of the observed data in each monitoring well, and the presence of a trend line where this occurs. a table summarizes the results and allows, via checkboxes, to decide the exclusion or not of the wells having a trend in their data.

- Block PS4 (Calculation of the median values)

As described for block CS1, relating to the component separation method, the software calculates the median values from the available concentration time series that have passed the previous steps, assigns them to the respective monitoring well, and displays them into a table.

- Block PS5 (Spatial outlier identification and management)

The same approaches and the same methods adopted in Block PS2 are now repeated at this stage to identify and manage the potential outliers which can be found among the median values.

- Block PS6 (Analysis of the statistical distribution of the dataset)

This fundamental step is presented at this point of the procedure, but it is performed by the software even before the study for the identification of both temporal and spatial outliers. the reason why this analysis recurs more than once lies in the fact that the applicability of certain methodologies, like those previously described for the outlier identification and those coming afterwards, for the NBL estimation, depends on the probability function that best approximates the available observed data. Moreover, given that the statistic sample can undergo the loss of data, precisely because of the potential exclusion of outliers, it is essential to have the possibility to repeat the analysis in this eventuality. This operation is carried out by applying appropriate tests, such as normal quantile–quantile (Q-Q) plots, Shapiro and Wilk [31], D'Agostino [32], and Lilliefors [33], all of which included and were automatically performed by the software.

- Block PS7 (Spatial and temporal consistency of the dataset)

This evaluation, which is also performed in the Component Separation method and it is already described in the Blocks CS5-CS7, distinguishes different levels of spatial and temporal "coverage" of the available data and guides the NBL estimation process. There are 4 distinct cases (A, B, C, and D), which are described below and are automatically identified by the software and proposed to the operator according to the spatial and temporal coverage of the data left over by the previous steps.

- Block PS8-PS12 (NBL determination for the datasets falling into cases a and B)

These datasets exhibit an adequate spatial dimension. Case A, unlike case B, also shows an adequate temporal coverage. There is no substantial difference in the NBLs estimation between these two types of dataset. the only distinction is reflected in a higher level of confidence to be attributed to the NBLs determined for the dataset of case A. the software assigns to the NBL value of the maximum observed median, provided that the dataset is normally distributed. If the dataset shows a non-normal distribution, the NBL is given by the 95th percentile of the identified PDF. in particular, the software automatically sets out whether the observed medians are best approximated by a log-normal or a gamma distribution or whether it is best to normalize the data or finally if a non-parametric distribution is the right solution. the parameters of the recognized probability density function are then estimated to fit the observed data within the same automatic calibration procedure described in Block CS3 and the NBL is calculated following the implementations defined in Block CS4. in the case of a distribution suitable for a normalization process, the data are transformed through the Box-Cox transformation [34] and then best fitted by a normal PDF to move forward with the same process described above. Non-parametric datasets (set of data that are not satisfactorily approximated by any distribution) are processed through a graphical method whereby the software draws the cumulative frequency curve of the data and identifies, as representative of the NBL value, the one corresponding to the 90th percentile. Finally, the software can also evaluate the NBL through well-known parameters such as the upper tolerance limit (UTL) and upper prediction limit (UPL) [35].

- Block PS13-PS14 (NBL determination for the datasets falling into case C)

This type of dataset shows an adequate temporal dimension but a poor spatial coverage. In this case, the procedures described for datasets a and B used to treat the medians of all monitoring wells, which are applied on the data of the single observation point. An NBL value is thus estimated for each of them. the final NBL representing the entire dataset is given by the maximum value among the estimated ones.

- Block PS15-PS16 (NBL determination for the datasets falling into case D)

When the data does not have a significant dimension in either time or space, it is expected that further data and information must be collected and, in the meantime, an estimate of a provisional NBL can be made if a total number of observations ≥ 10 is available. in this case, the NBL is equal to the 90th percentile of the whole dataset.

2.1.5. Final Report

The software automatically generates an output pdf file at the end of the estimation procedure, whether it has been carried out using the component separation or the pre-selection method. This file contains a detailed report in which all the operations and the choices, made by the operator to reach the final result, are recorded. All the graphic and numerical outcomes are also collected and displayed chronologically.

2.2. Study Area

GuEstNBL has been used in the proposed study to analyze the data recorded within the industrial area of the town of Lamezia Terme located in the Calabria Region, Italy. the site, which is part of the S. Eufemia plain, lies south of the town of Lamezia and overlooks the Tyrrhenian Sea, involving an area of approximately 12,000.00 m^2 (Figure 2).

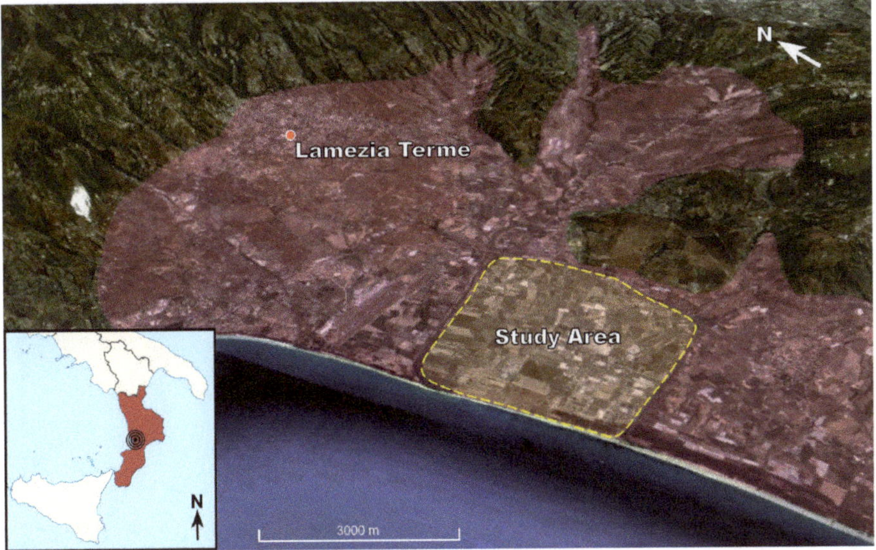

Figure 2. Aerial view of the Lamezia Terme District (pink background) and study area location (yellow background), Italy. Image taken from Google Earth.

The study area, with a roughly rectangular shape, was bounded in order to have a sufficiently homogeneous zone from the morphological, geological, and hydrogeological point of view. the shallow aquifer flowing at little depth from the ground surface is the subject of the investigation. Some monitoring wells, pertaining to the facilities operating within the aforementioned area, have been affected by anomalous values of As, Fe, and Mn concentration. the general purpose of the project was therefore to monitor the portion of the aquifer underlying the study area and to evaluate which part of the observed contamination is the result of natural phenomena, as well as which is attributable to industrial activities. the latter relate to the following sectors: treatment, transformation, and recovery of special and hazardous waste; oil refining and biomass production; wastewater treatment; carpentry and paintwork; farming activities.

In geological terms, the study area is characterized by gravelly-sandy alluvial deposits with silts and clays of continental and marine origin. the analysis of the available stratigraphies highlighted the presence of significant peat beds as a peculiar geological feature of the area. in general, the peat layers found during the drilling operations of the observation wells, most of which have been deepened up to 10 m from the ground level, have a thickness ranging from 0.5 to 3 m. It is likely to believe that the peat presence is due to an extensive swamp that anciently covered the coastal stretch of the plain.

In total, 17 monitoring wells (Figure 3) were used for groundwater periodic sampling and hydraulic levels recording by following the U.S. EPA indications [36]. the wells, owned by the facilities operating in the area, are part of the Integrated Environmental Authorization (IEA), an environmental administrative intervention procedure that must be carried out on certain industrial activities that are likely to affect the safety, health of people, or the environment. the IEA was established with the Council Directive 96/61/EC to comply with the principles of Integrated Pollution Prevention and Control (IPPC) dictated by the European Union since 1996. the hydro-geochemical investigation of the area concerned the analysis of heavy metals (As, Fe, Mn, Cu, Zn, Pb, Ni, Hg, Cd, Cr), cations (Na^+, K^+, Ca^{2+}, Mg^{2+}), anions (Cl^-, SO_4^{2-}, NO_3^-, NO_2^-), Ammonia Nitrogen (NH_4^-), Total Phosphorus (P), Phosphates (PO_4^{3-}), Hydrocarbons (C > 12), and pesticides.

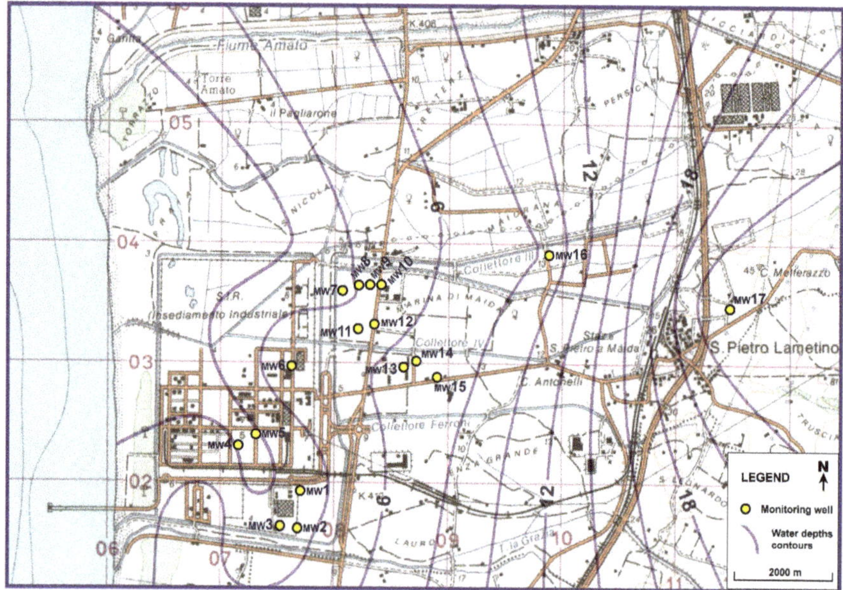

Figure 3. Map of the monitoring wells and two-dimensional groundwater level distribution (expressed in m above mean sea level).

Furthermore, the redox potential (Eh), the pH, and the temperature values were measured in place during the sampling procedures, as well as the electrical conductivity, the dissolved oxygen (O_2), and the chemical oxygen demand (COD). the sampling frequency was on a quarterly basis for a total period of 2 years. Eight withdrawals were hence collected and analyzed for each monitoring well, for a total of 136 samples.

3. Results

In this section, the results obtained by the interpretation of the acquired data is discussed. the qualitative trend of the water depths contours, obtained by means of an ordinary kriging performed on the median values of the observed hydraulic head values, showed a prevailing flow component that moves from east to west (Figure 3). the groundwater contour lines (this is even more evident at a larger scale) show a narrower arrangement in the innermost areas to the west. This trend highlights zones with shorter hydraulic paths and higher hydraulic gradients, which are typical of lower permeability geological formations. in these areas, the aquifer is fed by the rivers flowing through the plain. Moving to the east, toward the coast, the contour lines distance progressively increases, causing a consequent raise of the hydraulic paths and the decrease of hydraulic gradients. This suggests the occurrence of zones with higher permeability. the hydraulic heads distribution of these zones indicates that the aquifer feeds the rivers. in summary, the water-table trend suggests that the aquifer is recharged in the innermost areas while the coastal strip of the plain acts as a drainage zone. This evidence correlates with the hydro-geo-chemical observations since the innermost monitoring wells MW16 and MW17 have higher values of dissolved oxygen associated with positive values of redox potential and pH values between 7 and 7.4, attributable to an oxidizing environment and to a groundwater recharge area supplied by superficial waterbodies. the same parameters, monitored in the other wells closer to the shoreline, show instead conditions attributable to a reducing environment: negative redox potential, 6.7 < pH < 7.0, lower values of dissolved oxygen (Figure 4). the aforementioned conditions can be related to a drainage area characterized by a slow water flow with long residence times.

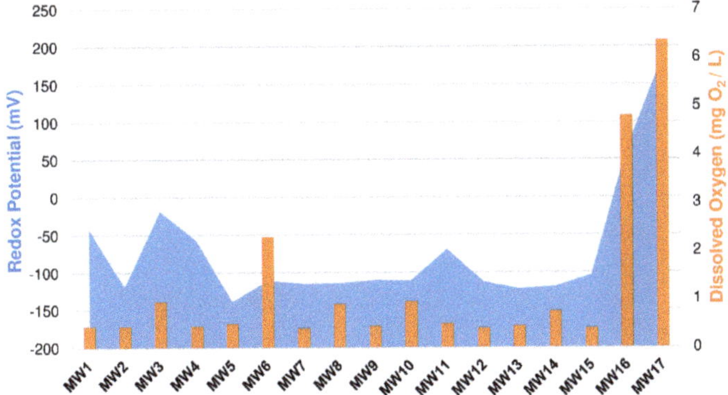

Figure 4. Correlation between the median values of dissolved oxygen and redox potential observed in each monitoring well.

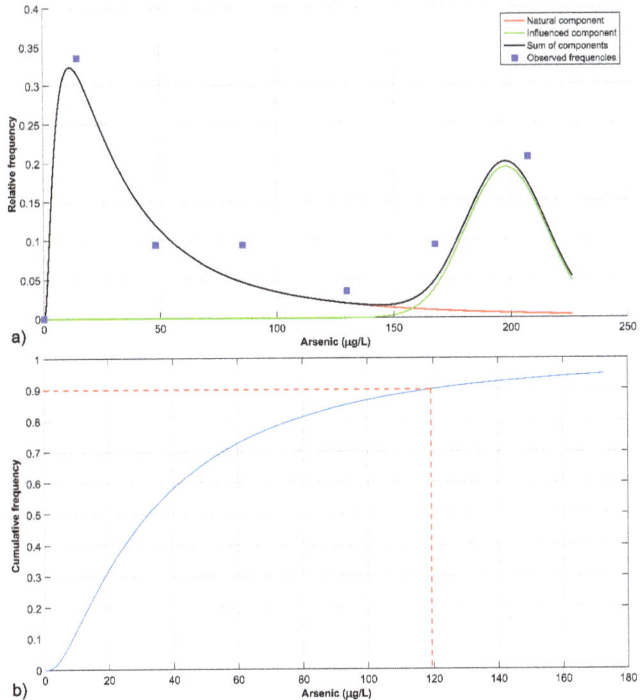

Figure 5. GuEstNBL output: (**a**) Arsenic empirical frequencies together with the estimated natural and anthropogenic components, and their weighted sum distribution; (**b**) cumulative distribution function (CDF) and 90th percentile of the identified log-normal PDF.

Based on the aforementioned findings, wells MW16 and MW17 have not been included in the dataset for the NBLs estimation. the remaining 15 observation points, falling within a system with rather homogeneous hydro-geochemical properties, and providing both an appropriate spatial and temporal coverage, are the points whose data have set up the input for the GuEstNBL software. the resulting NBLs are therefore assigned to a reduced aquifer portion and delimited by the sampling points involved in the guided estimation procedure. the available data are almost all affected by the presence of ammonia nitrogen (NH_4^-) with concentrations exceeding the limit of 0.1 mg/L. the pre-selection method is therefore not compatible with the dataset widely influenced by one of the principle anthropogenic markers (see Section 2.1.4). the application of the component separation (CS) method on the chemical species under consideration is examined below.

3.1. Natural Background Levels Estimation

3.1.1. Arsenic

The outcomes stemming from the application of the software guided procedure for arsenic are shown in Figure 5. the reference threshold value for this element is set at 10 µg/L by the Italian environmental laws [37].

Table 1. GuEstNBL output: Results of the guided estimation procedure for arsenic dataset.

Parameter	Natural Component (f_{nat})	Influenced Component (f_{inf})
Mean µ (µg/L)	32.28	197.86
Standard deviation σ (µg/L)	1.02	16.96
Mixture weight a	0.49	
NBL_{90} (µg/L)	119.24	

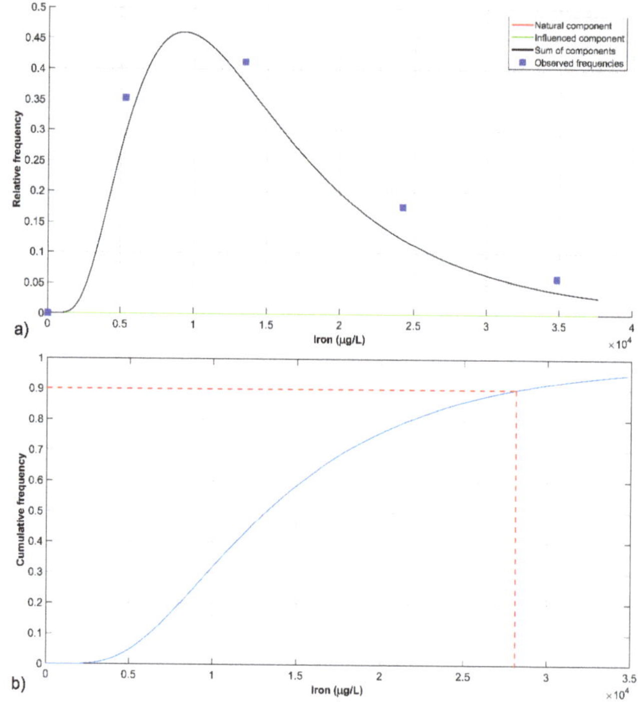

Figure 6. GuEstNBL output: (**a**) Iron empirical frequencies together with the estimated natural and anthropogenic components, and their weighted sum distribution; (**b**) CDF and 90th percentile of the identified log-normal PDF.

The distribution used to interpret the experimental data is characterized by a natural component that displays a sharp and narrow peak at about 15 µg/L. the log-normal (natural) component is prevalent for concentrations lower than 150 µg/L, while for larger concentration values the contribution of the Gaussian (anthropogenic) component takes hold and outclasses the former (Figure 5a).

the estimated NBL, namely NBL$_{90}$ corresponding to the 90th percentile (Figure 5b), together with the mixture weight a and the estimated standard deviations (σ_{nat}, σ_{inf}) and means (μ_{nat}, μ_{inf}) of the two PDFs, are listed in Table 1. the estimated background value is larger than the reference limit and hence, in line with what was proposed by Muller et al. (2006), a threshold value corresponding to the background level must be assigned to the area.

3.1.2. Iron

The curve fitting of the observed relative frequencies associated with the iron concentration values derived from the water samples collected in the monitoring wells, is shown in Figure 6a. the concentrations moving around values of about 10,000 µg/L are associated with high relative frequencies. the natural distribution exhibits a good agreement with all the experimental data (the log-normal PDF coincides with the sum distribution), while the influenced component is completely absent (the Gaussian PDF is flat and lies on the abscissa axis). Table 2 shows the estimated NBL, together with the mixture weight a and the estimated parameters for f_{nat} and f_{inf} distributions.

Table 2. GuEstNBL output: Results of the guided estimation procedure for iron dataset.

Parameter	Natural Component (f_{nat})	Influenced Component (f_{inf})
Mean µ (µg/L)	13,183.62	-
Standard deviation σ (µg/L)	0.59	-
Mixture weight a	1.0	
NBL$_{90}$ (µg/L)	27,960.52	

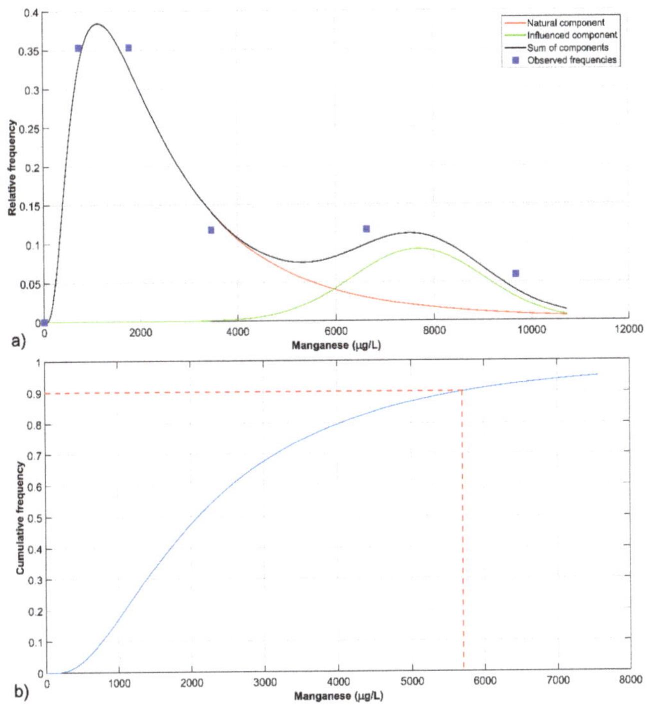

Figure 7. GuEstNBL output: (**a**) Manganese empirical frequencies together with the estimated natural and anthropogenic components, and their weighted sum distribution; (**b**) CDF and 90th percentile of the identified log-normal PDF.

The iron estimated NBL_{90} is extremely larger than the reference limit (200 µg/L [37]) and, hence, according to what was done for the arsenic results, a threshold value corresponding to the background level must be assigned to this element (Figure 6b). a natural background value larger than the current regulation limit indicates that the state of the investigated aquifer cannot be regarded as poor as long as the iron concentrations do not exceed this new threshold because of the occurrence of a specific natural signature typical of the examined context.

3.1.3. Manganese

The reference threshold value for manganese is 50 µg/L [37]. Figure 7a shows the results of the CS method for this metal. the lowest monitored concentrations (around 1000 µg/L) are associated with high relative frequencies. the observed frequencies are well interpreted by the log-normal (natural) distribution for concentrations up to 4000 µg/L; then, the Gaussian (anthropogenic) component become relevant.

The estimated NBLs together with the parameters resulting from the curve fitting are collected in Table 3. a natural background concentration (NBL_{90}) of 5718.20 µg/L was estimated, which is higher than the regulation limit and, thus, as in the previous cases, becomes the new threshold value (Figure 7b).

Table 3. GuEstNBL output: Results of the guided estimation procedure for manganese dataset.

Parameter	Natural Component (f_{nat})	Influenced Component (f_{inf})
Mean µ (µg/L)	2097.65	7671.01
Standard deviation σ (µg/L)	0.78	1350.7
Mixture weight a	0.65	
NBL_{90} (µg/L)	5718.20	

4. Discussion

The guided procedure offered by the software has comfortably led to NBL estimation. the three selected elements have important natural components and this can be understood by considering the observed site characteristics and correlation between the concentration values of the involved chemical species. Figure 8 shows a comparison between the median values of arsenic, iron, and manganese concentration recorded in each well during the two-year monitoring period. It can be noticed that iron and manganese concentrations have substantially the same pattern, as it is clearly evident from the two peaks registered for the first element that correspond to the peaks of the second one (monitoring wells MW2 and MW4) even if with significantly lower values. Arsenic basically shows high concentrations in correspondence to iron and manganese high concentrations, and low concentrations where the presence of this two elements is poor. the investigation of the origins of arsenic high concentrations in groundwater is a widely documented topic in the literature [38–45]. All authors highlighted the primary role of peat, which was widely present in the current study area, in the release of arsenic associated with high concentrations of iron and manganese in anoxic conditions. in the past century, the S. Eufemia plain was a large swamp reclaimed at the end of the 20 s. the study area is one of the sectors most affected by swamping. the presence of large marshlands, in which the deposition and decay of the vegetation occurred, is responsible for the creation of the peat deposits. the arsenic release mechanism, which is more accredited in the aforementioned literature, is based on the formation of iron and manganese oxides and hydroxides (naturally and richly present in the Calabrian soils) on which the arsenic is adsorbed. These complexes are stable in an oxidizing environment (the samples taken in wells MW16 and MW17 contain almost zero concentrations of the three elements), while they dissolve in a reducing environment where the bridging oxygen that binds arsenic to iron and manganese oxides and hydroxides fails, causing their simultaneous release into the water.

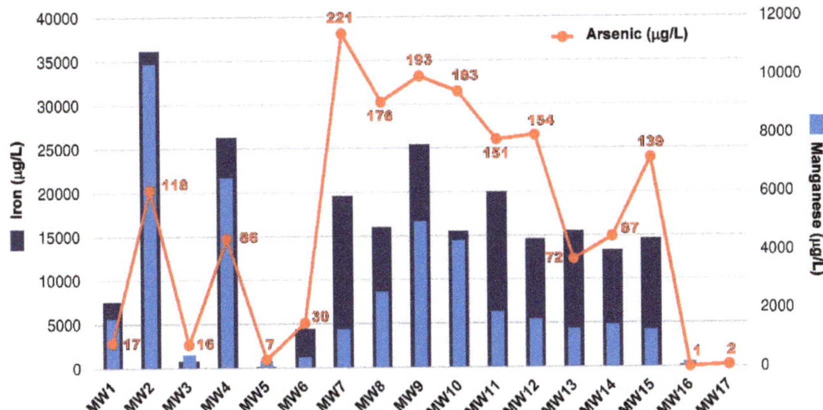

Figure 8. Arsenic (red line with dots), iron (dark blue bars) and manganese (light blue bars) median concentration values recorded in each monitoring well.

This can explain what happens in every other well pertaining to the reducing zone of the aquifer and having significant concentrations of all three species (see Section 3 and Figure 4). The review of the dataset, acquired in the two-year monitoring period, did not show significant concentrations of hydrocarbons and pesticides. On the other hand, a variable concentration of Ammonia Nitrogen (NH_4^-) and Nitrates (NO_3^-) was found throughout the study area. Nitrite (NO_2^-) concentration was always found below the detection limit of the adopted analytical method. the trend of nitrogenous compounds highlights concentrations of NH_4^- above the regulation limits in almost all sampling points. the presence of NO_3^- has been recorded in the upstream sector of the aquifer, more oxygenated, and with positive Eh values, while its concentrations become very low in the downstream sectors, scarcely oxygenated, and with negative Eh values. This suggests the generation of oxidizing and reducing processes typical of the nitrogen (N_2) cycle, according to which nitrogen compounds are transformed through aerobic (nitrification) and anaerobic (denitrification) processes, starting from the N_2 contained in organic substances. Commonly, these processes occur because of the decomposition of dejections and so their occurrence is used as an indicator of organic pollution of anthropogenic nature. However, nitrogen compound transformations can also be triggered by the decomposition of natural organic substances such as the peat found in the study area. the current analyses do not allow to definitively establish whether these compounds pertain to an anthropogenic or a natural pollution, this determination will be the subject of further studies. For this reason, NBL determination was not possible by means of the pre-selection method and the component separation method was used instead. the software therefore provides the possibility of using both approaches, depending on the anthropic impact on the study area, and therefore on the presence, or not, of anthropic markers with relevant concentrations within the observed dataset.

The implemented procedure is hence a versatile instrument but the reliability of its results depend on the compliance of the following general criteria:

- The input data must come from a single aquifer or a portion of it. the study area must have homogeneous characteristics in terms of geological and hydro-geo-chemical properties, since the procedure provides the estimation of a single NBL value assigned to the whole investigated area.
- The investigation scale depends on the spatial extension within which the aquifer homogeneous characteristics are maintained. This means that, depending on the case, the software can be used for local or regional studies. in practical applications, what provides the actual size of the area to be charged with the estimate result is the number and the distribution of the available observation points falling within the zone of homogeneity. the minimum number of observation points,

their distribution, and sampling frequencies are discussed in Section 2.1.3 (Blocks CS5-CS7 (NBL reliability)).

- The operator using the software is an integral part of it. He or she makes important decisions by systematizing the information coming from the study of the conceptual model and from the results gradually emerging from the guided procedure. This implies that not everyone is able to obtain reliable results.

5. Conclusions

The collaboration between the Department of Environmental Engineering of the University of Calabria and the Department of Environmental Policies of the Calabria Region resulted in the implementation of a useful procedure that relied on the joint application of a web-based geographic information system (GIS) and an ad hoc developed software (GuEstNBL). Thus, the onerous and complex issue of the Natural Background Levels estimation was more easily addressed. the environmental agencies responsible for the protection, management, and remediation of the territories can take advantage of such a tool, since it is equipped with a graphical wizard that guides the operators to implement the component separation and pre-selection methods to perform all computations required therein. the procedure can be further improved by adding additional statistical and graphical tests, and, in the present state, is going to be the official evaluation tool of the Calabria Region to identify groundwater thresholds.

In the present study, the developed calculation program was used for the NBL estimation of arsenic, iron, and manganese in a potentially contaminated industrial site located in Lamezia Terme, Italy. the outcomes of the software elaborations, together with the cross-reading of geological, hydrogeological, and hydro-geochemical data, have shown that the presence of the anomalous concentrations detected in the shallow aquifer may be due to natural phenomena linked to the degradation of peat deposits that drive the triggering of the redox reaction that generates a dissolution of the pollutants into the groundwater. the three investigated chemical species have in fact important natural components that result in natural background concentrations higher than the regulation limit. Thus, they can become the new threshold values. NBL estimation was obtained through the component separation method since the dataset was almost only affected by the presence of ammonia nitrogen (NH_4^-) with concentrations exceeding the limit of 0.1 mg/L. the pre-selection method was therefore incompatible with the dataset widely influenced by one of the principle anthropogenic markers.

The results of this study showed that groundwater systems are complex and require an accurate characterization to distinguish effective anthropogenic contamination from natural conditions. This is extremely important to define clean-up goals consistent with specific natural features of an analyzed water body, since the costs for aquifer management practices and remediation actions are dramatically high. the realized calculation framework has a great exploitation perspective since it can be used worldwide.

Author Contributions: Conceptualization, F.C. and M.D.B.; methodology, F.C. and M.D.B.; software, F.C. and M.D.B.; validation, F.C., M.D.B., A.C., E.C., and C.M.; formal analysis, F.C. and M.D.B.; investigation, A.C. and E.C.; resources, F.C., M.D.B., A.C., E.C., and C.M.; data curation, F.C., M.D.B., A.C., E.C., and C.M.; writing—original draft preparation, F.C.; writing—review and editing, F.C.; visualization, F.C. and M.D.B.; supervision, O.R., I.T., and S.S.; project administration, O.R., I.T., and S.S.; funding acquisition, O.R., I.T., and S.S. All authors have read and agreed to the published version of the manuscript.

Funding: This research was funded by Regione Calabria.

Acknowledgments: This research has been supported by the "AIM: Attraction and International Mobility", PON R&I 2014-2020 Calabria

Conflicts of Interest: The authors declare no conflict of interest. the funders had no role in the design of the study; in the collection, analyses, or interpretation of data, in the writing of the manuscript, or in the decision to publish the results.

References

1. Hinsby, K.; Condesso de Melo, M.T. Application and evaluation of a proposed methodology for derivation of groundwater threshold values-a case study summary report. *Report to the EU Project 'BRIDGE'*. 2006. Deliverable D22. Available online: http://nfp-at.eionet.europa.eu/Public/irc/eionet-circle/bridge/library?l=/deliverables/d22_final_reppdf/_EN_1.0_&a=d (accessed on 29 September 2020).

2. Edmunds, W.M.; Shand, P.; Hart, P.; Ward, R.S. the natural (baseline) quality of groundwater: a UK pilot study. *Sci. Total Environ.* **2003**, *310*, 25–35. [CrossRef]

3. European Commission. *Guidance on Groundwater Status and Trend Assessment, Guidance Document #18*; Technical Report; European Communities: Luxembourg, 2009; ISBN 978-92-79-11374-1.

4. Panno, S.V.; Kelly, W.R.; Martinsek, A.T.; Hackley, K.C. Estimating background and threshold nitrate concentrations using probability graphs. *GroundWater* **2006**, *44*, 697–709. [CrossRef] [PubMed]

5. Walter, T. Determining natural background values with probability plots. in Proceedings of the EU Groundwater Policy Developments Conference, (UNESCO), Paris, France, 13–15 November 2008.

6. Wendland, F.; Hannappel, S.; Kunkel, R.; Schenk, R.; Voigt, H.J.; Wolter, R. a procedure to define natural groundwater conditions of groundwater bodies in Germany. *Water Sci. Technol.* **2005**, *51*, 249–257. [CrossRef] [PubMed]

7. BRIDGE. Background cRiteria for the IDentification of Groundwater Thresholds. 2007. Available online: http://nfp-at.eionet.europa.eu/irc/eionet-circle/bridge/info/data/en/index.htm (accessed on 27 August 2020).

8. Edmunds, W.M.; Shand, P. *Natural Groundwater Chemistry*; Blackwell Publishing Ltd.: Oxford, UK, 2008; p. 469.

9. Hinsby, K.; Condesso de Melo, M.T.; Dahl, M. European case studies supporting the derivation of natural background levels and groundwater threshold values for the protection of dependent ecosystems and human health. *Sci. Total Environ.* **2008**, *401*, 1–20. [CrossRef]

10. Wendland, F.; Berthold, G.; Blum, A.; Elsass, P.; Fritsche, J.G.; Kunkel, R.; Wolter, R. Derivation of natural background levels and threshold values for groundwater in the Upper Rhine Valley (France, Switzerland and Germany). *Desalination* **2008**, *226*, 160–168. [CrossRef]

11. Kim, K.H.; Yun, S.T.; Kim, H.K.; Kim, J.W. Determination of natural backgrounds and thresholds of nitrate in South Korean groundwater using model-based statistical approaches. *J. Geochem. Explor.* **2015**, *148*, 196–205. [CrossRef]

12. Preziosi, E.; Giuliano, G.; Vivona, R. Natural background levels and threshold values derivation for naturally As, V and F rich groundwater bodies: a methodological case study in Central Italy. *Environ. Earth Sci.* **2010**, *61*, 885–897. [CrossRef]

13. Stan, C.O.; Buliga, I. Natural background levels and threshold values for phreatic waters from the Quaternary deposits of the Bahlui River basin. *Aui Geol.* **2013**, *59*, 51–60.

14. Sun, L.; Gui, H.; Peng, W.; Lin, M. Heavy Metals in Deep Seated Groundwater in Northern Anhui Province, China: Quality and Background. *Nat. Environ. Pollut. Technol.* **2013**, *12*, 533–536.

15. Molinari, A.; Chidichimo, F.; Straface, S.; Guadagnini, A. Assessment of natural background levels in potentially contaminated coastal aquifers. *Sci. Total Environ.* **2014**, *476–477*, 38–48. [CrossRef]

16. Preziosi, E.; Parrone, D.; Del Bon, A.; Ghergo, S. Natural background level assessment in groundwaters: Probability plot versus pre-selection method. *J. Geochem. Explor.* **2014**, *143*, 43–53. [CrossRef]

17. Chidichimo, F.; De Biase, M.; Straface, S. Groundwater pollution assessment in landfill areas: Is it only about the leachate? *Waste Manag.* **2020**, *102*, 655–666. [CrossRef]

18. ISPRA—Istituto Superiore per la Protezione e la Ricerca Ambientale. *Linee Guida per la Determinazione dei Valori di Fondo per i Suoli e per le Acque Sotterranee*; Manuali e Linee Guida 174/2018; ISPRA: Rome, Italy, 2018; ISBN 978-88-448-0880-8.

19. Tiedeken, E.J.; Tahar, A.; McHugh, B.; Rowan, N.J. Monitoring, sources, receptors, and control measures for three European Union watch list substances of emerging concern in receiving waters—A 20 year systematic review. *Sci. Total Environ.* **2017**, *574*, 1140–1163. [CrossRef]

20. Xiao, F. Emerging poly- and perfluoroalkyl substances in the aquatic environment: a review of current literature. *Water Res.* **2017**, *124*, 482–495. [CrossRef]

21. Tursi, A.; Chatzisymeon, E.; Chidichimo, F.; Beneduci, A.; Chidichimo, G. Removal of Endocrine Disrupting Chemicals from Water: Adsorption of Bisphenol-A by Biobased Hydrophobic Functionalized Cellulose. *Int. J. Environ. Res. Public Health* **2018**, *15*, 2419. [CrossRef]

22. Tursi, A.; De Vietro, N.; Beneduci, A.; Milella, A.; Chidichimo, F.; Fracassi, F.; Chidichimo, G. Low pressure plasma functionalized cellulose fiber for the remediation of petroleum hydrocarbons polluted water. *J. Hazard. Mater.* **2019**, *373*, 773–782. [CrossRef]

23. De Vietro, N.; Tursi, A.; Beneduci, A.; Chidichimo, F.; Milella, A.; Fracassi, F.; Chatzisymeon, E.; Chidichimo, G. Photocatalytic inactivation of Escherichia coli bacteria in water using low pressure plasma deposited TiO2 cellulose fabric. *Photoch. Photobio. Sci.* **2019**. [CrossRef]

24. Barnett, V.; Lewis, T. *Outliers in Statistical Data*, 3rd ed.; Wiley: Chichester, UK, 1994; 584p.

25. Huber, P.J. *Robust Statistics*; John Wiley and Sons: Hoboken, NY, USA, 1981.

26. Walsh, J.E. Large sample nonparametric rejection of outlying observations. *Ann. I. Stat. Math.* **1958**, *10*, 223–232. [CrossRef]

27. Dixon, W.J. Processing Data for Outliers. *Biometrics* **1953**, *9*, 74–89. [CrossRef]

28. Rosner, B. On the Detection of many outliers. *Technometrics* **1975**, *17*, 221–227. [CrossRef]

29. Mann, H.B. Nonparametric tests against trend. *Econometrica* **1945**, *13*, 245–259. [CrossRef]

30. Kendall, M.G. *Rank Correlation Methods*; Griffin: London, UK, 1975.

31. Shapiro, S.S.; Wilk, M.B. An analysis of variance test for normality (complete samples). *Biometrika* **1965**, *52*, 591–611. [CrossRef]

32. D'Agostino, R.B.; Stephens, M.A. *Goodness-of-Fit Techniques*; Marcel Dekker Inc.: New York, NY, USA, 1986.

33. Lilliefors, H.W. On the Kolmogorov-Smirnov Test for Normality with Mean and Variance Unknown. *J. Am. Stat. Assoc.* **1967**, *62*, 399–404. [CrossRef]

34. Box, G.E.P.; Cox, D.R. An analysis of transformations. *J. R. Stat. Soc. B* **1964**, *26*, 211–252. [CrossRef]

35. ISO 16269-6, Statistical Interpretation of Data, Part 6: Determination of Statistical Tolerance Intervals, Technical Committee ISO/TC 69, Applications of Statistical Methods. Available online: https://www.iso.org/obp/ui/#iso:std:iso:16269:-6:ed-2:v1:en (accessed on 27 August 2020).

36. Puls, R.W.; Barcelona, M.J. *Ground Water Issue: Low-Flow (Minimal Drawdown) Groundwater Sampling Procedures*; Technical Report EPA/540/S-95/504; US EPA: Washington, DC, USA, 1996.

37. D.Lgs. 152/06—Decreto Legislativo n. 152 del 3 aprile 2006. Norme in materia ambientale. Pubblicato nella Gazzetta Ufficiale n. 88 del 14 Aprile 2006—Supplemento Ordinario n. 96. Available online: https://www.gazzettaufficiale.it/dettaglio/codici/materiaAmbientale (accessed on 29 September 2020).

38. Mc Arthur, J.M.; Ravenscroft, P.; Safiulla, S.; Thirlwall, M.F. Arsenic in groundwater: Testing pollution mechanisms for sedimentary acquirers in Bangladesh. *Water Resour. Res.* **2001**, *37*, 109–117. [CrossRef]

39. Mc Arthur, J.M.; Banerjee, D.M.; Hudson Edwards, K.A.; Mishra, R.; Purohit, R.; Ravenscroft, P.; Cronin, A.; Howarth, R.J.; Chatterjee, A.; Talukder, T.; et al. Natural organic matter in sedimentary basins and its relation to arsenic in anoxic ground water: the example of West Bengal and its worldwide implications. *Appl. Geochem.* **2004**, *19*, 1255–1293. [CrossRef]

40. Peters, S.C.; Burkert, L. the occurrence and geochemistry of arsenic in groundwaters of the Newark basin of Pennsylvania. *Appl. Geochem.* **2008**, *23*, 85–98. [CrossRef]

41. Saunders, J.A.; Lee, M.K.; Shamsudduha, M.; Dhakal, P.; Uddin, A.; Chowdury, M.T.; Ahmed, K.M. Geochemistry and mineralogy of arsenic in (natural) anaerobic groundwaters. *Appl. Geochem.* **2008**, *23*, 3205–3214. [CrossRef]

42. Hung Hsu, C.; Tang Han, S.; Hsuan Kao, Y.; Wuing Liu, C. Redox characteristics and zonation of arsenic affected multy layer aquifers in the Choushui River alluvional fan, Taiwan. *J. Hydrol.* **2010**, *391*, 351–366.

43. Lawati, W.M.; Rizoulis, A.; Eiche, E.; Boothman, C.; Polya, D.A.; Lloyd, J.R.; Berg, M.; Vasquez-Aguilar, P.; Van Dongen, B.E. Characterization of organic matter and microbial communities in contrasting arsenic rich Holocene and arsenic-poor Pleistocene aquifers, Red River Delta, Vietnam. *Appl. Geochem.* **2012**, *27*, 315–325. [CrossRef]

44. Rotiroti, M.; Sacchi, L.; Fumagalli, L.; Bonomi, T. Origin of Arsenic in groundwater from multilayer aquifer in Cremona (Northern Italy). *Environ. Sci. Technol.* **2014**, *14*, 5395–5403. [CrossRef]
45. Carraro, A.; Fabbri, P.; Giaretta, A.; Peruzzo, L.; Tateo, F.; Tellini, F. Effect of redox conditions on the control of arsenic mobility in shallow alluvional acquifers on the Venezia Plain (Italy). *Sci. Total Environ.* **2015**, *532*, 581–594. [CrossRef]

MDPI
St. Alban-Anlage 66
4052 Basel
Switzerland
Tel. +41 61 683 77 34
Fax +41 61 302 89 18
www.mdpi.com

Water Editorial Office
E-mail: water@mdpi.com
www.mdpi.com/journal/water